I0112568

COSMIC FRAGMENTS

INTERSECTIONS: HISTORIES OF ENVIRONMENT, SCIENCE, AND TECHNOLOGY IN THE ANTHROPOCENE

Sarah Elkind and Finn Arne Jørgensen, Editors

COSMIC FRAGMENTS

DISLOCATION AND DISCONTENT
IN THE GLOBAL SPACE AGE

EDITED BY
ASIF A. SIDDIQI

UNIVERSITY OF PITTSBURGH PRESS

Published by the University of Pittsburgh Press, Pittsburgh, Pa., 15260
Copyright © 2025, University of Pittsburgh Press
All rights reserved
Manufactured in the United States of America
Printed on acid-free paper
10 9 8 7 6 5 4 3 2 1

Cataloging-in-Publication data is available from the Library of Congress

ISBN 13: 978-0-8229-4843-8
ISBN 10: 0-8229-4843-5

Cover photo: Courtesy of Blue Origin
Cover design: Alex Wolfe

To the Yanadi

CONTENTS

PART II. EMPIRE

PART III. WASTE

PART IV: RUPTURE

ACKNOWLEDGMENTS

The seed idea for this book can be traced back to my time as a long-term fellow at the Huntington Library in San Marino, California, in 2015–2016 when the library's then director of research, Steve Hindle, invited me to organize a conference on the global aspects of the history of space exploration. The idea was to bring together scholars from many different disciplines to weigh in on less-explored themes that might break new scholarly ground. My thinking about the conference was considerably enriched by a panel held at the annual meeting of the Society for the History of Technology in the fall of 2018 held in Saint Louis, Missouri, under the title, "Other Spaces: Displacement, Disruption, and Violence in the Space Age." I owe all the participants in that panel—Edward Jones-Imhotep, Ellen Power, Anna Reser, and Lisa Ruth Rand—an immense debt of gratitude for their brilliant contributions and for helping me sharpen my thoughts. Drawing from the ideas of this panel, the Huntington conference was planned for May 2020 under the title, "The Global Cosmos: Dislocation and Discontent in the Global Space Age." Unfortunately, a month before our planned conference, something truly global disrupted life across the planet, the COVID pandemic. The Huntington Library elected to permanently cancel the event rather than postpone it. In soldiering on, all except one of the participants in the original meeting, plus a number of new colleagues, finally convened online in a wonderfully generative workshop in September 2021. We discussed our papers, exchanged ideas, learned many new things, and came away with a renewed sense of purpose. The outcome of that workshop is, in essence, the book you have before you.

I want to thank the O'Connell Initiative for the Global History of Capitalism at Fordham University, which sponsored the online workshop in 2021. At Fordham, I also want to thank Audra Croke who helped organize the event and Javier Calderon Abullarade who took excellent notes for all the participants. At the University of Pittsburgh Press, I am grateful to Abby Collier who took on this project and graciously supported it to completion as well as to Kelly Lynn Thomas and Alex Mathews who helped shape it into being. I also want to thank the two coeditors of the "Intersections: Histories of Environment, Science, and Technology in the Anthropocene" series, Sarah S. Elkind and Finn Arne Jørgensen. A note of thanks to Esther Liberman Cuenca for her support and encouragement. I'm especially grateful to each of the contributors who gave their energy, ideas, and time to this project. Finally, many thanks to family in New York and Dhaka.

COSMIC FRAGMENTS

INTO THE COSMIC (AGAIN)

ASIF A. SIDDIQI

This book attempts to wrest the history of space exploration from its normative fetishization of machines, men, and manifest destiny. Our broader goal here is to enlarge the purview of what constitutes the history of space exploration, to foreground those narrative fragments normally consigned to the edges—environmental damage, Indigenous dispossession, infrastructural entanglements, failed pathways, and cultural registers of ambiguity and decline—to the center of space history, a field largely understood as a manifestation of Cold War tensions and subsequently, as a legacy of it. The chapters in this book suggest that those aspects of space history that are often seen as unencumbered by ambiguity and moral opprobrium, such as the landing of humans on the Moon or the use of space technologies to improve our daily lives, cannot be understood without appeal to ambiguous, extractive, and often violent phenomena.

A cursory consideration of the literature on the history of spaceflight shows a subfield concerned primarily with the fetishistic connection between nation and technology, manifest in several generations of scholarship on the Cold War "space race."[1] In part due to the centrality of the Apollo Moon landings in the history of technology (and undoubtedly shaped by the dominance of Thomas P. Hughes's heuristic notion of "technological systems"), the result has been an array of overdetermined narratives about a massive federally funded program that successfully marshaled American science, technology, and capitalism to achieve a singular goal.[2] The Anglophone literature has been dominated by explaining the techno-triumphalism of Apollo while also acknowledging without question that technology was as much an ideational factor enabled by deep-rooted notions of "frontier" as it was a material representation of American exceptionalism.[3] In all these early accounts, space history was unable to elude the vast shadow of Apollo or indeed the frame of the "space race." Critiques of American exceptionalism, techno-fetishism, and the frontier analogy were even rarer, despite (or perhaps because of) their deep historiographical and cultural roots in settler colonialism.[4]

The isolation of space history from the history of science was particularly evident as historians of science and medicine produced a vast canon of literature on colonial (and then, postcolonial) science and medicine in the 1990s and beyond.[5] Space history, for the most part, worked in isolation. As late as 2019, a cultural history of the intellectual genealogy of American views of the space "frontier" avoided any mention that nineteenth century visions of the American West frontier were produced on the effacement of the bodies, lives, and presence of Indigenous peoples in North America.[6] In the past decade, some historians as well as science and technology studies scholars from history-cognate fields such as anthropology, sociology, and geography have enabled a fundamental shift by producing studies on such topics as the making of "place" in scientific discourses of outer space, the use of popular media in making outer space "familiar," and most important, the colonial-like inequalities engendered by space systems on the Earth.[7] While much of this work owes more to social science than to the *history* of science, it has collectively steered space history away from the parochial concerns of space "programs" (e.g., the Apollo program, the Chinese program, the Venus exploration program, etc.) toward a number of conceptual priorities within the history of science, including those related to the human body, waste, infrastructure, social movements, tacit knowledge systems, transnational knowledge networks, cultural representations, philosophical archetypes, and global histories.[8]

Some historians of science and technology have thus moved the discussion of space history beyond the United States, compelled to fill in the web of activity that circled American space activities, whether in Western Europe or the Soviet Union.[9] The considerable work on cosmic enthusiasm in the Soviet context tapped a particularly rich vein of topics, including populist science under socialism, the central role of mysticism in cosmic thinking, and the paradoxes of Soviet science.[10] More recent cultural histories have provided rich insights on space activities, suggesting a kind of coproduction of imagination and engineering, often embodied in the biographies of important technoscientific figures during the Cold War.[11] Attempts to restore forgotten actors into the history of space have resulted in a number of popular works on women and people of color who have contributed to the work of NASA.[12] Now a critical mass of new scholarship seeks a fundamental repositioning of the history of spaceflight from its former mode.

FRAGMENTS

In bringing together some of this new work, as well as introducing other contributions, this book seeks to destabilize the master narrative of space exploration. The organizing heuristic here consists of the "fragments" that result from our efforts to deconstruct and dismantle the received wisdom. Fragments operate in multiple registers in this book. First, these accounts function as fragments at the edges of the stories we have normally been told about spaceflight—forgotten, ignored, or invisible until now, but existing at edge-sites where the unitary narratives no longer apply. Their location at the periphery of the master narrative, however, does not make them unimportant; in fact, as many of the authors explicitly argue, the cosmic fragments presented here are not simply additions and embellishments to the history of space travel but, on the contrary, *central* to it. They reveal unstable, uncomfortable, and unseen processes that were fundamentally symptomatic of the spirit, ideology, and practice of space exploration in all its dimensions in the past half century.

The heuristic of fragments also functions as a referent for a form of modernity, with all its contradictory impulses, emblematic of space exploration. In thinking through the relationship between modernity and fragmentation, the art historian Linda Nochlin, in her book, *The Body in Pieces: The Fragment as a Metaphor for Modernity*, describes "that sense of social, psychological, even metaphysical fragmentation that so seems to mark modern experience—a loss of wholeness, a shattering of connection, a destruction of disintegration of permanent value that is so universally [first] felt in the nineteenth century as to be often identified with modernity itself."[13] Many consider the space program as one of the primary symbolic referents for twentieth-century modernity and its futurist imaginaries; if so, as with the modernist impulse, space exploration too can be characterized by loss of connection, destruction, and disintegration. In one way or another, all the chapters presented in this book suggest this internal friction, between the utopian, upward moving, and positivist imperative of the cosmic imaginary, and the frisson manifested in more earthly dislocations and displacements, as if the optimistic parabolic arc of the rocket as it lifted into the heavens was pulled apart and distended by the stresses and pressures of earthly life. William Tronzo, writes in *Fragment* of "the contrasting modalities of the fragment" in the art historical tradition, but he might as well have been imagining how space exploration has imprinted itself onto the social order, as something that has been simultaneously "received and created, oppress[ed] and liberat[ed], past and future," with each of these registers possessing "the resonance of archetypes."[14]

These considerations, of the fragment as a symptom of modernity and its inherently "contrasting modalities," hover above all the chapters. The contradictions embedded in this most modernist human imperative of the late twentieth century—space exploration—are explored in this book through engagement with two broad historiographical traditions within the history of science and technology, one on the intersection between capitalism and knowledge production; and the other on science, technology, and the environment. These two traditions require some elaboration.

Although capitalism has fundamentally shaped the contours of much twentieth-century science and technology, especially after 1991, the links between science and capitalism remain woefully understudied.[15] While the converse—the relationship between science and socialism, particularly in the Soviet context—has been the subject of considerable work throughout and after the Cold War, in the West, science and technology have usually been understood as operating in a world without ideology—where the political economy is taken as a given rather than an externality (or a special case, as with socialism).[16] In other words, the universalist and normative position on science have often been conflated with mature forms of capitalism in the West, but without identifying any particular ideological foundation.[17] Yet, while state funding was the norm in both the Soviet Union and the United States in the early decades of the space era, Western space activities were very distinctly marked by capitalist relations, through public–private partnerships with a significant role played by large and small aerospace and defense contractors.[18] Long past the Apollo era now, and especially since the end of the Cold War, we find that capitalist modes leveraged on excess, exploitation, and extraction are poised to expand into the cosmos, perhaps in the same way that Lenin suggested that once capitalism had expended its markets at home, it would have to move outward as part of colonial and imperial projects.[19] As Peter Dickens notes, "Capitalism now has the cosmos in its sights, an outside which can be privately or publicly owned, made into a commodity, an entity for which nations and private companies can compete."[20] Many of this book's authors suggest that fragments of the cosmos have already been subordinated to the vicissitudes of capital.

Recent work on the Anthropocene has drawn attention to the possible mapping between the Anthropocene and capitalism—Donna Haraway has coined the term "capitalocene" to denote this juxtaposition—which frames the second broader theme discussed here, of the historical relationship between space exploration and the environment.[21] In a literal sense, the environmental history of space has been concerned with fragments—both the fragments that constitute orbital debris around the Earth and the detritus left behind on the terrestrial landscape. But how do we conceptually relate this cosmic detritus to the technological systems supporting space travel? One of the most important contributions of recent scholarship in environmental history has been to problematize the supposedly discrete "boundaries of social, technological, and environmental things and processes."[22] Extending Hughes's notion of sociotechnical systems, we might think of "envirotechnical" systems, which, in the words of Carl Zimring and Sara Pritchard, "[seek] to capture the ways in which objects, artifacts, and systems are both natural and technological."[23] In furthering our thinking about this interweave between nature and technology in the Canadian North, Edward Jones-Imhotep introduces us to an understanding of the "geography of natural hostility" that causes technological breakdowns, while producing a mutual understanding of the limits of both nature and technology. In the hostile environments beyond our terrestrial atmosphere, we might also think of a similar correspondence between nature and technology where, in Jones-Imhotep's striking turn of phrase, we come to think of nature as "technology made fallible."[24]

The environment—and nature more broadly—is also brought into sharper relief through the sense of "place" that dreams of spaceflight have fostered. For example, scholars have now explored the ways in which images of the full Earth from outer space, in all its beautiful fragility, may have given fuel to the environmental movement in the 1970s and onward.[25] This link between the cosmic and the terrestrial operates in many registers: Valerie Olson, for example, has shown how our conceptions of outer space contribute to the creation of what constitutes the natural and social environments.[26] We also find the converse—that our conception of "place" in outer space has been fundamentally shaped by our experience on Earth. The Earth—and its social, institutional, and environmental settings—remains a persistent and indelible analogue, not simply in the artistic depictions of far-off planetary landscapes but also in the kinds of questions planetary scientists ask about extraterrestrial objects such as exoplanets.[27]

LANDSCAPE, EMPIRE, WASTE, AND DECLINE

Drawing from this literature to illuminate the fragments of the history of the Space Age, the chapters in this book are organized around four themes: landscape, empire, waste, and decline. Each section contains three full-length chapters followed by a single "Fragment"—a smaller text highlighting a discrete empirical example that informs the larger theme of that section. The first section of the proposed volume, "Landscape," introduces readers to three chapters on the relationship between landscape and infrastructure in the history of spaceflight. Each of these chapters invites us to consider the accommodations made—with Indigenous people, with the natural environment, with the local political economy—with the emplacement of advanced infrastructure to support space activities. Anna Reser explores the physical, legal, and symbolic processes of "emptying" the land on which the Kennedy Space Center was constructed in Florida. Through this process, she argues, NASA mobilized a flexible slate of meanings to denote "emptying," meanings that could be turned and tweaked to make the spaceport seem both inevitable and desirable, while simultaneously concealing the various displacements and disruptions necessary to create such a "space place." Her work foregrounds the notion of "tropicality" in the ground infrastructure of space exploration, whereby the natural environment at the Kennedy Space Center can be seen as echoing the visual registers of empire and colonialism. Christine E. Evans and Lars Lundgren explore the history of two rival global satellite networks, the US-led INTELSAT and the Soviet-led Intersputnik, the latter mainly positioned in socialist countries, to highlight how a satellite earth station in Nicaragua contributed to unanticipated and unsettling transnational connections and substantial resistance to its construction. The authors offer four "perspectives" for thinking about the Managua ground earth station—as "layered" infrastructure, as artists' renderings on postage stamps, as part of Soviet efforts to sell their technology in the 1980s, and as sites of espionage. They find it impossible to render such space infrastructure—often ignored as "banal and unworthy of attention"—through clean and linear narratives of national "progress" or international cooperation in space. Instead, we are left with cycles of "repetition and mimesis" and "conflict and erasure," when such technological sites appear and disappear and reappear in larger narratives on the global Cold War. Finally, in the chapter on the construction of the Sriharikota launch-site infrastructure in the Indian state of Andhra Pradesh, I introduce the concept of the "logic of location" to describe the ratiocination behind the siting of technoscientific infrastructure in "empty" landscapes. In the context of Indian modernization in the 1960s and 1970s, this logic of location was starkly defined in terms—mathematical, rigorous, and devoid of ambiguity—suggesting that any opposition to it was either irrational or antimodernist. I show that Indian elite scientists were able to mobilize powerful narratives of national development and the emancipatory power of spaceflight to obscure the terrestrial realities around Sriharikota. The result was the violent displacement of a large number of the Indigenous (Adivasi) population known as Yanadis from Sriharikota, to be replaced by the gleaming monumental architecture of the launchpads as markers of the modern state. In the fragment by Eleanor S. Armstrong that closes out this section, landscape appears in a different register, as a mode to render outer space legible to publics in interactive spaces such as museums. She shows how pastoral themes that invoke "the wilderness, the garden, and the farm," often used in science media to naturalize technoscientific concepts to the general public, are underpinned by "American colonialism and dominance over the environment." She argues that because they reify certain "gendered, racialized, sexualized, and metropolitan" assumptions about space travel, more inclusive "ecocritical" orientations are needed.

The second section, "Empire," consists of three chapters and one fragment that focus on notions of empire. Alice Gorman begins with a contemplation on the centrality of colonialism within discourses of space exploration, both in the past and for the future. Her chapter traverses a vast landscape, both terrestrial and extraterrestrial, from the Woomera range in Australia to other ground space sites in rural Alabama, New Mexico, French Guiana, and ultimately to the Moon. She finds that the "myth of the empty land"—a myth handed down from colonial times—was a sine qua non for progress-oriented narratives of space exploration. This persistent myth is now deployed to support capitalist and extractive economies for mining resources on the Moon, whereby "to exploit the Moon," she argues, "we have to make it strange and unfamiliar,"—that is, erase millennia of cultural associations of the Moon, "relegating them to the realm of superstition." Next, Haris A. Durrani recovers the story of the first geostationary satellite launched by NASA in the 1960s and its role in a web of legal practices (including lawsuits) over what constituted "free space." He shows how lawyers conceptualized the legal geography of the United States in and beyond Earth to assert the expanded boundaries of the American administrative state. This notion was underpinned by a new concept of American extraterritoriality that foregrounded technological control as a basis for conceiving the United States as a new global empire. Ultimately, he argues for the inextricable relationship between empire and capital at the heart of the Space Age. Finally, Nelly Bekus, in her chapter on the afterlife of the Soviet Baikonur Cosmodrome, now in the independent nation-state of Kazakhstan, reminds us of a kind of internal colonization that marked the existence of the Soviet empire during most of the twentieth century. She introduces the concept of the "sacrifice zone" in thinking of Baikonur, where considerable damage to the environment (and thus, to the quality of life) was accepted as the cost of the futuristic Soviet (and later, Russian) space programs. In chronicling the work of anti-space activists in Kazakhstan in post-Soviet times, she finds the collapse of the utopian promise of space travel, now replaced by a deep ambivalence of the real costs of this project to Kazakhstan. In the fragment that closes out this section, Rebecca Charbonneau provides a brief schematic that situates the rhetoric and actors of the Cold War US and Soviet space programs as embodying expansionary and imperial ideologies, often rooted in past myths. She uses the example of the search for extraterrestrial intelligence (SETI) to foreground the ways in which these rhetorics of empire often translated from "abstract scientific theories" to actual practical ramifications, in terms of both the imperial frames that shape our expectations of possible extraterrestrial civilizations and the actual use of SETI equipment for military purposes during the Cold War.

In the third section, "Waste," the authors tackle not only waste but also how waste can be repurposed through reuse and repair. First, Julie Michelle Klinger, in her chapter on physical sites affected by space travel such as mines, launch sites, and asteroids, invites us to rethink how terms such as "wasteland" and the "frontier" are key concepts mobilized to reclassify remote places. As a result, these places are reconfigured as sites of imminent intervention and displacement by space activities underpinned by either profit or militarism. These sites are also rendered ahistorical, she argues, each without a past or an afterlife, existing only in the moment when they become legible to earthly extractive processes. She argues that "because future space travel takes off-Earth mining as a given, theorizing extractive practices on Earth can shed light on evolving engagements with outer space." Next, Réka Patrícia Gál explores the central role of reuse and repair in the history of space exploration, one whose received narrative has been largely obsessed with a fetishization of high technology and innovation. Revisiting an episode in the early history of human spaceflight—NASA's risky repair of the Skylab space station in 1973—she asks us to rethink the history of long-duration human spaceflight as emblematic of the "dual and constructive relationship of repair and innovation within the Anthropocene," where repair was both a resourceful act and a political position to negate the normative cycles of innovation and waste.[28] In the subsequent chapter, Subodhana Wijeyeratne finds a complicated story in postwar Japan, as the state sought to locate an appropriate site from which to launch rockets, abandoning some places and supporting others. In some areas, there was considerable local resistance to the siting of technoscientific infrastructure. In other launch places, such as Uchinoura and Tanegashima, there was "long-term decline of these areas in terms of both economy and population." Yet Wijeyeratne shows that at one abandoned launch station at Michikawa Beach, locals were able to repurpose the old site to function as a symbol of modern Japanese technical accomplishment. The suggestion here is that even in failure, the ruins of the Space Age have unexpected afterlives. In the concluding fragment, Lisa Ruth Rand writes about the ruins of the Space Age, the fragments of spent rockets on the ground, quiet satellites in orbit, and the assorted detritus of normal spacecraft operations on Earth, suggesting that, despite the prevailing reputation of space exploration as one of high-tech accomplishment and exploratory spirit, the Space Age, from the very moment of its arrival with the launch of the Soviet Sputnik satellite, can be equally understood as "moments of breaking," producing ruin and ruination.

In the final section, "Decline," all the chapters, in one way or another, find patterns of ambiguity and decline in episodes typically tied to triumphalism, jingoism, and the higher cause of space enthusiasm. Darina Volf explores the banner Soviet–American project at the height of détente, the Apollo–Soyuz Test Project from 1975, which has often been explored exclusively at the level of Cold War high politics. She recenters the story not on political leaders or astronauts but on the media, which, she argues, served as a central site for social negotiation over the importance of space exploration in both the US and the Soviet Union. Because the project was conceived entirely as a media spectacle—the emplacement of space into the theater of public opinion—when expectations were not met, both sides lost interest in similar future ventures. Apollo–Soyuz thus remained a kind of odd discontinuity in the Cold War rather than the first of many joint projects. In her essay on Hollywood space-themed cinema of the past quarter century, Esther Liberman Cuenca suggests that although the power of American exceptionalism retains a grip on cinema—especially through positivist narratives that valorize the exceptional technical acumen of NASA—recent movies have replaced the powerful nostalgia for the Apollo Moon landings with a certain feel-good liberal internationalism. From the late 1990s onward, she finds a reduced representation of Russia as a partner, which has been replaced by a growing coterie of other nations, such as China and Western Europe, in addition to an extremely heightened sense of the potential of global catastrophe due to the Anthropocene. Her chapter suggests that decentering the American experience masks a form of decline (regarding the reduced role of America) coupled with anxiety (about the increasingly apocalyptic outcomes resulting from human-caused climate change). On the other hand, Natalija Majsova explores the vibrant genre of Russian space films, which, in immediate post-Soviet times, confronted the material realities of a space program in decline with a form of deep irony and revelatory humor, thus highlighting both the inequalities that marked the Soviet space program and its normative collective memory. She shows that the resulting fragmentation of the master narrative of Soviet space exploration was then followed by a new generation of movies infused with neocolonial attempts to foreground uncritical mythmaking of a Great Russia. In the final fragment attached to this section, Alexander C. T. Geppert explores the multiple meanings of the term "space age"—its emergence, periodization, and residue—in late twentieth-century global culture. His careful exegesis of the concept reveals a fissure between the historical concept (often mobilized as an anticipatory metric) and its historiographical use (uncritically used to mark a discrete time period in the present and the past). In thinking about a possible "global" Space Age, Geppert juxtaposes multiple fragmented narratives from different spaces and times to advance a more useful concept of "planetization," which was originally offered in a different context after World War II by the French Jesuit thinker Pierre Teilhard de Chardin.

The book closes with a meditative epilogue by Edward Jones-Imhotep, in which he looks at the long history of space exploration not simply as a record of human endurance—a quest to imprint our histories with a form permanence—but also as an embodiment of entire economies that were expendable, used, and ultimately abandoned. In looking at the fragments of this history, he revisits some of the themes raised in the earlier chapters—emptiness and expendability, for example—and introduces some new ones, such as "estrangement" and "remains," to offer some final thoughts on the notion of that "the history of space activities is a history of things not meant to survive." This is not merely a disinterested academic judgment, he suggests, since "ruins are no more innocent than landscapes"—they have real-world consequences. In that sense, my hope is that future historians continue to reveal other cosmic fragments.

PART I

LANDSCAPE

CHAPTER 1

"LOOSE IN SOME REAL TROPICS"

Environmental Images of Florida's Space Coast

ANNA RESER

When America's "first spaceport" rose out of the palmetto scrub along Florida's Atlantic coast at the beginning of the 1960s, it seemed to many observers that a long-lived and stubborn wilderness had finally been tamed by technology. NASA's Kennedy Space Center (KSC), as it is now known, was to be the centerpiece of a vast spaceflight program that would reshape American culture and foreign policy into an image of total technological dominance. Boldly carved into the "wasteland" of Cape Canaveral, set astride ancient Indian burial mounds and delicate coastal ecologies, the spaceport would propel humanity to unknown cosmic shores, just as Europeans had landed on the same white beaches that now housed rocket gantries. It seemed as improbable as it was satisfying—the sight of white painted rockets rising above what had been just a few years earlier nothing but a mosquito-infested swamp. It was a symbol of progress, of colonial uplift of the most romantic kind, and a bold vision for the future.

This image of NASA's facilities in Florida became an enduring icon of the space age, but it was one constructed in the mien of a much older aesthetic tradition. In Florida, NASA and other observers deployed rhetorical and visual images of the area surrounding the launch facilities at Cape Canaveral that depicted the landscape as primitive, backward, wasted, and inhospitable or even uninhabitable. These environmental images served to naturalize the use of the land and to draw a pleasing aesthetic contrast between the high-technology project of the space program and what observers described as a primordial landscape of swamps, lagoons, estuaries, and marshland. The facilities of NASA, in its own self-image and in the public imagination, became a welcome developmental influence on a troublesome, intransigent landscape that had long resisted civilization. Distorted images of the human history of the area, especially those of Indigenous people and early European settlers, marked KSC as a site with a long history of colonization—a practice that the spaceport would triumphantly take to the stars.

These images generated a highly flexible set of meanings. NASA could, for instance, highlight the pleasures of working in a tropical landscape for its employees, often by emphasizing how women employees looked on the beach, while simultaneously laying claim to a rugged masculine history of struggle against a hostile nature over which technology ultimately prevailed. The Indigenous people of the area, particularly the Ais, could be portrayed at once as primitive squatters on land destined for greater uses, and as the innovative arrow-firing progenitors of the steely-eyed missile man. This flexibility was a productive tool for NASA, boosters, and outside observers who struggled to make meaning of the new and often fantastical space program and who fell back on familiar tropes drawn from the colonial gaze in their descriptions of the new wonders of the space age.

This analysis draws on NASA public affairs documents and the Kennedy Space Center employee newsletter *Spaceport News*, as well as journalistic accounts that reinforced the environmental meanings the agency was making about its new facilities in Florida. Rather than representing a historic break with Florida's long environmental and human history, the construction of Kennedy Space Center was aligned with the colonial imaginary that has shaped images of settlement and displacement in Florida since the first moments of European contact. This legacy of colonization, of forceful displacement and colonial uplift of "wasteland" and primitive landscapes, was a selling point for the US space program, as it was with other colonial space projects in the Americas. Critics of recent human spaceflight initiatives, especially commercial programs, have pointed out the imperialist and militarized aspirations of the aim of settling the solar system. But it is vital to understand that such aspirations and ideological commitments have been at work within America's civilian program since its inception, and the consequences of the colonial orientation and aesthetics of spaceflight have always been realized first here on Earth.

THE CREATION OF THE SPACE COAST

Kennedy Space Center is located on the Atlantic coast of Florida about forty-five miles east of Orlando. The center's main facilities were constructed in the early 1960s on Merritt Island, inside an initial land acquisition of 88,000 acres stretching from New Smyrna Beach in the north to Patrick Air Force Base to the south. The total land area of KSC is separated from the mainland to the west by the Indian River, and Merritt Island and Cape Canaveral are separated by the Banana River.

The region of Florida's east coast surrounding Cape Canaveral is among the oldest sites of Spanish contact in North America, having been encountered and named in the sixteenth century.[1] People had lived there for centuries before, among them the Ais people and their earlier ancestors who left behind burial mounds and shell middens that were ultimately encircled by KSC's large perimeter. Peter John Ferdinando has described the scope of Ais influence in this part of Florida, including their conflict and trade with the Spaniards and other European colonial forces well into the seventeenth century.[2] The Ais in particular are afforded a good deal of attention in NASA's own materials describing and promoting the history of Kennedy Space Center.

By the end of the nineteenth century, European settler families were growing citrus and sugarcane in the area and fishing in the rivers and lagoons. The mainland to the west saw a more significant population increase during this time, as it became a popular hunting and sporting destination, abetted by the arrival of the railroad in 1887.[3] Michele Currie Navakas's study of geography and settlement in Florida proposes that the liquidity of Florida's landscape—physically, metaphorically, and in terms of the region's resistance to European patterns of land use and settlement—marked it out as an especially troublesome part of the American imperial project.[4] Because Florida was mostly unsuited to plantation-style cultivation, because its geography itself was in question for so long (island or peninsula?), and because the Indigenous people of the area fought so fiercely against displacement, Florida became an object of concern for the architects of early America. "People have taken root in Florida for thousands of years," Navakas writes, "despite the fact that Florida's liquid landscape challenges crucial notions of land, space, and boundaries that underlie familiar British and Anglo-American forms and practices of founding."[5] Yet Florida was still considered "wild" and largely empty well into the postbellum period. Florida "had been ravaged repeatedly during three Seminole Wars, it lacked a basic infrastructure of roads and railroads, and it contained millions of acres of unsurveyed land at a time when no other state on the Eastern Seaboard contained any."[6]

In the twentieth century, the US military began building up its homeland defense ahead of its entry into World War II, which included the creation of the Banana River Naval Air Station.[7] Since the late 1940s, the US Air Force had used this area as part of what would be named the Bahamas Long-Range Proving Ground, which allowed for missile testing to be monitored by a series of tracking stations in the Atlantic.[8] The Air Force took over the Banana River Naval Air Station and redesignated it as Patrick Air Force Base in 1950.[9] The first missile launch from Cape Canaveral in 1950 was a modified V-2 rocket, technology recovered from Germany at the end of World War II. Over the course of the 1950s, the Cape was the site of more V-2-adapted missile tests as well as newly developed missiles such as the Army's Redstone and Jupiter missiles. The Air Force's Atlas missiles, first test-fired in 1957, were followed by Titan and Thor.[10]

Despite difficult conditions for cultivation and the strictures of Southern social order in the postbellum period, the area was home to historical communities of African Americans who made their living fishing and growing citrus. Since the 1890s, communities of freed people, like the Laughing Waters estate of the Campbell family that later became the town of Allenhurst, had been protected enclaves where Black people could live and work for themselves, pursue education for their children, and seek out self-determination for their families and communities.[11] Allenhurst was among the communities uprooted by NASA when the agency purchased the land to build its new facilities in the 1960s. Residents were displaced to other nearby communities, where some went on to work for NASA, or left the area altogether.

NASA announced in 1961 that it would be acquiring 88,000 acres of land on Merritt Island to build new facilities, most of which was to be a permanent launch operations installation.[12] In addition to the existing launch complex on the Cape and the new NASA installation to be built on Merritt Island, the major part of the installation would be unused land set aside as an exclusion zone. While some of the land for NASA's new spaceport was purchased from individual landowners by the Army Corps of Engineers, which managed the land acquisition, a good deal of it was acquired by eminent domain. Even when the corps successfully negotiated a sale, at least one family was forced to move multiple times as the scope of the land acquisition changed.[13]

NASA constructed an industrial area that housed administrative and engineering activities, including checkout and assembly of vehicles, and a new launch complex on Merritt Island for the massive Saturn series of rockets that would launch Apollo to the Moon. The site also featured a modernist E-shaped headquarters building, an operations and checkout building with a high bay with an enormous door, and the Vehicle Assembly Building (VAB) in which the stages of Saturn rockets were assembled. Adjacent to the VAB was the Launch Control Center, with its reinforced firing-room windows facing the launch area.

EMPTY, TROPICAL, AND PRIMITIVE

As the center came into existence, NASA deployed this long human and environmental history of Cape Canaveral to create meaning around its new facilities and to generate desirable images of its technological encounter with untamed nature. Among the many themes used by NASA as well as outside observers to generate these images, three stand out as emblematic of the colonial gaze with which the agency's own public relations efforts regarded the new center and its environs. First, the land that would become Kennedy Space Center was consistently characterized as empty, and where it was not empty, it was made so. Second, the location of the new spaceport was distinctly tropical, with the requisite climate and biome as well as many of the social relations embedded in Western ideas of the tropics. And third, the area around Kennedy Space Center was portrayed as primitive and underdeveloped, and the human history of the area was relegated to a deep past where the harms of its colonization could be more easily dismissed. These three meanings—empty, tropical, and primitive—were expressed in visual and rhetorical images meant for internal consumption and meaning making at NASA and in public accounts that shaped Americans' understanding of the new site and its connection to the larger ideals of the space age.

The land acquired for the new center was frequently described as empty, uninhabited—even uninhabitable—wild, virgin, or untamed. This formulation obviously deflects attention from the displacement that NASA's land use required, but it also serves an aesthetic purpose that draws on familiar visual tropes of technology in nature. Emptiness is the prerequisite condition for a landscape to achieve an aesthetic interruption that Leo Marx has argued captures a particularly American way of understanding nature and industrialization.[14] This was new theory in the 1960s, being developed by historians and theorists of technology like Marx who sought to explain how technology functions in American culture and national self-image. In 1965 Bruce Mazlish published an edited volume, commissioned and funded by NASA, that explored the potential social and cultural impact of the space program through analogy to the coming of the railroad.[15] Marx, drawing on his then recently published *The Machine in the Garden*, analyzed the nascent space program through his formulation of the pastoral ideal and offered a set of frameworks for thinking about the relationship between the image of the locomotive in the landscape and the rocket similarly situated. Marx noted that it is the landscape that dictates the affective and symbolic power of the machine in the garden. He wrote that "the affecting object was not so much the machine itself, but rather the spectacle of the machine in the natural landscape."[16] Writing only a few years after construction had started on NASA's new spaceport, Marx argued that it was simply too early to make an accurate assessment of the space program's social and cultural impacts. Rather, he wrote, "It is necessary to discover how collective images of space exploration comport with existing modes of belief."[17] One of these existing modes was the colonial vision of a tropical Florida prime for uplift by the civilizing high-technology forces of the space program.

Consider, for example, the account of NASA's first director of public affairs, Gordon Harris, in a NASA publication, *The Kennedy Space Center Story*. In describing how the space center came to be, Harris was careful to address the aspects of community life that were displaced by NASA's arrival and provide explanations for how the agency was able to replace them or accommodate them. Of the citrus growers, for instance, Harris reported that "within the Federal reservation are 185,000 citrus trees planted on 3,306 acres. The groves were leased back to their former owners by the Government. They care for the trees and harvest the annual crops of fruit. In return for this privilege, they pay annual lease fees to the U.S. Treasury."[18] Harris noted that recreational areas were available for hunting and fishing and that Brevard County maintained a stretch of seashore for public use.[19] However, Harris also reiterated that the land was mostly wild, and he listed the various animals and plants that could be found there. He closed the first chapter of his history with a description that would become commonplace in accounts of the Kennedy Space Center: "This is the unique environment of almost virgin wild land contrasting sharply with Space Age facilities serving the needs of the national program today and in the future."[20] For Harris, as it would be for many others, one of the center's distinctive features was the contrast between the most advanced technology taking root in the most unlikely landscape. But emptiness is not enough for this contrast to be potent, to become iconic. The tropical character of the "virgin wild land" on which the center is built is just as important.

In Harris's chapter 1, "A National Resource," he opened with a description of a space center defined by its unique relationship to the environment and to the history of Florida:

> Uniquely a creation of the Space Age, the Center presents sharp contrasts between its physical setting, early history and the gargantuan engineering

achievements which transformed palmetto scrub, marshland and citrus groves into the first operational Spaceport. Archeologists found traces of human activity before the Christian era, Indian burial mounds and refuse piles of later times, and indications of French and Spanish explorations before the birth of the Republic. Professor Charles Fairbanks of the University of Florida observed that the site was one of the places where Western civilization came to the New World; now it is destined to become the place from which our civilization goes out to other worlds.[21]

In 1963, with the first phase of land acquisition complete and construction well underway, NASA executed an agreement with the US Fish and Wildlife Service to redesignate all of the "empty" land around the spaceport—most of which formed the protective exclusion zone—as a wildlife refuge.[22] This redesignation ensured that the "wilderness" around the center could not be further developed and would therefore remain an integral part of the identity of KSC. It was also a tidy solution for NASA that maintained the emptiness it needed as an exclusion zone, but conveniently "filled" that emptiness with now-desirable wildlife and undeveloped, "pristine" wilderness in need of preservation. A new introduction to the volume printed in the 1980s began with a description of KSC that suggests that the spaceport represented a favorable fusion of nature and technology:

> Located on the east coast of Florida approximately midway between Jacksonville and Miami, the 56,700 hectares (140,000 acres) controlled by the Center represent a melding of technology and nature. Wildlife thrives here, alongside the immense steel-and-concrete structures of the nation's major launch base. KSC is a national wildlife refuge, and part of its coastal area is a national seashore by agreement between the National Aeronautics and Space Administration and the Department of the Interior. Over 200 species of birds live here year-round, and in the colder months large flocks of migratory waterfowl arrive from the North and stay for the winter. Many species of endangered wildlife are native to this area: the Southern bald eagle, dusky seaside sparrow, brown pelican, manatee, peregrine falcon, green sea turtle, and Kemp's Ridley sea turtle.[23]

In this description, KSC was not merely coextensive with the Merritt Island National Wildlife Refuge established in 1963; "KSC *is* a national wildlife refuge." The identification of KSC so closely with the surrounding environment and the specific characterization of nature that catalyzes the association offers an important path toward analysis of the colonial orientation and aesthetics of spaceflight. Once the emptiness of the land had been physically and metaphorically secured, by the removal of residents and then the designation of the wildlife refuge, meaning could be created from the environment developed along familiar pathways carved by NASA's use of the land.

David Arnold's formulation of "tropicality" is helpful in framing these images and discourses about nature at KSC in terms of the values historically assigned to tropical landscapes in colonial contexts. "Tropicality" can be thought of in this context as a particular register of the visuality of the space program and of KSC in particular, one that it shares with the visuality of empire. While explicit accounts of displacement, dislocation, and removal are prevalent in my sources, this colonial register is often more subtly conveyed by specific environmental conventions of representation that frame Cape Canaveral and the surrounding area as tropical. Arnold writes: "Calling a part of the globe 'the tropics' (or by some equivalent term, such as the 'equatorial region' or 'torrid zone') became, over the centuries, a Western way of defining something culturally alien, as well as environmentally distinctive, from Europe (especially northern Europe) and other parts of the temperate zone. The tropics existed only in mental juxtaposition to something else—the perceived normality of the temperate lands. Tropicality was the experience of northern whites moving into an alien world—alien in climate, vegetation, people and disease."[24] Descriptions of the space center often featured accounts of the tropical flora and fauna, complaints about mosquitoes and heat, and ruminations on the wildness of the surrounding landscape. Emptiness, primitivity, and tropicality are not exclusive to accounts of the Florida site, and in fact comprise a consistent pattern of representation of space ground installations in the Western Hemisphere. Geographical constraints privilege sites for rocket launches with access to an equatorial orbit, meaning that many launch sites share similar climates and biomes.

Sean Mitchell's study of a failed satellite launch facility and space program in Brazil similarly grapples with the way that space projects introduce and sharpen inequality and challenge the beneficent ideal with which space projects are pitched to their various publics. In Brazil, planners' visions of competing with other nations on the international stage of spaceflight in the 1980s were met with unexpected resistance from people living near the launch site. When the people of Alcântara were displaced from their ancestral homes and their self-sustaining lifeways, in violation of constitutional mandates designed to protect them, they organized to oppose further impoverishment that came with development of the site.[25]

The same patterns of representation found in Florida trouble the meanings of spaceflight infrastructure wherever it confronts tropicality.

The anthropologist Peter Redfield has also considered the question of tropicality in relation to the history of spaceflight in Kourou, French Guiana, where the European Space Agency currently operates a vast infrastructure to launch rockets. The surface installations of space programs there are an explicitly colonial project, Redfield argues, to make a troublesome colony valuable. Located near the equator and along a coast facing open sea to the east, like the Florida installation, Kourou became a desirable location for a spaceport even though this same geography partly accounted for earlier failures by the French to develop the colony.[26] Redfield identifies an environmental narrative similar to those written about how KSC made the wild land of the Cape useful, noting that "wilderness can have its uses, even for high technology. Or, more pointedly, space technology did not erase wilderness but found parts of it useful once it was properly redefined."[27] Redfield suggests that this particular redefinition of wilderness is specific to space technologies. He writes about a "technological irony of rocketry: the more remote a location, the better suited it is for explosive experiments. Thus, when seeking to leave the globe, wasteland becomes valuable, and underdevelopment can appear a virtue. The same tropics that in the nineteenth century bore a sinister reputation for disease and disrepair beckon a key technology of the twentieth century."[28] As with the colonial project in Guiana, the technological redefinition of wilderness is central to the identity and meaning of the Florida site that would become the new spaceport. These mobile environmental conditions were used by NASA and by outside observers to construct a unique identity for KSC among the palmetto and lagoons, one that envisioned spaceflight technology as a miraculous counter to a wild landscape and a civilizing force in a region that has often resisted assimilation.

The historian Neil Mahr argues that "in many ways it was nature that first lured NASA to the Cape." The geography of the region provided a favorable orientation to orbit for launches; waterways provided transport for massive rocket components; and the Atlantic Ocean provided a safe zone for the splashdown of falling vehicles.[29] And despite the frictions NASA experienced with the nascent environmental movement, nature became a core part of the identity of American spaceflight as a result. Accounts of the environment are a staple of the center's internal employee newspaper *Spaceport News*. First published in December 1962, *Spaceport News* provided NASA employees with useful news and updates about work and life at the Cape.

The sixth issue of *Spaceport News* contained a short piece about a prank played on new employees that involved convincing them that the Cape lighthouse was a rocket with a very long countdown. The piece gave a brief history of the lighthouse and the Cape, emphasizing the emptiness and bleakness of a landscape as yet untouched by the space program: "The only features to break the total isolation of the present Cape area in 1868 were a few scattered houses on the north beach and a pier and old hotel on the south shore. Clouds of mosquitos and horseflies swarmed over an area inhabited mainly by snakes, scorpions and the occasional alligator which came waddling across from the Banana River."[30] Later that month the paper interviewed an employee who talked to *Spaceport News* about what it was like working on the Cape before NASA began building the center. The rocket program veteran described a battle with an undeveloped landscape, saying that "the only buildings on the Cape then were Central Control and a cafeteria."[31] There were, in fact, many buildings on the Cape before NASA arrived, including private homes and businesses that were seized, destroyed, or repurposed by the agency as NASA facilities, with space for astronaut classrooms.[32]

Another feature from that same spring returned to the nineteenth-century history of the area, detailing the family history of a NASA employee whose family settled on the East Coast of Florida in 1883. Mosquitoes, vectors of tropical disease, featured heavily in remembered descriptions of the land and environment that older family members related to *Spaceport News*. "We used to say," the employee's father recalled in the interview, "that when mosquitoes were out, you could strain a pint cup through the air and catch a quart of them."[33] Snakes were also a memorable part of living in the area, according to the same interviewee, who confirmed that "they were all over the place! It's a wonder to me more people weren't bit. But nobody paid much attention to them."[34] Not mentioned in this tale of settler grit are the Black communities that also called the Cape area home until the establishment of the center. This is in keeping with the general tone of the newspaper as a whole, which reported through the 1960s almost exclusively on the activities of white people and occasionally of the many white women who worked for NASA.

These environmental tropes were used to describe KSC well into the Apollo program. In 1968, the October issue of *Spaceport News* looked back on the preceding decade of spaceflight as the center prepared for the launch of Apollo 7, the first crewed Apollo launch. The issue included a condensed history of KSC titled "KSC Story: From Marshland to Spaceport."[35] Using language and style similar to Harris in *The Kennedy Space Center* Story, the piece opened with a proclamation that "what is now KSC was virtually semi-wilderness when Pioneer I, the first U.S. deep space probe was launched from Cape Kennedy on October 11, 1958," and that "the former virgin lowlands adjacent to Cape

Kennedy became the nation's first operational Spaceport."[36] The same issue featured a story about the early history of rocket launches from the Cape before NASA was even formed. It employed the same language to describe the area, stating that "the Cape was still an untamed spit of land when the first Redstone missile cut a smoky trail through the sky on August 20, 1953."[37] The primitive, untamed, virgin past of the environment was not a fixed point in time but an infinitely mobile environmental condition that was conjured to heighten the sublimity of high-technology activities.

Spaceport News also helped create environmental meanings for the center on a much smaller scale. For instance, the paper reported on the "gator-in-residence" at KSC in 1969, which was "one of two placed in the pond in front of the Headquarters Building by the U.S. Fish and Wildlife Service about a month ago as part of the program to restore its natural ecology."[38] By running a contest to name the alligator and calling it spaceport personnel's "unofficial pet," the paper mobilized the wildlife of the area in the formation of the image of the tropical spaceport.

A piece from March 1969 detailed the landscaping projects undertaken around various buildings and facilities on the site. The piece was illustrated with a photograph of the employee Gail Richards in front of the Visitor Information Center and featured a characteristically pin-up-style caption that read "Tropical Plants and Gail Richards . . . add beauty to the Visitors Information Center at KSC."[39] The writer detailed some of the native plants that were used to decorate the spaceport, noting that "to the northern newcomer, they may appear to be just some more of the kookie Florida vegetation but they're plants that have been proven for the area and can be recommended for home landscaping projects. For that's where many of them stood originally—around the private homes dotted around the property bought by NASA for KSC."[40] The knowing aside about "northern newcomers" helped create a sense of unity for KSC employees that was explicitly about living and working in what was perceived as a unique or "kookie" environment, in addition to marking the site as Southern. The *Spaceport News* piece gave more detail about the physical processes of acquiring these properties and their plants than most accounts of land acquisition either from the 1960s or later historical accounts: "They were gently dug out of their former homes and moved to a holding area off the Kennedy Parkway and held for use in such projects as the VIC [Visitor Information Center], saving much in landscaping expense." The plants were then placed in front of the Visitor Information Center, the central hub for public access to the spaceport, signaling the center's enmeshment in the landscape and all that had been accomplished in taming it.

While tropicality was often a negative attribute, denoting a hot climate or tropical disease that NASA and its projects sought to mitigate, NASA also harnessed it as a lure for employees and an aesthetic attraction for visitors and supporters. Henry Knight Lozano's studies of tropicality in the southern United States show that promoters and developers in Florida preferred to characterize the state as "*tropical* rather than *southern*," and that Black Floridians forced into menial service roles after Reconstruction were leveraged by boosters to heighten the sense of nostalgia for a simpler colonial past that elites saw as a relief from the frenetic modernity of the Gilded Age.[41] The aim was to create a fantasy of the tropics that was easily accessible to American elites—all the pleasures of a tropical climate and slow pace of life, without a taxing visit to the colonies. The image making surrounding Kennedy Space Center in the 1960s owes a great deal to this imagining of Florida.

The plant relocation piece, featuring a woman employee alongside the plants as a beautifying influence on the site, also points to another aspect of the tropicalization of the site. Nicole Cox has written about the way that Florida development boosters deployed gendered imagery of the tropics to encourage relocation to a beautiful and Edenic tropical zone where the flora was as beautiful as the women. Advertising images centered visions of feminine and natural beauty as a special feature of Florida. These images often explicitly "used the presence of Seminole Indians in Florida to market the state's environment and women as exotic and mythic."[42] Tropicality was invoked through the image of a racialized feminine other. *Spaceport News* deployed similarly gendered tactics, but with its mostly white women workers. Gail Richards posing with the relocated plants was one of the less explicit pinup images that the paper used in its run. Women workers in bikinis splashing in the surf, published for no reason other than to titillate male readers and mark out women workers as workplace attractions, were common.

Observers drew on the presence of a racialized other, typically the Indigenous people of the distant past, to evoke primitivity in tandem with the hallmarks of tropicality.[43] Mark David Spence has demonstrated how the creation of the most prominent national parks in the United States relied on emptying the "wilderness" of Indigenous people in the name of preserving the natural environment. National parks, he argues, do not protect or delineate "remnants of a priori Nature," but "enshrine recently dispossessed landscapes."[44] When NASA and the federal government agreed in 1963 to create the Merritt Island National Wildlife Refuge, they transformed a landscape with a rich history of centuries of human life, recently emptied of its inhabitants, into a protected "wilderness." Furthermore, KSC's Public Relations Office and Visitor's Center created and used images of the antiquity of the spaceport and of the Indigenous people who lived in the area until the nineteenth century to naturalize the presence of the spaceport in the "wilderness" of the coast by situating KSC at the apex of technological progress.

The racialized other in the case of KSC was not the citrus farmers who were displaced in the 1960s, but the Indigenous people of the site's more distant past. The capture of the cultural significance and the physical remains of the Ais and other Indigenous tribes by NASA when it took control of the land heightened the sense that its presence was part of a long history of progressive colonization and the technological apex of the various peoples who had inhabited the land before first contact. The meaning of the Kennedy Space Center site drew on a complex set of interlocking histories that naturalized the new spaceflight infrastructure. In this sense, the histories that are not deployed, like that of Allenhurst and the Black communities in the area, are as telling as those that are. The agency generally ignored the immediate past of the site by reaching back to a deeper past to flavor its own history. The colonial imagination of the tropics is completed by an imagining of a native people suited to the climate and environment of the tropics in ways that white settlers and visitors were not. At KSC, the Ais and other groups were variously portrayed as the primitive natives of an American tropics and as the ancestors of the technological heritage that NASA claimed to inherit.

Spaceport News published a number of items about the archaeological sites on KSC property and their investigation by anthropologists and archaeologists as a matter of interest for employees. A piece from 1968 began, "The first missiles—with chipped flint nosecones—were launched from the land now owned by KSC some 3,000 years ago by primitive Indians."[45] The writer naturalized the presence of such a high-tech endeavor in a wild landscape by placing the spaceport in a progressive history of technology from the arrowhead to the rocket. In describing the period from 800 to 1,000 AD, the writer claimed that "from the Spaceport south was one of the few areas in the world where people maintained a fairly civilized standard of living," suggesting that KSC had a history of innovation, into which the spaceport rightly fit.

The writer referred to "the natives of the Spaceport area," the Ais who descended from the people who made the middens and burial mounds that archaeologists were studying on the spaceport site, as though they were the ones who had settled on NASA land, not the other way around.[46] The paper wove a very short but brutal history of the Ais in which they failed to be civilized by contact with the Spanish, and were ultimately wiped out in conflict with other tribes. The precontact ancestors of the Ais were described in archaeological terms, with descriptions of what they left behind paraphrased from an archaeological report about an excavation in 1963.

The subject of this report, a site called Ross Hammock, was brought to the attention of KSC staff by the Florida Anthropological Society (FAS) in the 1960s.[47] The archaeological report on the site, furnished to NASA and the National Park Service, documented the value of the site and made recommendations for its preservation. Fieldwork on the shore of the Intercoastal Waterway revealed "a complex of two very large sand burial mounds and a fairly extensive, but not extremely large, shell midden village area."[48] Among the excavated objects were pot sherds, patterned with "check stamp" indentation; human bones including skulls; shell beads; vessels; and tools made from shells and stone. The report concluded that the site contained valuable data that should be protected by NASA and studied.[49]

In *Spaceport News*, the archaeological sites near the spaceport as well as the antiquity of the area were meant to be consumed by employees as an interesting feature of their workplace. Within the agency, however, they presented a challenge to NASA's control of the land. In 1964, the Department of the Interior forwarded correspondence about the FAS's interest in preserving the site to Harris, noting that "we have no idea as to whether the proposal of the Society to study and develop the site fits into your program of land use."[50]

The Ross Hammock site is located near the northern boundary of KSC on the mainland above Mosquito Lagoon. NASA's facilities, then nearing completion, were clustered on Merritt Island and the Cape about twenty miles south. Thus, it was unlikely that NASA would need to utilize the site for anything other than the exclusion zone, but NASA's program of land use was firmly focused on the future. In response to the FAS's campaign to preserve the site, the agency drew up a set of restrictions that made it clear that while the site should be studied, NASA reserved all rights to the land for its own purposes, in view of uncertainty about future land use needs.

The restrictions called for NASA's right to "construct such roads, buildings or other facilities of a permanent or temporary nature, and perform other such work on or across lands within the area covered by the permit as KSC may from time to time determine to be necessary or desirable in the interest of the United States."[51] This meant that NASA reserved the right to designate the site as an important archaeological find in need of preservation or as a site for activities of national importance at its own discretion.

NASA's control over the archaeology sites allowed it to capture the cultural meaning that could be gleaned from them. The restrictions on the permit indicated that any items that had been recovered would be handed over to the Florida State Museum but only *after* they were made available to KSC for "temporary or permanent retention and public display in the Visitor Information Center or other repository at KSC."[52] The restrictions also called for periodic updates to be made to the Public Affairs Office so that KSC could use that information for education and outreach. KSC also reserved the right to photograph the site and distribute images as it saw fit. This area was a resource for NASA in the multiple cultural and media products that could be extracted from the site.

NASA's control over sites like Ross Hammock allowed the agency to dictate the terms under which they were integrated into the public image of KSC and to create a narrative about the past that naturalized the presence of the new spaceport within the tropical landscape. The archaeological sites were, like the endemic wildlife, presented as unique features of an already extraordinary place and ones that placed the space center in a lengthy imagined lineage of technological development. The agency's actions were structured by the colonial attitude to the tropics as resource-bearing land, shrouded in the difficulty and danger of the tropics, and marked by the presence of racialized others who were relegated to an imagined primitive past.

We have seen how NASA fashioned its own identity for the Kennedy Space Center site by leveraging colonial tropes about the tropics and distributing those meanings in internal documentation of its own activities. But it is also worth noting, if briefly, how these meanings were relayed to NASA's larger publics by contemporary media accounts. The transport of these images of emptiness, tropicality, and primitivity into wider public discourse helped to cement the center's image and played a part in the continuing association of spaceflight with the colonial project.

Among the more widely read accounts of the environment around KSC were those written by outside observers, who were sent to document the momentous events taking place on the Florida coast. Norman Mailer's account of Apollo 11, first serialized in *Life Magazine* and appearing later as *Of A Fire on the Moon*, introduced readers to the spaceport area in explicit contrast to the more sterile confines of the Manned Spacecraft Center in Houston, which more closely resembled a suburban engineering campus than the tropical installation of the spaceport.[53] To Mailer's mind, the Florida coast was much more suited to the surreality of the space program precisely because the environment contrasted so sharply with the high-tech work of KSC. Finally "loose in some real tropics," he observed, "it is country beaten by the wind and water, not dissimilar to Hatteras, Chincoteague and the National Seashore on Cape Cod, unspectacular country, uninhabited by men in normal times and normal occupations, for there are few trees and only occasional palms as ravaged and scabby as the matted backside of a monkey, a flat land of heat and water and birds. . . . It is country for hunting, for fishing, and for men who seek mosquitoes; it was next to uninhabited before the war. Now, first Spaceport—think on it! first *Spaceport*."[54] The spectacle of the rocket in the marshland captured something essential about the space program that for Mailer had no equal in the terrazzo-floored offices of the Manned Spacecraft Center.

Tom Wolfe, in his novelization of the history of Project Mercury, *The Right Stuff*, took up the subject of the site of KSC in a similar fashion. He observed that the land on which the spaceport was built was the same kind of land as Edwards Air Force Base in the Mojave Desert, where many of the first astronauts had come up as jet test pilots. Of the beach in Florida, Wolfe wrote: "It was one of those bleached, sandy, bare-boned stretches where the land that any sane man wants runs out . . . and the government takes it over for the testing of hot and dangerous machines, and the kings of the resulting rat-shack kingdom are those who test them."[55] In understanding that the land around the Cape was similar to other "empty" spaces that the government used for the building and testing of "hot and dangerous machines," Wolfe represented the Cape as a literal evolutionary backwater, describing it as "the sort of hopeless stone boondock spit where the vertebrates give up and the slugs and the No See'um bugs take over."[56] Wolfe also cast the landscape as primitive, prehistoric, backward, and at fundamental aesthetic odds with the high technology activity that was taking it over in the early 1960s.

The environmental images in these accounts were not merely literary devices. They employ the same language as the stories NASA itself used to fashion a new identity for its tropical facilities. This imagery became the norm for describing Kennedy Space Center and was also present in journalistic accounts of KSC. In 1964 the *New York Times* published an article titled "Visit to the Three Cape Kennedys" by Robert Whalen. The "Three Capes" that Whalen referred to are "the launch area, where the space story up to now has unfolded"; the new facilities that NASA was purpose-building for its own activities; and the communities in the surrounding area. Whalen expressed some disbelief that activities like the construction of the world's largest building and the assembly and launch of rockets to the moon would take place in "an improbable set-ting of sand, water and scrub growth."[57] Earlier that same year, another journalist described the construction projects on the site as "the free world's greatest rocket center . . . rising rapidly on once-useless Florida swampland to support America's boldest adventure into space."[58] These descriptions of the Cape posit that the area was empty—of actual human life, or of what the writers consider valuable human activity—and thus suited to being appropriated by NASA for its high-tech purposes.

While the specific technology of spaceflight, and perhaps the scale and speed of the development were new and even shocking to Americans, the playbook for this kind of land use had already been written. Across tiny islands turned into mile-wide craters by nuclear tests, on the pocked soil of proving grounds at home and abroad, on the grounds of military installations erected on territories and protectorates across the globe, the United States had long since proved its power to displace and dislocate in the name of technological progress.

BOCA CHICA AND BEYOND

Playing up the contrast between the high-technology activities of NASA and the "primitive" environment surrounding KSC was a representational strategy that was incorporated into the spaceport's identity and readily taken up by influential observers whose accounts were widely read. NASA's use of land in Florida generated meanings about spaceflight that were connected not to some larger cosmic purpose but to the very tangible, earthly concerns of the environment, resources, people's homes, and modes and legacies of colonization and displacement.

Tropicality remains a fixture in the way spaceflight projects are conceived and executed. The long-awaited launch in 2021 of the James Webb Telescope from the Kourou spaceport in French Guiana was heralded by NASA's TV commentator Rob Navias as "liftoff from a tropical rainforest to the edge of time itself."[59] Tropicality is part of the aesthetic and affective register of spaceflight.

Now, in the era of nascent commercial spaceflight, the question of the orientation of space projects to the land they use and consume remains crucial. Geographies of spaceflight still require launch sites near the equator, and the effect of the vision of tropicality remains a threat to the equitable development of space technologies. As long as Westerners still conceive of nontemperate locales as primitive, backward, and marked by racialized others—and they do—the land use practices of empires can continue to be used and naturalized. This process is already underway in Texas, where SpaceX is replicating the patterns of displacement and disruption that characterized the construction of KSC's spaceport.[60]

Amid the heady talk of establishing human settlements on Mars and extending humanity's reach into the solar system, critics have raised strong objections to structuring future space initiatives within the framework of colonization. But this framework also animates the much closer work of spaceflight here on earth. The logic of colonization, the tools and images of empire, are already fundamental to the terrestrial infrastructures of spaceflight projects. To set humanity's sights on Mars is also to turn a conquering gaze on the land here on Earth that will be needed to make the dream of space colonization a reality.

CHAPTER 2

FRAGMENTED HISTORIES OF A SATELLITE COMMUNICATIONS EARTH STATION

CHRISTINE E. EVANS AND LARS LUNDGREN

In the summer of 1986 Salman Rushdie accepted an invitation to tour Nicaragua as the guest of the Sandinista government, on the occasion of the seventh anniversary of the revolutionary seizure of power from the Somoza dictatorship.[1] On July 17, the revolutionary anniversary, Rushdie found himself accompanying the Nicaraguan president Daniel Ortega to a ceremony that celebrated not the revolutionary events themselves, but the opening of a new satellite communications Earth station that would link Nicaragua to the Soviet-led Intersputnik satellite network.[2] Intersputnik, a multilateral international organization to facilitate global satellite communications, was created by the USSR and its socialist bloc allies in response to the US-led creation, in 1964, of the International Telecommunications Satellite Organization, or Intelsat, which was expanding rapidly around the world in the late 1960s and 1970s.[3] In this context of global rivalry and overlapping networks, the Intersputnik earth station in Managua was immediately the subject of conflicting interpretations, as Rushdie observed. "We arrived at the dish antenna," Rushdie recounted, "which sat in the Managua hills not far from the wooden FSLN [Sandinista National Liberation Front] sign, and listened to speeches from Soviet dignitaries. The new installation had been paid for by the USSR," Rushdie noted, "and the US was already calling it a spy base." Rushdie, a sympathetic, if critical, observer of Sandinista-era Nicaragua, was skeptical of these US claims.⁴ "It looked," he noted drily, "like a telephone system to me."⁵ Rushdie's comments about the Earth station's opening and the response it drew from US observers suggests at least two points—first, that what observers saw in the construction outside Managua of a new satellite antenna was highly dependent on their point of view, and second, that communications infrastructure—even when it involves hypermodern space technology—is frequently experienced or interpreted by those who use it as banal and unworthy of attention.

The Managua Earth station, located on a forested hillside above a volcanic lake, the Laguna de Nejapa, has evolved considerably over time, serving different purposes and being understood in different ways. The historical evidence available for investigating this space infrastructural installation is itself fragmented and compromised from the beginning by the divided perspectives of the creators of that evidence. As a contested installation reshaped by the overlapping incursions of sometimes violent and always self-interested superpowers, a great deal of the evidence available to us about Nicaragua's satellite Earth station was produced by outsiders. That evidence is characterized by its erasure of local experiences and fractured by the conflicts that led to the construction, reconstruction, and redefinition of that Earth station over the course of several decades up to the present. We thus employ the term "perspectives" here intentionally. Like a drawing that uses perspective to represent a three-dimensional object in two dimensions, these sources depict the Earth station from a series of fixed and distant points—the views of American and Soviet diplomats and telecommunications officials, as well as technical experts for capitalist and socialist world communications satellite corporations. The perspectives of Nicaraguan politicians appear occasionally—those of nonelite Nicaraguans, almost never in the sources we were able to access from our own distant and limited position.

At the same time, the banality of Nicaragua's satellite communications Earth station—its status as "just a telephone system"—and its failure to reflect clear-cut Cold War boundaries also makes it "unthinkable," to borrow Michel-Rolph Trouillot's term. As Trouillot has written, "When reality does not coincide with deeply held beliefs," humans impose "interpretations that force reality within the scope of these beliefs."⁶ By failing to be either a significant site for espionage or a material embodiment of a binary Cold War alliance, despite its construction in the midst of revolution and Cold War proxy conflict, the Managua satellite Earth station is largely invisible in conventional historical narratives. It becomes briefly thinkable and visible again and appears in a handful of fragmented sources and newspaper accounts, only at moments like an opening ceremony, when it briefly seems able to affirm existing beliefs and stories. Here, we have assembled some of these fragments of evidence about the Managua Earth station's life as "technology-in-use" and considered what they can tell us about quotidian space technologies that fall well outside of our Cold War expectations.[7] Taken together, these fragments suggest when and why satellite Earth stations become visible, and to whom. These sources present the Managua Earth station as a technical object celebrated at opening ceremonies and on postage stamps, as an achievement—American or Soviet—in the Cold War competition to sell aerospace technology globally, and as a site of espionage.

We conclude with our own argument, limited by our specific position, that the Managua Earth station suggests how the layering of new infrastructures atop old ones can unsettle our assumptions about how the Cold War shaped global communications infrastructures. On the hillside outside Managua, this layering is quite literal: the Soviet and Nicaraguan officials at the Intersputnik station opening ceremony in July 1986 had failed to mention that the new Earth station had been installed directly next to another satellite communications Earth station, one that had been linked up to the US-led satellite Intelsat network since 1972.[8] While historical sources that capture the layered infrastructures of the Managua Earth station are fragmented and partial, they nonetheless reveal how entangled, even indistinguishable, ostensibly rival Cold War space technology networks could be on the ground, far from Washington and Moscow.

THE VIEW FROM THE FIRST OPENING CEREMONY

In a telegram of June 1, 1973, the US ambassador to Nicaragua, Turner Shelton, reported on the official inauguration, on May 27, of the first satellite communications Earth station in Central America. This new Intelsat satellite Earth station allowed Nicaragua's state telecommunications agency to send and receive telephone, telex, and television traffic to and from the US-led network's Atlantic satellites.[9] Ambassador Shelton's terse description of the ceremony noted the participation of Nicaragua's National Governing Council along with the former (and future) president, General Anastasio Somoza Debayle. Somoza Debayle was present in his capacity as director of the General Office of Telecommunications and Postal Services; he had stepped down as Nicaraguan president in 1972, but effectively remained in power. Also attending the inauguration was the president of Communications Satellite Corporation (COMSAT), John Johnson, as well as the Japanese ambassador to Nicaragua. During the inauguration, Ambassador Shelton delivered a congratulatory message from President Nixon.[10] Shelton noted that the inauguration ceremony was broadcast on television throughout Nicaragua and pointed out that the ceremony itself was somewhat belated. The station had actually begun operation in early December 1972, just weeks before a devastating earthquake on December 23 destroyed large parts of Managua. In light of the devastation of the city and the need to provide emergency services, the formal inauguration was delayed until May 27, 1973.[11] However, Ambassador Shelton noted, the new Earth station had not been damaged in the earthquake and "served as a vital link with the outside world during a critical emergency period."[12]

During the wave of Earth station openings and Intelsat membership expansion around the world in the early 1970s, US officials were eager to emphasize that each new Earth station was the product of a strong bilateral relationship between the US and the country operating the new Earth station. As new Intelsat-certified Earth stations began to open, the standard rhetoric that accompanied their opening emphasized the power of communications to bring two countries together.[13] Shelton's telegram, however, gives no indication of such rhetoric, despite the fact that Shelton was the subject of criticism within the State Department for his uncritical embrace of the dictatorial Somoza Debayle.[14] Indeed, as noted above, the Japanese ambassador to Nicaragua also attended the opening ceremony, presumably in part because two Japanese firms, Mitsubishi Shoji Kaisha Limited and Mitsui and Company Incorporated, had built the Earth station.[15]

Instead, the first and longest section of Shelton's report focused on the novel joint venture, NICATELSAT, between COMSAT, the US corporation that served as the US representative within Intelsat, and the Nicaraguan Telecommunications Corporation, or TELCOR. NICATELSAT represented a significant innovation for both COMSAT and the Somoza regime in Nicaragua, reformulating a relationship of direct, monopolistic exploitation by US telecommunications firms to one that allowed the Nicaraguan government more control while still ensuring a significant return on COMSAT's investment. The dominant player in early telecommunications infrastructure in Nicaragua was a US subsidiary of AT&T, which had a complete and highly profitable monopoly on all telephone and telegraph service in Nicaragua by the early twentieth century.[16] By the 1960s, the Nicaraguan state had shifted, like many other developing countries, toward seeking greater state control of telecommunications systems, due to both strategic considerations and dissatisfaction with the limited service provided by foreign firms. As would be the case with the creation of NICATELSAT, telephone and other communications technology and infrastructure were obtained from foreign firms via loans and barter but were owned and operated by Nicaragua's state telecommunications ministry.[17]

Under the terms of the joint venture, COMSAT would provide the capital needed for construction of the Earth station, which would then be operated by NICATELSAT. The government of Nicaragua was to be the majority owner, with TELCOR owning 51 percent of the shares in NICATELSAT and COMSAT owning 49 percent. COMSAT's investment was secured by mortgage bonds and stocks in the corporation, which TELCOR was required to pay off first with any profits from the Earth station's operation. After COMSAT's investment had been repaid with interest, it would continue to enjoy 49 percent of the profits for at least five years.[18] COMSAT also retained significant control over NICATELSAT's conclusion of contracts and selection of leadership.[19]

From Shelton's perspective in June 1973, the Managua satellite Earth station represented an achievement of US diplomacy, which could, he believed, successfully marry the financial interests of COMSAT to the economic and infrastructural development needs of the Nicaraguan government. The Earth station had connected Nicaragua to the outside world at a moment of crisis and had done so via a novel joint-venture arrangement that gave Nicaragua an Earth station and COMSAT a guaranteed profit on its investment for at least a decade. Not apparent in this upbeat report was the turbulent economic and political context. In the wake of the December 23, 1972, earthquake, which killed between five thousand and ten thousand people and reduced much of Managua's built infrastructure to rubble, Somoza Debayle, ruling through puppets and in his capacity as head of the national guard, oversaw the theft of millions in international relief aid and used the crisis to seize land and businesses for his family. As David Johnson Lee has argued, the failed reconstruction of the city after the earthquake helped catalyze resistance to the Somoza dictatorship.[20] From the perspective of Ambassador Shelton in reporting on its first opening ceremony, however, the Managua Earth station appears as a success story about US foreign policy and economic assistance to Nicaragua. Shelton's depiction of the promise of infrastructural modernization (a satellite Earth station that kept working through a devastating earthquake) and the successful new way to preserve US corporate profits within Nicaragua's telecommunications sector offered his readers in the State Department a picture of the Managua Intelsat Earth station as a stable, forward-looking point in Nicaragua's shaken landscape.

THE VIEW FROM A POSTAGE STAMP

What happens to an Earth station after it is "inaugurated" and when the political context in which it was created is no more? Despite the well-worn claim, first articulated by Susan Leigh Star, that infrastructures are "by definition invisible" and only "become visible on breakdown," in fact, as Brian Larkin has argued, infrastructures can be invisible or spectacular, or anywhere between those two poles.[21] We might add that they can also move in and out of visibility over time. The moment of highest visibility is the moment of construction of new technical objects, and finding images of satellite ground stations constructed *after* the first years of this new technology can be quite difficult. Google Earth offers the most immediate access to viewing these infrastructures from orbit, though the decision to engage in this kind of archeology from orbit both offers new ways to study these objects and problematically enmeshes the researcher in global surveillance networks.[22] Before widespread access to satellite images of the whole globe, however, there were images of Earth stations nestled within another media system that circles the globe: postage stamps.[23]

Somewhat unexpectedly, a set of stamps issued in 1981 offers a link across the conjuncture of the 1979 Sandinista revolution. In 1981, two years after the Sandinista victory, the government of Nicaragua issued a series of stamps featuring communications satellite technology. The stamps were likely designed by, or at least in collaboration with Intelsat, since each included Intelsat's logo prominently on the stamp.[24] These stamps, one of which depicts the Managua Earth station, demonstrate that, on some occasions, Intelsat's presence was actively promoted by the Sandinista-era state. In this instance, the history of the Managua Earth station across significant Cold War events appears entangled rather than fragmented and divided.

The celebration of space technology, including communications satellite technology, as a symbol of modernization was widespread in the early decades of space exploration and communications. In the 1960s–1980s, national telecommunications agencies routinely issued stamps celebrating the opening of a satellite communications Earth station and their country's corresponding entry into one of two global satellite communications networks, Intelsat and Intersputnik.[25] These stamps tended to focus very narrowly on the new Earth station itself and its futuristic modern architecture, often set in a characteristic national landscape—barren hills in Iceland, deserts in Iraq, and, most frequently, lush tropical scenes for countries around the equator. These stamps follow an established pattern in space technology imagery that stresses the

2.1. Postage stamp, Nicaragua, 1981. Personal collection of the author.

contrast between the hypermodern technical object and the "tropicality" of the site. While the landscapes depicted in Earth station stamps were not always tropical, they were always empty and wild. This pattern often served, as Anna Reser shows in her chapter on the construction of the Kennedy Space Center, to present the sites of space technology installations as primitive and empty and thus available for, even in need of, transformation via the construction of space infrastructure, concealing the fact that these sites were already home to both human residents and delicate ecosystems.[26]

The Nicaraguan stamp series issued during the early years of the Sandinista era celebrated space technology and the satellite age in ways that reflected the complex geopolitical relationships that space technology created there. This stamp series may already have been in production before the Sandinista revolution, reflecting the fact that media and infrastructural changes are often out of sync with political ones. The decision to still issue this series highlights Nicaraguan officials' own agency in refusing to be limited in their foreign policy choices by US–Soviet competition. Like politicians in other nonaligned countries, Nicaraguan officials sought to forge ties with Intersputnik and the Soviet Union while still not fully abandoning ongoing infrastructural ties with Intelsat. In the case of Nicaragua and many other postcolonial countries, the presence of the Intelsat logo on the stamp may also have served to emphasize the internationalism of satellite communications institutions, creating a shield against charges of allowing former colonial powers to meddle in their internal affairs. The rhetoric of scientific internationalism was useful not only for superpowers concealing their Cold War machinations but also for postcolonial governments seeking to avoid the appearance of giving up hard-won sovereignty.

Most stamps in the series feature Intelsat satellites against a starry sky, juxtaposing the technical object with the remote, wild landscape of space. One stamp in the series, however, does depict the Intelsat Earth station outside Managua. This stamp fits neatly within the visual vocabulary, outlined above, that was typical of many satellite Earth station stamps in this period. We see the Earth station with its large antenna, set on a forested hillside in the lower half of the stamp, with a closeup of the antenna against the backdrop of the Nicaraguan flag in the right-upper quadrant. In the left-upper quadrant, an Intelsat logo takes the place of a satellite, with its signal represented abstractly by colored beams (see fig. 2.1).

While this Intelsat stamp's depiction of the Managua Earth station's forest setting is not entirely inaccurate, based on its similarity to contemporary photographs of the Earth station on google maps, its decontextualized appearance serves to obscure, if not erase entirely, several salient facts about the Earth station's location, including its position not on some remote stretch of one of Nicaragua's two ocean coasts, but above a small lake in the middle of populous, urban Managua. The image's perspective also positions the viewer above the Earth station, as if they are seeing it from a surveillance airplane, or perhaps a position somewhat higher up the opposite hill. One such building, on the hill directly across the Laguna de Nejapa from the Earth station, is the US Embassy. Like most other Earth station stamps, Nicaragua's does not contain any reference to the process of construction of the Earth station, or the web of international political and monetary exchanges that the construction of this new technical object entailed.[27] In other words, while the stamp highlighted Nicaragua's continued integration into the Intelsat network even during Sandinista rule, at the same time, it revealed only a fragment of the Earth station's complex history.

THE VIEW FROM THE SOVIET MINISTRY OF COMMUNICATIONS

As the 1981 Intelsat stamp series demonstrates, Nicaragua did not end its relationship with Intelsat after the 1979 Sandinista revolution. Instead, it simply added another ground station and large satellite antenna—a Soviet TNA-77 like those that had been installed in Intersputnik Earth stations in countries like Laos, Vietnam, and Afghanistan in the preceding years—right alongside the existing Intelsat station. This new satellite dish would point at the Soviet/Intersputnik telecommunications satellite over the Atlantic. What were ostensibly separate Cold War satellite communications networks, corresponding to rival Cold War blocs, were in fact deeply entangled on the ground in Cold War hotspots like Managua and were characterized as much by cooperation and coexistence and as they were by competition, suspicion, and hostility.

Two years before the Earth station's opening, in 1984, experts from the Soviet Ministry of Communications evaluated the plans for this new Intersputnik station outside Managua. The plans, the group of evaluators determined, "deserved an outstanding grade."[28] What made this project outstanding, the report suggested, was not its quite modest capacity, which was limited to twenty-four phone channels (with the possibility of expansion to sixty in the future) and the ability to receive one black-and-white or color TV channel (with the possibility to distribute television broadcasts from Nicaragua in the future).[29] Instead, what was remarkable about the new Earth station was its supremely efficient manufacturing and design—the whole Earth station was largely prefabricated and packed into just six shipping containers, for easy and inexpensive assembly on-site.

Perhaps the greatest evidence of this project's cost efficiency, however, was its "maximal" reliance on the infrastructure already in place for another Earth station located just 150 meters away—the Intelsat earth station in place since 1972. Since the Intelsat station was nearby, the new Intersputnik Earth station could employ not only the existing station's electrical power source and backup generator but also the personnel already working at the Intelsat station, and even the employee break room.[30]

Accompanying the report was a hand-drawn diagram of the planned Earth station, depicting two identical Earth stations nestled on the same hillside, with arrows and an angle indicating their antennae's respective orientations toward the Intelsat and Intersputnik satellites over the Atlantic.[31] This sketch reduced the installation of an Earth station to its barest features. It depicts the topography of the hills descending into the basin of the Laguna de Nejapa—both the hill with the Earth station and the opposite slope appear—in order to illustrate how each antenna will access the sky.[32] A sketch like this could be made of many different hillsides, anywhere on the globe.

Looking forward to the successful installation of the Intersputnik antenna and control equipment alongside the existing Intelsat facility, Soviet telecommunications officials saw a bright future for Soviet leadership and the generation of revenue in the work of installing lower-cost satellite communications technology around the developing world. The handful of documents relating to Nicaragua, in the archive of the Ministry of Communications in Moscow, erase the people and landscape of the Laguna de Nejapa—with the exception of the convenient presence of an existing Earth station—reducing the construction of an Intersputnik station to an achievement of low-cost, prefab architecture that could be replicated in other, vastly different places. This focus on profit and expansion into global markets from a socialist country might seem surprising; as historians have shown, however, the Soviet state consistently sought hard currency revenue from a variety of forms of international business activity and trade, even as it publicly justified those activities as profoundly different from their profit-seeking, capitalist counterparts.[33]

This prefab approach to satellite communications Earth station architecture was relatively new. Intersputnik Earth stations were typically constructed on-site from components made chiefly in the USSR and installed by teams of Soviet engineers. The standardized nature of Earth station construction was reflected in a commemorative volume published by the Special Construction Bureau of the Moscow Energetics Institute (OKB MEI), the scientific institute responsible for design, construction, and installation of satellite antennae and Earth stations for Intersputnik.[34] In this volume, the Managua Earth station appears only as a place name in a long list of installation destinations, designed to convey the prestige and wide travels of the engineers profiled in the commemorative volume.[35]

This was not always the approach taken by Soviet technicians and architects abroad. The socialist-world architects who designed and helped build housing and landmark public buildings across the Global South in the 1970s and 1980s took a more original and customized approach to each site. Those projects, as Lukasz Stanek and others have documented, involved a great deal of collaboration with local officials and architects and resulted in extensive customization and adaptation of materials to local climatic conditions and cultural preferences.[36]

Only a single mention of the experience of Earth station installation trips appears in the OKB MEI volume, as part of the biography of S. I. Dorn, an administrator who was, his biography notes, especially good at working around the rules to organize "a 'collective-farm-like' everyday existence in the place of arrival. That meant," the author continued, "sending, together with the equipment, some amount of everyday necessities and food supplies. At that time," the author noted, "that kind of work was officially forbidden, but in reality was practiced widely everywhere."[37] It seems unlikely that this pleasurable memory of bending rules and surviving on imported provisions in remote and undeveloped settings actually fits the case of the Managua Earth station, which was in a city and next to an existing Earth station, which had facilities including an employee break room. In Soviet antennae engineers' commemorative texts, however, the reality of work in urban settings and with local technical professionals was replaced by a generalized story of building space technology in supposedly undeveloped wilderness.

If the Soviet Ministry of Communications' view of the Managua Earth station does not tell us much about the people who worked there and lived nearby, it does show us the energy with which the USSR was pursuing global profits from its space technology sector, supporting recent work that highlights Soviet participation in the processes of capitalist globalization.[38] Other documents in the Ministry of Communications' archives suggest that Soviet officials were having fifteen to twenty meetings a month with multinational aerospace corporations and other capitalist world actors by the 1980s.[39] As Artemy Kalinovsky, James Mark, and Steffi Marung point out, "The idea of Western capitalism as the only engine of globalization [has] bequeathed a distorted view of socialist and postcolonial states as inward looking, isolated, and cut off from global trends."[40] However invisible Nicaraguans' own interests and responses remain in the Soviet archives, the Intersputnik and Intelsat stations collocated in the Managua Earth station remind us that Soviet and Nicaraguan citizens were active participants in the expansion of global flows of signal and money.

Indeed, Nicaraguan officials' views on the significance of the new Intersputnik station *were* expressed at the station's opening ceremony, as described already in our introduction. The Nicaraguan president Daniel Ortega and the Soviet minister of communications, V. A. Shamshin, a newly elected member of the Soviet Communist Party Central Committee and a radio electronics specialist who had helped build the Soviet domestic satellite television network, gave speeches. From the perspective of the Soviet participants in the inauguration ceremony, new signal flows connecting socialist-world and other "progressive" countries were the main occasion for celebrating the new Earth station.[41] An article in *Pravda* published two days later, on July 19, described Ortega's remarks, which emphasized something slightly different: unlimited, global access, albeit facilitated by Soviet largesse. "This is a historic event in the life of the Nicaraguan people," Ortega said. "Our country has received access to the Intersputnik system," he continued, "which allows us to expand our communications with the whole world."[42] Ortega also used the new satellite connection to make a phone call to the Nicaraguan Embassy in Moscow, an action that reinforced the ways in which the new satellite Earth station would extend the Nicaraguan government's own global reach.[43]

Ortega's remarks, as recorded in *Pravda*, understandably did not mention Nicaragua's continued membership in the Intelsat network explicitly, but they present the Intersputnik station as expanding Nicaragua's global connections, rather than displacing previous connections and networks. In some sense this is not remarkable: Nicaragua, with its collocated and infrastructurally integrated Intelsat and Intersputnik stations, was not the first country to operate Earth stations belonging to both systems. Indeed, the USSR itself had operated Intelsat stations since 1973.[44] But the opening of Managua's Intersputnik station did mark a turning point: the end of US and Intelsat efforts to prevent Intelsat members from belonging to both systems. While both the USSR and Cuba were users of Intelsat stations, they were not Intelsat members, unlike Nicaragua. A year before the opening of the Managua Intersputnik station, in October 1985, Intelsat's board of directors passed measures to equalize the status of members and nonmember users like the USSR and to approve requests from Nicaragua, Iraq, and Algeria to use the Intersputnik system while retaining Intelsat membership.[45] The Soviet press covered this agreement with Intelsat as a victory for the Soviet Union's "Star Peace" program, designed to contrast with the US "Star Wars" and founded on "equal, mutually advantageous cooperation on a global basis."[46] Perhaps it was that, but it was also an achievement for nonaligned countries wishing to avoid having to choose one Cold War satellite communications network over another.

UNAUTHORIZED VIEWS AND ESPIONAGE

In addition to the contested meanings of the new Intersputnik Earth station's activities at the opening ceremonies, other, secret signal flows were taking shape around and above this new network node. On the evening of the inauguration of the Intersputnik Earth station in Managua, Captain Ricardo Wheelock, the chief of intelligence of the Sandinista Popular Army and the brother of Jaime Wheelock, a top Sandinista general, appeared on Radio Sandino to give an intelligence report concerning US activities in the country.[47] The broadcast outlined in great detail the large number of espionage flights carried out by the CIA in Nicaraguan airspace. Wheelock informed his listeners that the US SR-135 aircraft had "carried out 47 flights in the past six months," and that the aircraft "gathers . . . [like] a vacuum cleaner, all the electromagnetic spectrum of Nicaragua." Spy planes overflying the Earth station could listen to radio broadcasts and telephone calls, and intercept telexes and all other communications using the electromagnetic spectrum. "It permits them, practically on a daily basis, to have an X-ray of the deployment of the troops of the Popular Sandinista Army," Wheelock complained. "And," he continued, "obviously that X-ray of the deployment of our troops is passed on" to the Contras, a US-backed and -funded right-wing insurgency against the Sandinista government.[48]

In addition to monitoring the electromagnetic spectrum, the US used U-2 and TR-1 aircraft to photograph important installations including not only the Intersputnik Earth station but also ports and airports. Illustrating the high resolution capacity of these photography operations, Wheelock went on to explain that "if I get out of this building while one of these U-2 aircraft passes overhead to observe the Intelsat satellite, excuse me, I mean the Intersputnik communications center, the picture taken by this aircraft's cameras can tell whether I have shaved or not."[49] Here, as Wheelock's slip of the tongue suggests, what mattered was not whether the Laguna de Nejapa Earth station was the Intelsat or the Intersputnik station (since it was, from that day forward, both). The really significant signal flows from this network node were the Nicaraguan communications being intercepted by US spy planes.

The use of aircraft for capturing radio communications or taking high-resolution photographs also serves as a reminder of the many ways intelligence operations could be undertaken, all without requiring any dual use of the Laguna de Nejapa Earth station's equipment.[50] Indeed, as whistleblowers and intelligence reporters have demonstrated, for both political and economic espionage purposes, a long-standing US and UK signal intelligence program, known by the code name ECHELON, has collected and monitored *all* the state, corporate, and other private data flowing through Intelsat satellites around the globe.[51] ECHELON, however, has its own, dedicated Earth stations to capture and analyze Intelsat and other international satellite traffic.[52]

Despite pervasive US signal intelligence collection, decontextualized allegations that the Laguna de Nejapa Earth station could be used for Russian espionage continued to circulate long after the station's opening ceremony. In 2017 Laureano Ortega, one of President Ortega's sons, cut the ribbon to open the first Central American node in the Russian Global Navigation Satellite System (GLONASS) on the same hillside above the Laguna de Nejapa.[53] Just as, three decades earlier, Daniel Ortega had welcomed the inauguration of the new Intersputnik station as a "new page in history," the new GLONASS station was greeted by Nicaraguan officials as an infrastructural promise to modernize maritime transportation, fight drug trafficking, prevent natural disasters, and monitor climate change.[54] News coverage by BBC Mundo and the *Washington Post*, by contrast, echoed the allegations of dual use that Salman Rushdie had noted at the 1986 inauguration: "But is it also," journalists asked, "an intelligence base intended to surveil the Americans?"[55]

The reporting on the GLONASS inauguration did not mention the longer history of the site, but reader comments made the connection. Responding to a Cuban online news article about the opening of the GLONASS station, a Cuban doctor who lived across the lake from the ground station in the 1980s while working at the Lenin Fonseca Hospital nearby, pointed out that the station had been there at least since the late 1980s. From their rented house, the Cuban doctor recalled, he and his roommates could see both the Intelsat and Intersputnik antennae. The doctor and his friends lost their view of the Managua Earth station, however, after a wild party during which another Cuban doctor fired a gun at the Earth station and they were all evicted the next day.[56]

At the time of writing this chapter, anxieties about the Managua Earth station had resurfaced again on the eve of the Russian invasion of Ukraine on February 24, 2022.[57] The satellite Earth station above the Laguna de Nejapa has once again become visible as part of a larger story of Russian military aggression. For even if the Managua GLONASS station is engaged in dual-use activities—which is certainly likely—the past layers of satellite communications infrastructure on which it has been built and the many other forms of signal intelligence collection that take place around it remain invisible.

THE MANAGUA EARTH STATION AS LAYERED INFRASTRUCTURE

A wave of new research, including much that is gathered in this book, has emphasized the importance of investigating the specific sites in which space activity takes place on Earth—in the range of the places from which humans access and propel technical objects into space, to the more quotidian ground stations and communications and surveillance network "nodes" where space activity is tracked and messages sent, received, and intercepted.[58] These sites provide the opportunity to access a more nuanced, dynamic, and detailed picture of the histories and materialities of space exploration as well as of global communication networks.[59]

Much of this work, however, has focused on the initial selection and construction of these sites, emphasizing the entangled colonial and postcolonial political logics, technical requirements, and processes of violence, racism, and erasure that underlay their initial acquisition and construction. This emphasis makes sense both because of the moral significance of documenting those foundational moments of infrastructural violence—often as part of a reassertion of violent control by former colonial powers in their former colonies—and because of the emphasis, throughout the historiography of human space activity, on various "firsts." At the same time, however, scholars of media infrastructures have demonstrated that most new media infrastructures are built along the traces and pathways of older ones—conforming to a pattern that Lisa Parks and Nicole Starosielski describe as the "layering of an emergent system upon an existing one."[60] If these earthly sites for space activity are to serve, as Asif Siddiqi has argued, as a "possible entry point into rethinking the normative history of space exploration," we must consider not only the initial creation of these sites but also their ongoing existence and significance as nodes within larger, layered and changing networks of space, media, and technical infrastructures.[61]

Nicaragua's international satellite communications Earth station is certainly an example of this kind of infrastructural layering. With its two rival Earth stations located right next to one another and sharing fundamental infrastructures, the Earth station at the Laguna de Nejapa reminds us of the way that space ground infrastructures can make visible the novel geographies, relationships, and aspirations produced by the encounter between, on the one hand, local scientists and technical workers, government officials, and publics in the postcolonial world and, on the other, competing capitalist and socialist world technical specialists and officials who worked with them to plan, build, and utilize space infrastructures around the globe.[62] As Elizabeth Banks, Robyn d'Avignon, and Asif Siddiqi point out in their elaboration of what they call the "African-Soviet Modern," the entangled relationships and imaginaries generated by transnational technical and political relationships between the socialist and postcolonial worlds of the 1960s–1980s offer "a mode of thinking that resists the pull of teleology, one that frames this particular historical moment unencumbered by the knowledge that the Soviet Union collapsed and that a socialist pan-Africanist worldview no longer animates its future gaze."[63] This mode of thinking without recourse to teleology is especially evident in the case of the Managua Earth station, where material manifestations of past and ongoing relationships with socialist and capitalist superpowers overlap one another, are fundamentally entangled, and continue to coexist long after the historical ruptures of both 1979 and 1989–1991. Indeed, as Banks, d'Avignon, and Siddiqi likewise point out, existing national or regional chronologies, focused, for example, on the Cold War or decolonization, lack the explanatory power to make sense of the layered history of this specific space infrastructural site.

As we argue above, researching a network node like the Managua Earth station poses epistemological and methodological problems for the historian that are as layered and complex as this infrastructural node itself. Vanishingly few of the available sources depict the Managua Earth station in all its layered complexity.

As Siddiqi has argued, studying the history of science and technology raises problems "of power and agency . . . in postcolonial contexts lacking a substantive ethnographic or evidentiary record," especially from the formerly colonized side.[64] While remaining cognizant of the significant absences in this record and limitations on our own ability to access what may be available, what can we learn by assembling these fragmented and often unsatisfying glimpses of this single satellite communications Earth station? First, we see evidence that, from the 1970s onward, socialist and nonaligned countries like Nicaragua were as active in their pursuit of globalization in the space industry as the United States was by the 1970s and 1980s. If we wish, as Alexander C. T. Geppert suggests, to understand the contribution of space activity, and satellites in particular, to globalization in the 1970s, we need to include socialist bloc and nonaligned countries in our analysis.[65] Second, the consistency of the suspicions and allegations of dual use and espionage over the years, likely strengthened by the military heritage of space technology, gives us a sense of how Cold War stereotypes are reinforced and superimposed over other possible pasts. The infrastructural layering, to return to Parks and Starosielski's concept, and simple coexistence of Intelsat and Intersputnik antennae and ground equipment in a single Earth station is another form of "dual use," rendered invisible by recurring charges of espionage.

Our exploration of the history of this single Earth station is inevitably incomplete, limited by our own status as outsiders and the limits on research travel during a pandemic. Yet the partial and fragmented sources we have been able to access reveal through lines of infrastructural layering and entanglement as well as erasure and ongoing contestation. Brought together, these sources offer us multiple, often incommensurate, perspectives on a single satellite communications Earth station. These different ways of looking at an Earth station help us to see the history of human space activity on Earth as a cyclical story of repetition and mimesis, conflict and erasure, rather than as a triumphant series of "firsts" organized into a story of national progress or bilateral "international co-operation in space." We hope that this chapter has made the complicated and ambivalent nature of space technological sites on Earth somewhat more thinkable.

"PRACTICALLY NO HABITATION"

The Logic of Location and Indian Space Dreams
on the Bay of Bengal

ASIF A. SIDDIQI

On June 30, 2014, a little over a month after taking over his new post, the Indian prime minister Narendra Modi visited the principal Indian space launch site at Sriharikota Island on the eastern coast of India on the Bay of Bengal. Here, sprawled across an area of nearly 150 square kilometers was a vast network of advanced infrastructure, monumental in scale, including two giant launch pads, assembly and test facilities, tracking antennae, manufacturing and storage areas for rocket fuel, weather stations, and office buildings. The site had been under construction since about 1970, and Indian scientists had launched modest satellites into orbit at a regular pace from Sriharikota since the 1980s, as the island became the hub of a large and richly funded state program of space activities.

On this summer day in 2014, the Indian Space Research Organization (ISRO) was about to launch its reliable Polar Satellite Launch Vehicle (PSLV) rocket on a paid contract to deliver a French remote sensing satellite into orbit. As one of Modi's first public acts as prime minister, the visit underscored that the Indian space program would be a priority for the new government. After the launch went off successfully, in a surprising and rare move, Modi spoke to the press briefly in English instead of Hindi, in a gesture to ISRO scientists who were largely from non-Hindi-speaking states in South India. But speaking in English was also a signal to the world beyond India. In an expansive mood, Modi called upon Indian space scientists to develop a satellite "that we can dedicate to our neighborhood as a gift from India."[1] The word "neighborhood" was undoubtedly chosen carefully to convey a sense of the shared space India occupies in South Asia and beyond, but it held a particular resonance for many in the actual neighborhood of Sriharikota who had just experienced the violence of police action.

As contemporary reports noted, several days before Modi's visit, villagers from the Yanadi Indigenous community who lived around Sriharikota Island, were "rounded up as a protective measure for the dignitaries."[2] In anticipation of Modi's arrival at Sriharikota, the local Marine Police "launched combing operations along the 169-kilometer long coastal line."[3] As a result, the police "remov[ed] villagers from their communities . . . [and] then occupied the villages. The police also began removing Yanadi who live in the forests surrounding the island, taking them to shelters in Sullurpeta, a town about 16 [kilometers] . . . inland from the spaceport."[4] A journalist added, "The tribals will not be allowed to go back to their houses/villages till [Modi's departure]."[5] As if to add insult to injury, a week before their forced removal, the Yanadi people had met with local politicians campaigning in elections, pleading for help with an array of problems, including lack of water for irrigation and drinking, and absence of durable housing. Instead of help from the government, they had been pulled out of their houses and put in "detention camps" by the thousands.[6] Briefly in the news during Modi's visit, the Yanadi people soon disappeared from media encomia for the Indian space program, continuing more than four decades of erasures at the physical, social, and epistemic level. They barely appear, for example, in any account or history of the Indian space program.[7]

Instead, the dominant narrative of the Indian space program has been one of self-reliance, unbridled confidence, and unmitigated success, all couched in revolutionary terms. In 2013, K. Radhakrishna, ISRO's then chief, noted that "a great revolution . . . has taken place over these last 50 years in the country by a meagre expenditure that has been put into the space programme."[8] The material accomplishments of the Indian space program have indeed been extremely impressive: India remains one of only about a dozen countries with the capability to regularly launch satellites into space. And while most of its work is directed toward modest goals such as communications and Earth observation, it has also managed to send small probes to the Moon and Mars, impressive achievements that have brought global attention to its highly talented engineering force.[9] Occasional failures have brought attention to the space program from the broader media but, even then, hard questions about the return on investment are rarely raised in this context because the space program functions as an obvious source of pride for many who see it as a metric of India's arrival as a global technological power. And although satellites provide important data, for example, on India's national resources, a recent economic study found no conclusive evidence that indicated a positive "overall social rate of return for the Indian [space] programme."[10] In other words, while the space program undoubtedly represents a site of national pride, its material return measured by financial investments is harder to quantify. Modi's tenure as prime minister, meanwhile, has ensured the space program's alignment with a more nationalist agenda. In a perfect alignment between science, religion, and the cult of Modi, for example, in early 2021, ISRO launched into space, with great fanfare, both a photograph of Modi and a copy of the Bhagavad Gita.[11]

The launch site of Sriharikota, the most vivid terrestrial monument to India's space program is the subject of this chapter. I track its origins but reframe the story with the fortunes of the Yanadi people at its center. My story begins in the late 1960s and early 1970s when, to make way for a new space center in the coastal state of Andhra Pradesh, the government of India forcibly removed and resettled about ten thousand people from Sriharikota, some of whom belonging to the Yanadi people, one of roughly seven hundred Adivasi Indigenous communities—or "scheduled tribes" in the legal lexicon of the Indian bureaucratic state.[12] The forcible displacement of the Yanadi community enacted profound and permanent disruptions to their lifeways and political economies, resulting in deep precarity that is still felt more than fifty years later.

The chapter draws on Dennis Rodgers and Bruce O'Neill's notion of "infrastructural violence," originally employed to describe "how structural forms of violence often flow through material infrastructural forms" in everyday life in modern cities. In reappropriating the term here, I use it to describe social displacements imposed on marginal populations by technoscientific infrastructure in modernizing states.[13] Infrastructures, of course, produce many kinds of reconfigurations of the social, material, and natural worlds, but *scientific* infrastructure mobilizes positivist imaginaries in a distinctive manner. Scientific infrastructure—such as laboratories, telescopes, space launch sites, seismic stations, remote research stations, and so on—are naturalized as modernizing forces in locales across the globe, even as the knowledge that they produce usually have no immediate social benefit to local communities, but in fact, as I show here, have often resulted in enormous displacements. Spaceflight infrastructure on the ground but facing upward into the heavens, are coded, in Peter Redfield's evocative formulation, as "placeless space[s] of modernity"—antiseptic, standardized, and absent any markers that might provide a sense of location.[14]

The placement of such infrastructure, usually in the tropics and often in spaces denoted "empty" by the forces of modernization followed from a rationalizing framework, a "logic of location." Here, the logic can contain multiple, overlapping, often contradictory rationales, but it usually functions as a rhetoric of demarcation, dividing what is relevant from what is not. In siting technoscientific infrastructure—especially related to sensing or accessing the world(s) beyond the terrestrial—scientific and geographical rationales are always "relevant," legible, modern, and most often, obdurate. From this reasoning, these "scientific" rationales cannot be changed or imagined to be different since they are drawn from nature itself. Other relevant considerations are mutable and malleable—such as political arrangements with local authorities. Most important, the logic of location casts the aspirations and voices of local populations as less "relevant" because their predicaments are always seen as amenable to alteration through disenfranchisement, displacement, or deracination. The logic of location naturalizes these exclusionary practices by rendering opposition to them as being antimodern, ahistorical, and against the greater good.[15]

This logic, invoked by the US, the Soviet Union, and Western European countries during the Cold War, justified placing important space infrastructure in former colonial spaces under the guise of scientific imperatives. Collectively, these countries placed over two dozen stations across an incredibly diverse array of geographies to support their own space activities, including the Apollo missions to the Moon.[16] In the case of India, the violent logic of location was free, at least in literal historical terms, from the burden of colonial violence. This was not a case of Western science seeding technological systems into the remaining fragments of empire. We find here a more complicated story of elite Indian scientists who, having secured a national commitment for space research from the government, put into action a series of scientific projects that echoed colonial practices, but now within the borders of India.

The plight of the Yanadi people was not uncommon: developmentalist-minded policies of modernization adopted by India and other postcolonial countries frequently went hand in hand with state-directed and capitalist modes of extraction, especially during the Cold War. In this new setting, the modern scientific nation was the beneficiary of science but only by excluding some from the protections it promised to grant all its citizens. Citizenship required a commitment to modern science that, as we will see, exceeded any attachment to the guarantee of an undisturbed home. I argue that the violence of displacement, dislocation, and damage were not appendixes to the history of spaceflight activities but uniformly encoded into its entire history. Such violence functioned squarely within the modernist aspirations of individual states, the international scientific community, and often, ordinary people, who were all activated by the desires and promise that space exploration invoked.

INDIA IN SPACE

As with much of what has been written about the history of Indian science and technology, the history of the Indian space program has typically been framed as a story of "great men" in the noble pursuit of some higher ideals. Working under the shackles of colonial rule, their biographies are presented as surrogates for the voyage of the nation itself, a teleological journey from colonial subordination to anticolonial sensibility to independent action. The lives of the principal figures in the birth of the Indian space program, Homi Bhabha and Vikram Sarabhai, follow this formulation. Both Bhabha and Sarabhai came from extremely wealthy families who were well-connected to the economic, scientific, and intellectual elites of colonial India. Both attended Cambridge University although they had different interests within physics, with Bhabha being the theorist and Sarabhai the experimenter. After independence, both men populated multiple organizational strata in India spanning the government, the industrial sector, the burgeoning scientific establishment, and most important, prestigious international scientific networks that included some of the most important physicists of the time such as Paul Dirac, John Wheeler, Bruno Rossi, and Albert Einstein.[17]

Both Bhabha and Sarabhai were close to Jawaharlal Nehru, the first prime minister of India, a fact that undoubtedly helped prioritize their particular visions of science and technology within the discourse of national development. From the 1950s on, their prolific writings on the role of science and technology in India betrayed a deeply technocratic vision typical of many prominent Indian scientists of the period. In 1958, for example, Sarabhai noted, "Our national goals involve leapfrogging from a state of economic backwardness and social disabilities—attempting to achieve in a few decades a change which has historically taken centuries in other lands. This involves innovation at all levels."[18] This innovation would focus not only on those industries that already had a home in India but also on technologies that emerged in the postwar era as harbingers of Western modernity, such as nuclear power, media, and space.[19]

Space was not an obvious priority for newly independent India, but scientific elites such as Bhabha and Sarabhai employed two rhetorical strategies to justify the space program in its early days, rationales that have remained sutured to both popular and academic discussions on the place of space in modern India. These included the principle of self-reliance ("indigenous" was and continues to be the word of choice, ironic given its later resonance) and, more important, the potential for space activities to be directed to the most urgent areas of national development, including education, poverty alleviation, weather forecasting, and agricultural planning. In many ways, the coupling of these narratives—that is, an indigenous space and only for national development—produced a powerful effect that shielded the relatively expensive space program from potential critiques leveled by those who saw spending on space as frivolous given the range of other problems in postcolonial India. Who would oppose a national initiative identified with the cutting edge of science that could rank India with the leading advanced countries in the world, especially if those investments were made to support the creativity of local talent and directed to alleviating many of the problems of development?

In practice, the emerging Indian space program was less homegrown than coproduced, drawing from a range of practices and artifacts from countries such as the United States, France, the Soviet Union, Japan, and West Germany that circulated across global networks.[20] Space as a mode of national development also helped distinguish India's space program as an alternative to both the Soviet Union and the United States, aligning perfectly with India's nonaligned status. Sarabhai would make much of the fact that India was not interested in self-aggrandizing space "spectaculars" where space was a battleground for an ideological competition. Instead, Indians would use space as an instrument for national development.

Bhabha's and Sarabhai's technological enthusiasm did not emerge out of a vacuum but was enabled and justified by a number of Western thinkers including Walter Rostow, the Harvard economist and author of *The Stages of Economic Growth: A Non-Communist Manifesto*, published in 1960, just on the cusp of the formation of the Indian space program.[21] Rostow's belief in the proper path of development—linear, with a discrete point of "takeoff," and one measured against a normative Western model—was deeply influential, as modernization theory extended its grips into international philanthropic and economic institutions, many of which, such as the Ford and Rockefeller Foundations, had firm links with the South Asian elite.[22] Accepting without question the portability of technology, Bhabha noted late in his life that "what the developed countries have and the underdeveloped lack is modern science and an economy based on modern technology."[23]

With these ideas in mind, Bhabha and Sarabhai pushed through the creation of the institutional seed of the Indian space program, a small office called the Indian National Committee for Space Research (INCOSPAR) within India's Atomic Energy Commission (which Bhabha headed).[24] At the apogee of their respective careers in the early 1960s, the two men orchestrated an agreement with NASA to launch American rockets from a remote fishing village near Trivandrum (now Thiruvananthapuram) in the southern Indian state of Kerala. These modest rockets, later augmented by ones from France and the Soviet Union, were not powerful enough to deliver satellites into orbit but they did train a generation of young Indian scientists in a panoply of scientific and engineering disciplines.[25] Bhabha's unexpected death in a plane crash in 1966 did little to deter the momentum of the Indian space program. In 1969 Sarabhai engineered the formation of the ISRO, which laid out a ten-year plan of development of space technology, one that rejected the space exploration model of the superpowers in favor of fostering indigenous technological competence and services useful to India's population at large, such as agriculture and communications.[26] To do that, he imagined an idealized scientific citizen of India. In his article "Space Activity for Developing Countries," he noted, "Clearly the development of a nation is intimately linked with the understanding and application of science and technology. . . . History has demonstrated that the real social and economic fruits of technology go to those who apply them through understanding. Therefore, a significant number of citizens of every developing country must understand the ways of modern science and of the technology that flows from it."[27] Here we see his articulation of the ideal scientific subject who would internalize the language and grammar of space exploration as an obligatory participant in Indian economic development.

By the late 1960s, Sarabhai, with the sympathetic ear of Prime Minister Indira Gandhi, began planning a significant expansion of India's still seedling efforts. In a major report to the government of India, he suggested that the "most important task" was the need to "develop indigenously a satellite launch capability" so that India would not have to depend on others to carry out its space activities.[28] To launch the rocket, however, Indian scientists needed a launch site. Since the early 1960s, as I have noted, Indian space scientists had already operated a small site on the coast of Kerala in South India, which, while effective for small rockets, was not suitable for launching satellites into orbit. It had reached its limit with these small rockets, not for reasons of lack of infrastructure or problems of bureaucracy, but because of its precise location in relation to broader geographical, celestial, and environmental constraints, a kind of "logic of location" that determined precise sites ideal for space launches.

THE (VIOLENT) LOGIC OF LOCATION

Long before the beginning of space activities, scientists had known that launching a rocket in an easterly direction would grant significant benefits since the rocket's velocity could benefit from the eastward rotation of the Earth itself. With this additional cumulative velocity, rockets could launch heavier satellites into orbit. In seeking to implement Sarabhai's plan for the Indian space program, Indian scientists considered the old launch site at Kerala, but promptly dismissed it: if a rocket flew in an easterly direction from that location, it would, in fact, fly over a densely packed landmass: consequently, spent rocket stages—essentially toxic trash—would rain down over villages, towns, and cities. A more "optimal" solution was to locate a site on the *east* coast of India, where, as the rocket was rising into space, its spent trash rocket stages could fall into the Bay of Bengal without endangering people or property. In other words, at this stage, the logic of location determined that dumping toxic material into the ocean was a more acceptable outcome than potentially destroying villages. Besides directionality and the creation of trash, a third geographical requirement necessitated that the site be as close to the equator as possible, thus enabling the rocket to pick up some extra velocity on the way into orbit. In collating these conditions, Sarabhai's scientists suggested a location on the eastern coast of India, but as far south on the coastline as possible to be close to the equator. One suggestion was to locate the site at the very southern tip of the Indian landmass. But here there was another danger: at the southern tip, the launch site would be too close to Sri Lanka, and wayward rockets might therefore crash into a foreign country.

All these conditions, included as part of the logic of location, limited the optimal spaces for a new space launch site to a single stretch on the eastern coast of India in the South Indian state of Tamil Nadu. Sarabhai had one of his senior engineers, R. M. Vasagam, survey the Tamil coast to identify at least two potential spots for a launch site, one at the southern tip of the subcontinent at Kanyakumari and the other just south of the city of Nagapattinam, one of the most important fishing sites on the Indian coast. Both were considered ideal from a scientific point of view but the decision between the two required further intervention, this time from the local government of Tamil Nadu.[29]

It was at this stage that scientific and bureaucratic considerations were derailed by Sarabhai's bruised ego, setting in motion a series of events that would land the space program at the feet of the Yanadi people. To decide between the two possible sites in Tamil Nadu as well as to negotiate the terms of the site, Sarabhai visited Tamil Nadu (at the time known as Madras State) to meet with the state's chief minister, C. N. Annadurai. At the last minute, however, Annadurai was not feeling well and instead passed the responsibility to one of his ministers, K. A. Mathiazhagan, who was also the cofounder of the Dravidian Progressive Federation (Dravida Munnetra Kazhagam, or the DMK), the leading Tamil nationalist party. Sarabhai would not have expected a warm welcome given his own allegiance to the ruling Congress Party in India, and in particular, Prime Minister Indira Gandhi, who had aggressively sought to impose the Hindi language on South Indians. A witness remembers that when Sarabhai arrived to meet Mathiazhagan, he "was kept waiting" for a lengthy period before the meeting. And when Mathiazhagan finally agreed to meet the increasingly irritated Sarabhai, the Tamil minister supposedly advanced "impossible demands" for Sarabhai.[30] These demands undoubtedly included significant financial recompense from New Delhi to the government of Tamil Nadu at a time of great friction between the Congress Party and the Tamil DMK, especially over the imposition of Hindi on Tamil speakers.[31] All this was too much for Sarabhai. Deeply offended by Mathiazhagan's behavior, Sarabhai decided not to work in Tamil Nadu at all, and instead went to the state just north of Tamil Nadu, Andhra Pradesh, which, in December 1967, had also offered a spot on its coast.[32] Here, again the logic of location determined that the most rational spot would be somewhere at the very southern tip of Andhra Pradesh—that is, as close to the equator without encroaching into no-longer-welcome Tamil Nadu. This precise spot—and there was only one spot left after all these considerations— happened to be an island named Sriharikota.

Sarabhai sent one of his young engineers, Pramod Kale, to survey the Andhra coast, in particular, Sriharikota; he found it "satisfactory" and submitted a report to Sarabhai in August 1968.[33] The tone of the report was optimistic and anticipated "no obstacles."[34] Soon, the Andhra Pradesh government, whose chief minister, K. Brahmananda Reddy, belonged to Gandhi's Congress Party, agreed to donate 12,340 hectares (about 30,000 acres) in and around Sriharikota Island free of charge to the Indian Atomic Energy Commission (which supervised the space program), seeing this as an opportunity to invite infrastructural development projects into the state as a motor of economic growth.[35] An agreement between the Andhra government and the Atomic Energy Commission formalized this arrangement, which was publicly announced in April 1969.[36]

Thus, the logic of location that led to Sriharikota Island was shaped by a confluence of the geographical, the political, and the personal. The geographical and celestial requirements of spaceflight called for a location on the Indian coast with open water on the east to dispose of toxic rocket detritus not on populated areas but into the Bay of Bengal. The launch site also had to be as close to the equator as possible to make use of the rotation of the Earth. Sarabhai's bruised ego and unwillingness to acquiesce to demands from Tamil Nadu undoubtedly played a role. These demands left only a single option: a site at the southern tip of Andhra, whose government, friendly to the center, was eager to hand over the site to the government of India. The logic of location left little wiggle room for planners, or so they claimed: it was either Sriharikota or there would be no space program—this was, in fact, the exact manner in which Sarabhai presented his proposal to the government of India for final approval.

THE YANADI

A long land fragment that functions as a "barrier" to the Indian mainland, the island of Sriharikota is about one hundred kilometers north of Chennai (Madras) and part of the Nellore district in Andhra. The name itself is a neologism, describing the presence of half (*arc*) a crore (*cotti*) of Siva Lingams on the island itself.[37] Between the island and the mainland lies the Pulicat Lake, one of the largest marsh lagoons in India that made the island largely inaccessible for centuries until the advent of modern bridges. Government documents from 1969 and 1970 prepared by Sarabhai's team describe the island, as well as Pulicat Lake and the many islands in its environs, as remnants of an idyllic, untouched, underdeveloped "paradise." This is also how Sriharikota is described in an endless series of history books on the Indian space program as well as innumerable journalistic accounts: a remote idyllic paradise of natural beauty, a place of "rare and beautiful birds, including pelicans, kingfishers and pink flamingos."[38]

In the late 1960s, there were no roads on the island proper, no infrastructure, and no permanent dwellings on the island. In an official report issued at the time by the Indian Atomic Energy Commission, it was described as "a forest area *without any permanent habitation* or cultivation except eucalyptus and casuarina trees used mainly as firewood."[39] Another report a year later, in 1970, qualified that statement with a suggestion that there was "*practically* no habitation."[40] The Indian media reproduced such characterizations in their exact language: the *Times of India*, for example, noted: "The object of locating the station at Sriharikota is to provide a suitable range for launching scientific and technical satellites using multi-stage rockets. This work cannot be undertaken by the existing range at Thumba since it is surrounded by thickly populated villages. Sriharikota, in Nellore district in Andhra Pradesh, is an island and has practically no habitation."[41] Such accounts introduced the logic of location into discussions of the siting of Sriharikota while at the same time rendering invisible the actual population in and around Sriharikota, the Yanadi.

Records of a Yanadi presence in Sriharikota date back to the sixth or seventh century and there is some evidence of even older habitation. At different points in the past millennium, they had been in contact with various outsiders, but their lot changed considerably subsequent contact with the British who gained rights to the island after the independent Carnatic Sultanate, one of the major kingdoms of South India, came under the nominal control of the British East India Company in 1801. From the 1830s, the British reorganized their administrative jurisdiction over Sriharikota, incorporating it as part of its revenue administration—the British employed the Yanadi to collect jungle produce and firewood. As a result, the British reorganized the nearly fifty villages in Sriharikota into eleven larger ones.[42] The population of the island at the time constituted several Hindu castes, a Muslim community, and two "tribal" populations, one of them being the Yanadi, who were the "original inhabitants of the island."[43] The Yanadi population on the island constituted a small portion of a much larger community spread across the eastern areas of Andhra Pradesh, representing one of the largest Indigenous populations in South India.[44]

An American geographer's study from 1917 provides a window into Yanadi culture even as it reveals the deep biases inherent in the observer's eye. In speaking of the Yanadi, Sumner Webster Cushing wrote: "There is no record of the date of their arrival on the island. It has preserved them as a pure blooded tribe in all their primitiveness. Waves of conquest, waves of caste, waves of religion and of culture have swept over India and left these people untouched. For centuries currents of commerce on both sea and land have flowed past the island even within sight, but without effect on these people."[45] Yet he evinced grudging admiration for their deep knowledge of their environs:

> The low stage of culture of the Yanadis has one contrasting feature. They have a wide and practical knowledge of their jungle. Every kind of tree, herb and vine has a name. The properties of each are as well known to them as the letters of our alphabet to us. They gather roots, fruits, leaves and honey for their food and know well the best season for each variety and where it grows best. They have learned by experience the properties and uses of herbs and roots in medicine. Every ache and pain has its supposed remedy in the jungle flora.[46]

Their expertise and knowledge about the Andhra coast was, in fact, a key determinant of their fate. In 1835, when British authorities purchased the island in an auction, they, through Indian interlocutors, "employ[ed] the Yanadis in collecting minor forest products" since they knew so much about the island's natural environment, and "in return for their labor the government . . . paid them in clothing, grain and money."[47] Increasing contact with the authorities of the British Raj, mostly with Indians in the employ of the government, slowly reshaped their original "animist" beliefs as the Yanadi population was reordered into several fixed Hindu castes, although some resolutely retained their animist cosmologies. By the mid-twentieth century, their set of beliefs included a deity, Chenchamma, as well as other beings who "controlled human activities and nature" and who took the form of "supernatural beings such as nature spirits, ancestral spirits, village deities and malevolent spirits."[48] Perhaps the most defining characteristic of these beings was their connection to the land on which the Yanadi people lived—that is, their cosmologies were intimately tied to location.

INFRASTRUCTURAL VIOLENCE

When Sarabhai's team came to Sriharikota in 1969, the first order of business was to inform the Yanadi population that they would no longer be able to live on the island or, indeed, around it. One of Sarabhai's colleagues, R. Aravamudan, remembers that on Sarabhai's very first visit to the island in May 1969, he clearly and directly addressed the Yanadi after a big reception organized by the local Andhra government: "After the feast, Sarabhai addressed the local Yanadi tribals in English, the local collector providing the translation. The tribals, who had not ventured much outside their island, believed Sarabhai was the Raja who was going to change their lives."[49] And their lives did indeed change, although not without their considerable resistance: at first, barring a few exceptions, the population of several thousand inhabitants refused to move. The local government decided that some form of monetary compensation might persuade those who remained intransigent. Based on in-house evaluations carried out by the Collectorate of the Nellore district of the Andhra government, the authorities estimated that the total population of the island (which included many non-Yanadis) were entitled to 87.5 lakh rupees (or 8.75 million rupees). In fact, it is estimated that each Yanadi family received on average 115 rupees (about $15.30 in 1969).[50]

Displaced families were also given the option of "rehabilitation"—that is, the possibility of relocating to government colonies (or housing) with added monetary incentives.

A total of 2,085 families (about 10,000 people) were forcibly relocated in three phases between 1969 and 1972.[51] Although the Yanadi
made up a small proportion of the total—about 15 percent—they had
the distinction of being the only population displaced who were indigenous to the island. The movement of this large population was neither
entirely voluntary nor without incident. Some left out of fear, apparently
alarmed by a rumor that may have been circulated by ISRO employees
that the government was planning to test atomic bombs in Sriharikota.
The anthropologist S. Sudhakara Reddy, who interviewed relocated Yanadi people in the late 1970s found that "rumours . . . were spread that
no life would exist after the blast of the first bomb. . . . Many Yanadi and
non-Yanadi who were illiterate and poor feared very much that there was
danger to human life . . . and left the island without being noticed by the
Government."[52] The Andhra government, which was legally required to
find land for those who had been displaced, acquired plots of land from
a powerful landowning family in Nellore district, the Reddys. As soon
as the Yanadis arrived to take possession of this land, the landowners
reneged on their agreement. Under pressure from the local government,
the family only agreed to hire them as part of their large (and grossly
underpaid) labor force.[53] In a study from the 1980s, a group of anthropologists noted: "The new 'villages' where the Yanadi were rehabilitat-
ed in 1970 were poorly managed. There was an acute shortage of water.
The Yanadi were given land for cultivation but they did not know how
[to cultivate in this new ecology]. The compensation [money] [ran out]
within a week in the new place (mostly because of cheating by liquor
sellers). Several Yanadi suffered from diseases and died. Many starved
for several days."[54]

Not all the Yanadis left Sriharikota in 1970–1971. As part of the
arrangement that Sarabhai had orchestrated, nine hundred Yanadis remained on the island proper in eight "labor camp" villages to support
ISRO activities, but they too left in a slow dribble until 1977 when only
about five hundred remained.[55] They were joined by others sporadically
over the years. Despite tight security around the newly emerging space
infrastructure at Sriharikota, by 1975, many Yanadi had returned to
their former homeland, although by then, they had no legal claim to
the land. They were driven to return by the abysmal conditions of their
relocation sites but also by their own cosmology. Reddy notes that "pri-
or to displacement, their belief system was closely associated with the
environment of Sriharikota Island. The Island was the original place of
their deities who were propitiated for the tribal well-being. . . . Frequent
deaths in the [displacement] colonies were attributed to the wrath of
deities of the old habitat [for leaving the island]."[56] The returnees expe-

rienced significant violence at the hands of personnel hired by ISRO to ensure security in and around the island. They were treated like pariahs and often detained for extended periods without charges.[57]

Sarabhai, having engineered the original dispossession of the Yanadis in 1970 was not witness to this stage of the events—he passed away unexpectedly in December 1971, only fifty-two years old. He was succeeded as head of ISRO by Satish Dhawan, a highly respected mathematician who trained at Caltech in the 1950s before returning to India to head the elite Indian Institute of Science in Bangalore. On a personal request from Prime Minister Indira Gandhi, Dhawan reluctantly agreed to take over the helm of the Indian Space Research Organization.[58] Dhawan's primary remit was to build India's first rocket capable of launching a satellite into orbit, but he also expressed occasional concern for the fate of the Yanadi, spurred apparently by an inquiry from an anthropologist, P. Sudhakara Reddy, from the University of Hyderabad who wanted to study the conditions of the remaining Yanadi community in Sriharikota. Dhawan rejected that offer but hired two *other* anthropologists and put them on the ISRO payroll to study the surviving Yanadi population on Sriharikota as part of an "action anthropology project" geared toward proactive work.[59] The goal was to offer suggestions on what could be done for them as part of the ISRO economy on the island.

In the report, belatedly finished in 1985 and published in-house by the Indian Space Research Organization, the ISRO-employed social scientists claimed that "the Yanadi today is confident and feels secure on the island. They are far removed from the perpetual state of anxiety and helplessness." The positive sheen barely obscured their continued precarity: "The Yanadi today are able to have at least two square meals a day from their earnings. The number of starvation days in the Yanadi life have decreased so much so that a Yanadi does not go to sleep without food of some kind or other."[60] Unfortunately, by the time the report appeared, Dhawan had already retired, thus robbing the Yanadi of even a modicum of curiosity on the part of ISRO. Since then, for decades, they have continued to live in a state of permanent precarity, now spread out in many different areas of Andhra Pradesh.

INDIA, ADIVASIS, DEVELOPMENT

The Yanadi story was not unique in modern India; in fact, a similar dispossession had occurred in the space program nearly a decade before when Homi Bhabha and Vikram Sarabhai had set up the rocket firing range on the western coast of India in the state of Kerala. Seeking a site to study a certain cosmic ray phenomenon only observable near the southern tip of India, Sarabhai's engineers—with considerable help from NASA advisers—found a spot in a fishing village known as Thumba. In the process, the nascent space program forcibly removed the five hundred fishers and their families who lived in Thumba (and whose forebears had lived there since the fifteenth century). As had the Yanadi community, most of the population refused to move or give up their houses. The Kerala government promised compensation for each relocated family, but the problem was exacerbated because the funds would not be available immediately. Dazzled by the promise of a space program in Kerala, the local government reneged on promises of reemployment for the displaced. It required the intervention of a local bishop who convinced the fishers to surrender their land without any promises from the government.[61]

In India, such displacements—justified within the discursive apparatus of national development—have a long history, with violence and displacement inflicted particularly on the most vulnerable populations, Indigenous populations. Although Adivasis (or Scheduled Castes) were guaranteed certain rights by more than two hundred articles in the 1950 Constitution of India, there were continual struggles over the government's encroachment on their property and livelihood in the decades after independence. These struggles began during the colonial era, when the British, for example, forcibly introduced conceptions of private property and positioned them in a social order in which they were naturalized as a subordinate "other" to colonial subjects, relegated to the bottom of the social ladder. The British as well as many Indians used Hindu scriptures to legitimize this hierarchy.[62] After independence, three major acts—the Forestry Policy of 1952, the Wildlife Protection Act of 1952, and the Forest Conservation Act of 1980—"downgraded [Adivasis'] privileges to State concessions," in the words of an Adivasi activist.[63] In incorporating laws governing Adivasis as part of those ostensibly concerning forestry and wildlife, the government essentially reduced the claims of Adivasis to ones relevant to the management of natural resources.

Ignoring or downgrading Adivasi rights at the cost of national development passed into a more intense phase with the arrival of the era of liberalization in India in the 1990s, to the point that Adivasis remain a fundamentally disenfranchised population in India. Ironically, this is despite many acts signed by state governments guaranteeing the rights of Adivasis; these acts in fact serve to shield continuing dispossession through massive numbers of court cases that rule against the Adivasis.[64]

In *Churning the Earth*, Ashish Kothari and Aseem Shrivastava frame these forced displacements in India as part of a larger process they call "sub-colonialism," invoking not only its extractive nature but also its not-quite-colonial properties, now in a postcolonial context. They note that "both nature and rural communities across the country have been displaced, exploited, and been 'diverted' for mining, dams, industries, and so on, with the rate of diversion significantly rising since the 1990s [with] over 60 million people physically uprooted for the same."[65] They link subcolonialism to India's frontiers, especially in the twenty-first century, suggesting that "now, the Indian state is eyeing the 'frontiers' of the country that have so far been relatively less impacted by its development sub-colonialism—regions in its far north, northeast, and the islands off its coasts."[66] In many of these "frontiers," the most vulnerable populations have been the Adivasis who have borne the brunt of repeated state violence, and whose experiences have been easily ignored or trivialized by both nationalists and liberal elites, who prefer to accept such costs as part of India's dramatic economic growth in the past few decades.

In Andhra Pradesh, large development-minded projects, such as the Polavaram Dam, a venture given national project status, inflicted violence on the Indigenous populations of the Godavari district, which, by some estimates, displaced nearly three hundred entire villages comprising a largely Adivasi population.[67] We also see this pattern in the state of Orissa where, with the opening of the Bauxite mining sector for private investment in the 1990s, huge populations of Adivasi areas have become vulnerable to encroachment by multinational corporations seeking to make a huge return on aluminum, useful for household goods but also for extremely lethal incendiary weapons used in many modern battlefields.[68] Such practices have exacerbated the already existing precarity of heterogeneous and dispersed Adivasi communities in India, robbing them of forested lands and thus their livelihoods, to the point that Adivasis remain perhaps the most disenfranchised population in India, with much higher poverty and illiteracy rates than any other minority.[69]

The predicament of the Yanadi people continued to deteriorate through the late 1980s and into the 1990s, by which time they depended not on ISRO or the Andhra Pradesh government for largesse but on local nongovernmental organizations such as the Yanadi Education Society and the Association for Rural Development (ARD). By the 2010s, despite their considerable efforts (as well as a number of longitudinal granular ethnographic studies on the Yanadis), the socioeconomic condition of the Yanadi population in and around Sriharikota remained extremely tenuous, with continued and often violent conflicts with the Reddy landowning family and ISRO's security apparatus.[70]

Like the larger population of Yanadis in the state of Andhra Pradesh, those in Sriharikota continue to live in conditions redolent of labor bondage and squalor, with those around the space center subjected to the worst forms of violence and dispossession.[71] In addition to their economic precarity, they harbor a deep resentment toward how, in the words of one of their spokespersons, "a premier institution like ISRO was able to impact . . . the life of the Yanadi community," and hope for an opportunity "to let the global world [*sic*] . . . learn about this community and help them in bringing positive change."[72]

NETWORKED VIOLENCE

As an episode on social and natural disruptions wrought by infrastructure, we can locate the story of the Yanadi people among many other similar narratives of the detritus left behind in the rush to national development. But the Yanadi story also remains encoded in contradictions. It is a story at once erased from the narrative of the Indian space program while at the same time lodged deep within it. The twining of multiple threads in this story—the violence of infrastructure, postcolonial nation-building, and the demands of the space program—were woven into a clash of outcomes: a successful space program that was the pride of and inspiration for many Indians, and the near-erasure of a considerable portion of an Adivasi community.

The story of the displaced Yanadi population tells us that at its very roots, the Indian space program was conceived in a moment of great violence and dispossession. This, I argue, was not simply a regrettable outcome of the cosmic aspirations of a group of scientific elites intent on "making India modern." This mode of infrastructural violence was fundamental to the production of the space program, driven as it was not just by the "logic of location" but the *violent* logic of location—a logic that took into account geographical considerations, political alliances, scientific priorities, and personal whims, but not the aspirations and dreams of those who lived in those spaces. Driven by the violent logic of location, the technocrats who wished to use Sriharikota as a location for their space program saw the demands of science as obdurate and inflexible. In contrast, nature—which now included not only Sriharikota Island ("beautiful," "pristine") but also the people who lived on it ("primitive," "untouched")—was defined as open for violent transformation into a modern site of science. In arraying science against nature, Sarabhai and his supporters fought for the former to prevail. For that to happen, the latter had to be disciplined in service of the inspiring and emancipatory imaginary of space travel. Instead of the violence of these infrastructural irruptions, these "placeless spaces" of India's space program were reconfigured as symbols of modernity, monumentalism, and momentum.

This relationship between technology and nature also produced new configurations between citizenship and science in India. In his famous meditation on the past, present, and future of modern India, *The Discovery of India* (1946), Nehru idealized the notion of a "scientific temper" as a goal for future Indians, even as he sought to cleave a delicate balance between a universalized version of science and one that was more suited for India, a future where colonial technoscientific achievements were shorn of their violent legacy and reformulated for Indian development.[73] Echoing Nehru (whom he knew), Vikram Sarabhai, the architect behind the space program, too spoke repeatedly about cultivating ideal scientific subjects in newly independent India. In his speeches in the late 1960s, at the very same time that he personally signed off on the dispossession of the Yanadi people, Sarabhai consistently lectured to various domestic and international audiences of the need for a "complete scientific education" so that Indians could be mobilized for the future. The ideal subject would understand the ideal of science. Yet, for people like the Yanadis, who were liminal or marginal national subjects exposed to infrastructural violence, there was no such subjecthood. Instead, they suffered double erasures, the first time involving physical dispossession, and the second time, in their total absence from the memory of the space program. Yet, they were as much actors in that story as any of the engineers or scientists involved in the history of the Indian space program.

If the experience of the Yanadi underscores how the government of India (as well as its many state governments) has seen certain populations as disposable or at least as inconveniences in the government's pursuit of the project of socioeconomic development, the particular case of the space program raises some specific and, I think, interesting questions about how certain scientific projects—especially those that appeal to a national imagination aimed at some uncertain future horizon—can be mobilized for universal appeal even as they are deeply complicit in practices haunted by colonial modes of exploitation, extraction, and displacement. In the past sixty years of spaceflight activities, an entire culture of space enthusiasm has been built around seductive narratives—and dreams—of discovery, exploration, and wonder. The so-called revolutionary potential of space exploration has had powerful appeal to many, from the Soviets who linked it explicitly to Bolshevik millenarianism to the Americans who found a "new frontier" having exhausted the terrestrial one.[74] Such imaginaries have been very effective in shielding space activities from criticism. Who would condemn curiosity and the thirst for knowledge?

Recent liberal concerns about billionaires throwing away money at space often miss the networked violence of space activities on the ground. Originating during the Cold War in infrastructure built by the established superpowers across the globe, space programs remain linked to a considerable extractive economy all over the planet, all activated by the violent logic of location.[75] Both the Russians and Elon Musk, for example, have recently sought to position their space operations, with all of their expelled toxic fuels, massive amounts of trash, and extractive microeconomies, in places such as Papua New Guinea, to take advantage of proximity to the equator.[76] The case of the Yanadi and many others during the Cold War, in sites such as Kenya, Algeria, Kazakhstan, Brazil, Mexico, and Madagascar underscore that these practices long predated the arrival and incrustation of neoliberal extractive political economies across the globe.[77] Our dreams of space, unleashed, have produced over the past many decades a kind of networked violence across the tropical landscape of the planet.

The Yanadi too had dreams. In the 1985 anthropology report sponsored by ISRO to "actualize" their "development," the authors note: "In general, the Yanadi are exposed to various new ideas as a result of continuous interaction with the non-Yanadi and their ways of living. This includes exposure to mass media also. The acquaintance and mere seeing of the ISRO activities through rockets, electronic gadgets, and other equipments [sic] directly or indirectly have started influencing the Yanadi-world view and life. Observations have indicated that these have lead [sic] to change in dream patterns of the Yanadi. They have started dreaming about the sound and thunder of rocket firing."[78] We could interpret this in two ways. The first, a slightly generous reading, is that the Yanadis have been co-opted into the Indian national imagination of space and have thus begun to accept that the dispossession imposed on them was for a larger cause. But in a less sympathetic reading, the "sound and thunder of rocket firing" suggests a kind of violence, a fear, and a terror that perfectly distills their dispossession into the singular image of the glorious yet frightening form of a rocket screaming out of their former homeland and into the cosmos.

GROWING NEW NARRATIVES IN SCIENCE COMMUNICATION IN OUTER SPACE

From the Pastoral to the Ecocritical

ELEANOR S. ARMSTRONG

Food and drink are essential components of spaceflight. They often appear as subjects of comic, "relatable" anecdotes about astronauts who did not like the culinary offerings of those organizing their time in space. This included the British astronaut Tim Peake's rejecting of NASA's tea offerings, instead taking specially prepared Yorkshire tea (a limited edition, called Peake's Tea) to the International Space Station (ISS), a version of which now exists in the collections at the National Space Centre in the United Kingdom. Samantha Cristoforetti, the European Space Agency astronaut from Italy, drank Lavazza coffee from the specially designed ISSpresso machine. Certain local, culturally specific items are precious to astronauts—the Swedish astronaut Christer Fuglesang attempted to bring reindeer jerky on his shuttle mission. Reports on food for the Indian Space Research Organization (ISRO) Gaganyaan mission frequently mention a range of dishes such as *moong dal halwa*, *idli sambhar*, vegetable *pulao*, and *upma*, which have their origins in multiple states across India. Food for Chinese astronauts on Tianhe skews toward Sichuan-style foods based on astronauts' reception of foods on previous missions. Even before getting into space, food has been an important component in popular narratives about space. The ritualization of breakfasts of the early American space program astronauts—begun by Alan

Shepard's orange juice, steak, bacon, and eggs; and continued by others in the program—played into the characterization of white masculinity that was "required" for the program. There is ongoing popular interest in where and how people eat in space. Videos from the ISS often feature astronauts eating and drinking under zero-gravity conditions—astronauts celebrated the successful of growing chili plants on the ISS by eating the chilis in tacos.[1]

Such anecdotes tell us as much about individual astronauts or space programs as they do about how nationalist tendencies bleed into travel to outer space. These stories exist as narratives in the public consciousness, through circulation in news and social media posts and images; they are solidified through their depiction in spaces like museum displays or movies. For example, a glass display case in the Exploring Space gallery at the Science Museum in London contains a cabinet with a selection of foods that could be eaten on the International Space Station: a can of Coke, unlabeled tins to be opened, Extra gum, cereal cubes, chicken salad, sachets of Heinz ketchup, mayonnaise, and some seasoned scrambled eggs. While some of the foods displayed in the cabinet are exported around the world, many are culturally specific to white settlers of the United States. Coke, Wrigley (the makers of Extra gum), and Heinz, for instance, are US-based brands founded by white industrialists of the 1800s; whereas many of the other foods, such as the chicken salad, are simply popular US dishes.

Everyday science learning practices—which dominate ideas about informal education and science communication—encourage relating science to the audience in communication exercises. Such practices, however, inherently create some groups for whom the relation lands, and others for whom it does not. Relatability is touted as actions such as reframing distances from standard units to distances between well-known local landmarks or giving areas in units that relate to locally popular sports. Culturally dominant groups and ideas tend to guide what is deemed to be "relatable." In a similar vein, foods on display in a museum are often expected to have this relatability function—operating as being similar to "everyday" items that visitors will recognize but representing a subsection of the wide diversity of types of food that go into space worldwide.

Science communication and learning practices related to outer space are dominated by tropes, imagery, and rhetoric to make outer space understandable on Earth, because outer space is unknowable and has yet to be experienced by many in the audiences. Cultural shorthands employed in these practices are especially important to the understanding of which visions are being conjured to members of the public. While communication efforts more obviously draw on nationalist narratives, in the context of food and vegetation they also draw extensively on pastoral themes and ideas of terrestrial landscape—such as the wilderness, the garden, and the farm.[2] These are employed in learning contexts as a way of navigating the tensions between the technoscientific requirements of living beyond Earth and the "naturalness" of Earth, and the fact than many learners have not experienced both. However, I demonstrate that this prominence of the pastoral and its associated implications preclude other engagements with cosmic environments that could open new relations to living or thinking beyond Earth. In this fragment, I argue that emerging engagement with themes, tropes, and ideas from ecocriticism can pluralize the ways publics learn about outer space as well as inspire new relations with the cosmic.

To develop this argument, I begin by describing how the pastoral (and particularly the American Pastoral) is theorized and how its inclusion in space science narratives has been documented to date, theorizing an Outer Space Pastoral genre. I then show how it appears across three examples of science communication—a placard at a science museum, a postcard issued by a NASA center, and a blog post by a NASA astronaut—highlighting which tropes appear from the genre and how this shapes the perception of outer space. I select these objects through purposeful sampling from the data of three of my research projects on outer space science communication that span the period 2017–2023 to illuminate the key elements of the themes tackled in this fragment. They are, in this way, fragments that, in my research over time, have connected together and across projects to crystalize motifs of landscape in cosmic science communication. Finally, I demonstrate how other formulations of nature through forms of "ecocriticism," shown here in a self-published zine on space and the movie *WALL-E*, can support pluralized learning and communication orientations to outer space.

THE PASTORAL, THE AMERICAN PASTORAL, AND THE OUTER SPACE PASTORAL

The pastoral, or the bucolic, is a pervasive and long-standing rhetoric, replete with tropes, imagery, motifs, structures, and ideals that support its comprehension in public settings. Generally used as a mechanism for framing landscapes and nature, the pastoral's emphasis on the rural and the nostalgic sets it in opposition to motifs such as the metropolitan. Importantly, as William Empson notes, the pastoral as a rhetoric relies on cultural, textual, image, and linguistic shorthands that compress meaning into emblematic formats.[3] This is crucial, as it means that where the pastoral is invoked in the context of outer space science communication and learning, the learner is required to fill in the shorthand with their knowledge of the expectations of the genre from other contexts. The pastoral also has nationally specific sets of ideas. For example, the British Pastoral genre emphasizes class, invisiblizing rural labor, thus generating a strong tradition of Marxist critiques against the framing of work as unseen and unacknowledged. This contrasts with the American Pastoral, which has stronger masculine-colonial, agrarian worker connotations, thus producing feminist- and Indigenous-led critique.

The American Pastoral genre developed to accommodate and reify the masculine dimensions of colonialism, propagandizing the westward expansion of the United States across the continent, the "civilization" of nature, and the arcadian vision of the landscape by European colonists. This rhetoric is underpinned by utopian longings for new beginnings and draws most strongly on the agrarian tradition, where landowning farming is framed as part of a healthy nation. Greg Garrad proposes that different historical genres also create nuance in the pastoral.[4] He links the Romantic Pastoral with the American Pastoral, demonstrating that the Romantic Pastoral's emphasis on political engagement with nature, the vastness and endurance of nature, and its ties to progress are visible in the American Pastoral, where the agrarian work is seen as politically valuable to the nation. Elsewhere the Sublime Pastoral, also linked to the American Pastoral, dominates in visions of vast open spaces—such as in the photography of Ansel Adams—and the intertwining of technology and nature usurps the dichotomous spatial separation that happens in other pastoral contexts. The American Pastoral is a wide and continuously bifurcating motif with nuanced meanings, spanning texts such as Laura Ingalls Wilder's *Little House on the Prairie* children's series, Herman Melville's archetypal American *Moby-Dick*, to Kara Walker's subversion of the trope in *Kara Walker: My Complement, My Enemy, My Oppressor, My Love.*

The American Pastoral has been recognized as a fundamental part of the narratives underpinning communicating outer space. This includes rhetorical emphasis on the new, utopian beginning that is possible in space, which, as De Witt Douglas Kilgore analyzes in *Astrofuturism*, is pervasive in science fiction: a setting that requires "technological forces to tame natural worlds for human colonization and exploitation."[5] The dominating agrarian motif and the combination of technical and rural seen in the American Pastoral is identified by Chris Pak in the lineages of science fiction terraforming. Pak argues in a chapter titled "American Pastoral and the Conquest of Outer Space" that "aligning [science fiction] with the agricultural idyll [allow authors to bring] terraforming into contact with the familiar domain of agriculture."[6] The American

Pastoral's inclusion of the Sublime Pastoral is most efficiently shown in the rhetorical framing of the "wilderness" in the American West. This invocation of both the sublime and the wilderness, Elizabeth Kessler argues, makes its way through images of outer space into what she calls the "pretty sublime" in Hubble Images and their references to the famous Ansel Adams images of the American West.[7]

While the pastoral's general temporal relation is guided by nostalgia by creating the idealized past and the fallen present, the motif also makes space for the elegy to this idealized past, for the idyll of the present, or for a utopia of the future. When employed in narrating outer space, it is this temporal pluralism of the pastoral that often sees the past-present-future as entangled together. Kat Deerfield has argued that "the relationship between outer space and time is not straightforward . . . the present of space exploration is conceptually behind the future we imagined in the past, and that future is now an artefact of our history."[8] Because visions of outer space are temporally entangled, this creates, as Martin Parker and David Bell argue, "a future that never happened, or a history that seems not to connect with our present."[9] This iterates the pastoral rhetoric within narratives of outer space, creating recursive imagery, tropes, and ideals that are returned to by publics to make sense of the otherworldly; I argue that these are distinct and constitute an Outer Space Pastoral.

These different elements of the pastoral, specifically the American Pastoral, make their way into many different examples of science communication about outer space. Thus, cultural shorthands are specific to a particular rhetoric, shaping in turn how publics are primed to interact with cosmic landscape environments.

THE PASTORAL IN SCIENCE COMMUNICATION ABOUT OUTER SPACE

I have purposefully selected three cultural texts that demonstrate some of the characteristics of the invocation of the pastoral in their construction of food and landscape in relation to outer space: the representation of Food in Space in the Science Museum, London's Exploring Space Gallery (n.d., henceforth *Food in Space*); a postcard from the Studio at NASA/JPL, titled "Earth Your Oasis in Space" (henceforth *Earth*); and the blog post *Diary of a Space Zucchini* from *Letters to Earth: Astronaut Don Pettit*, an online blog curated by NASA (2012, henceforth, *Space Zucchini*). My goal here is to explicate some of the ways that these examples (and others) demonstrate pastoral tropes in science communication, and what these imply for learners when encountering outer space.

I selected the context of food in space to demonstrate how the public engages with meaning-making displays, especially in contexts beyond the United States, showing how motifs originating in the US might shape expectations and understandings of outer space in publics in the rest of the world. To characterize some of the ways that cultivating plants in space is narrated, I selected *Space Zucchini* as written by an astronaut. The two examples that describe experiments to grow dwarf wheat and zucchini are organized toward a future of farming in space where food is not packaged and shipped to space (as were the coffee, tea, jerky, or dinners that introduced this fragment) but grown there. I include *Earth* to exemplify some of the ways in which the pastoral informs this vision of the distant future, demonstrating the chronopolitics of the pastoral motif in outer space communication. I offer these cases to characterize the impacts of pastoral-informed representations of outer space in science communication.

The Living in Space section of Exploring Space at London's Science Museum has a subsection on *Food in Space*.[10] London's Exploring Space Gallery has been open to the public since the 1980s, with periodic updates over the intervening forty years, which include the section on Living in Space. Entry to the gallery is free for members of the public year-round and is the first ground-floor gallery of the museum with high foot traffic. Eschewing images of astronauts, the panel is covered with a large picture of many bananas packed tightly together, with white panel text writing. It has a photograph of a woman with curly hair wearing a white shirt, identified as Lisa Ruffa, who is in action with her hands manipulating the stalks of the wheat sample. The image is from a NASA series depicting people working on a project to cultivate dwarf wheat as a crop to grow in the confines of the ISS. Next to the photo is a glass panel that allows visitors to look into a display containing a silvery-metallic cabinet in which a series of space food items are laid out. Pictured on Earth and not under microgravity, Ruffa is identified as a "research technician . . . at the Kennedy Space Center." She is at odds with the rest of the people featured in the section, which focuses on astronauts' lives on the ISS in space—from their technical solutions to going to the toilet in space, to how they exercise, to photos of astronauts resting in sleeping bags. No other technicians appear in the display: no people training the astronauts before going into space or helping them to prepare for their flights. Ruffa is instead surrounded by male astronauts in space who are conducting space walks, sleeping, exercising, and reflecting on their work in space. She creates a fragmentation of the narrative of the section and the gallery at large: one of the only women named in the Exploring Space gallery, one of the only technicians featured, and the only person working on production of food or in relation to plants in the exhibition.

This representation of food (and the reproductive labor of creating it) is the only time it appears in the gallery. Elisabeth L'orange Fürst describes cultural scripts around food and cooking as tasks that fall into gendered patterns primarily done by women in the household, calling them "women's work"—important expressions of identity, closely tied to the construction of racialized femininities, and the use-value of different tasks in the home.[11] Food is intimately tied to femininity and acts of being feminine—to fail at food, Kate Cairns and Josée Johnston argue, "also means failing at femininity."[12] Therefore, while some postfeminist movements maintain that food is not closely associated with femininity or women, the choice of a woman—and a technician rather than (as elsewhere in this portion of the gallery) astronauts—harks back to the idea of a woman as the natural, nurturing caregiver and sustenance provider, which is tied to the reproductive labor of transforming farmed produce into food in the pastoral context.

The racialization of this food production is also important. While Lisa Ruffa is featured in this context, tying in with American pastoralist visions of white people participating in sanctified agrarian work in the expansion westward, it is not the case that these were the only participants in work with food and plants in the context of outer space. Racialization in the American Pastoral plays out as both an absence of Black people and the paternalistic treatment of Indigenous peoples. Charnell Chasten Long has documented how Black American women such as Julie Stewart and Sara Thompson were part of teams that planned and prepared meals for Apollo 11 astronauts, and while they were profiled in contemporary newspapers, these culinary workers were written out of canonical histories of outer space as demonstrated by their absence in official archives and publicity materials.[13] Much like Black women "computers" of the Mercury program, recuperated in Margot Lee Shetterly's *Hidden Figures*—and now made into a popular movie—knowledge about Black communities in the US that contributed to (or resisted) space programs in often pivotal but historically underrecognized positions are absent in the gallery's narration.[14]

Similarly, in online publicity in 2007 for this dwarf wheat as part of the space experiment from which the photo of Ruffa is drawn, NASA links the project directly with the sanitized Thanksgiving story of pioneers and Powhatans exchanging food. Their online content specifically links the corn that grew in the precolonial Americas to the US visions of continuing colonial-pioneer futures in space.[15] In the words of the publicity material, "the settlers established more friendly relations with the Virginia Indians who shared their knowledge about food, water and shelter." Here, drawing on settler pastoralism and the tropes of the "noble savage," the description of this project by NASA performs a well-rehearsed erasure of genocide, replacing it with narratives about friendship and helping, setting up a eulogy of the idealized past in the beginnings of the United States.

The second example is the *Earth* postcard, part of a series of retro-futurist space travel advertisements "Visions of the Future," aimed to "sell" various parts of the cosmos in the style of the travel posters for the American West of the early to mid-twentieth century.[16] The series—started in 2016 but continually expanded—is created by the NASA Jet Propulsion Laboratory (JPL) creative team, The Studio, in collaboration with JPL scientists and engineers. The images are available online to download for use without prior permission. The tagline of the postcard runs along the bottom of the image in an art-deco font, reading: "Earth Your Oasis in Space: where the air is free and breathing is easy." The image is dominated by an unspecified wilderness landscape—white-capped mountains in the distance, with lower bluish-green hills, green forests, and a blue lake. This vista is framed by two people in space suits, outlined as viewed from the back—ostensibly a man and woman—with the man's arm around the woman as they look out from their bench beneath a tree in the foreground with red apples hanging above their heads. The image shows a range of animals: birds in the clear blue sky, three deer (one antlered, one not, and one fawn) drinking down at the lake, butterflies near some flowers in the foreground.

The image draws on tropes of both the garden and the wilderness within the pastoral. The garden is most visible in the image's use of the apple on the tree, which both evokes the Garden of Eden before the fall and Isaac Newton's scientific discovery of gravity. This evocation of the idyllic, innocent past of Eden as part of the future of humankind speaks to the use of entangled time that characterizes the pastoral in communication of outer space. Lisa Messeri argues that this is not just characteristic of the rhetorical construction of our planet, Earth, but placed in reference to visions of other exoplanets, it suggests that nature on Earth is somehow both preserved and something to imitate elsewhere.[17] The postcard of exoplanet TRAPPIST-1e in the same series, for example, has a garden with a picket fence that gives us the vision of the suburban pastoral dream of lawns and domesticated nature, continuing the trope of the rural as an idealized utopia in space.

Themes of the wilderness are also at play here in the landscape of the image, with a dramatic mountain in the background of the picture, a trope that characterizes themes of the vastness, beauty, and endurance of nature in the pastoral. The treatment of spaces in the American West (to which this is an homage) as natural, nostalgic, and without maintenance erases the roles of Indigenous communities in the sustaining work with land that preceded settler (ab)use and their dispossession. In *Earth* invoking the national park and the "naturalness" of the space performs the same erasure of Indigenous communities that long maintained the land enclosed in national parks and were dispossessed in the parks' creation.

These are not the only times that this colonial element of pastoralism appears in outer space science communication. For example, Natalie Koch details how the Biosphere 2 project strategically intertwines "science and spectacle," using technoscientific solutions to create a rural idyll in a desert landscape.[18] Positioning Biosphere 2 as a stunt that poses a solution to white, western anxieties about the desert's extreme scarcity (barren, sparse, depopulated, otherworldly) as the marker of eco-apocalypse, Koch shows how the project both naturalizes the colonial taking of Indigenous lands and pulls in local Indigenous performance at ceremonies as justification of the practice.

The heteronormativity of outer space is also visible here, with both the couple in the foreground and the approximation of a nuclear family of deer down near the water. Within pastoral analysis, while homoerotic visions are permitted, the masculinity and heterosexualization of the pastoral as an idyll of the present is positioned in relation to the pejoratively framed queerness of the urban metropole. This anxiety about queerness in outer space is noted elsewhere. In *Queer Universes*, Wendy Pearson, Veronica Hollinger, and Joan Gordon note that while technoscientific capabilities are endlessly plural within science fiction, gender roles and sexualities are dominated by conservatively cis-heterosexual possibilities in most hegemonic science fictions.[19]

Finally, I consider *Space Zucchini*, a blog post that is part of a larger series of "Letters to Earth" written by the astronaut Don Pettit and hosted on NASA's blogsite.[20] The blog is written from the first-person perspective of the zucchini grown on the space station, with dated entries from January 5, 2012, through February, 16, 2012 (which match the dates of Expedition 30/31 that Petit participated in to the ISS in 2012), accompanied by several images of the zucchini growing on the space station. The blog describes various elements of growing the zucchini, from the setup of the bag that supported the plant's growth, to the details of how the plant was given liquids and nutrition, its development and blooming of flowers, and its interaction with astronauts on the space station.

The blog is dominated by technoscientific themes. Envisaged as a tool for science communication to engage publics with science research being done on the ISS, the expected knowledge of the reader about technical aspects of growing plants in space is high. The blog describes how light "makes my cotyledons a happy vibrant green" through "photosynthetic nourishment"; how the plant grows in a bag that is "not hermetically sealed" and is "definitely not hydroponics," but is instead held in place by "spongy material called pigmat. Used for absorbent packing for spacecraft supplies." Throughout the blog the zucchini places itself in relation to other members of the space station—describing (we assume) Don Pettit as "my gardener" and the ground crew as "my gardener has a gardener too." It notices how the gardener: "Takes pleasure in my earthy green smell. There is nothing like the smell of living green in this forest of engineered machinery. I see the resultant smile."[21] This creates the idyll of the human-in-nature, showing a pastoral evocation of an imagined preferred location away from the technoscience of the space station. David Valentine's consideration of *Space Zucchini* suggests that the construct of "nature" should be problematized in the beyond-terrestrial context of the ISS, as the "machine," "human," and "plant" have no way of being distinct from each other.[22] However, within the rhetorical context of *Space Zucchini*, work is being done to distinguish the human from the plant, creating distinct roles that fit the trope of the pastoral. When describing the life-support systems, and particularly the lights, *Space Zucchini* demonstrates that the conditions that are good for human crew members do not benefit the plant. This sets apart the human crew and their benefits from the plant, demonstrating the pastoral operationalization of nature in service of the human, rather than surmounting the divide between nature and the human.

Throughout *Space Zucchini* distinctly erotic descriptions of the plant are used. Opening with the plant being "thrust into this world," and reconciling itself as "naked to the universe," the erotic is characterized as a function of the plant itself. The text contains gendered descriptions ("being part of this all man-crew it was fitting for me to make only male flowers," making boys "gag") that are underpinned with heteronormative implications ("my gardener is behaving like an expecting father"). This characterization makes visible Louise Hutchings Westling's observation in *The Green Breast of the New World* of how the American Pastoral, in particular, encourages both the eroticization of and misogyny toward the landscape to consolidate the masculine and imperialist gaze on a feminized nature.[23]

Situating these three examples—the text in a museum, a postcard, and an astronaut's blog post—within larger discussions of science communication on outer space, I have demonstrated that the tropes employed are underpinned by a fundamentally American version of pastoral rhetoric, continuing ideas of masculine colonialism into narratives about outer space, and nature in service of a capitalist society. All contain the figure of the pastoral "shepherd," human in nature. Both Lisa Ruffa and Don Pettit occupy the idyll of the agricultural labor worker; and the couple on *Earth* occupy a classical pastoral vision of stewarding while *in* nature. Collectively, they demonstrate gendered, heterosexual, colonial visions of nature and plants. These tropes of the American Pastoral bolster ideas of American hegemony in space beyond the United States themselves. Ideas critiquing US hegemony—along with narratives of the space race that Asif Siddiqi describes in his introduction to this book—are strikingly absent as important spaces where ideas about space exploration are communicated to publics. Rather than continuing to rely on the American Pastoral, turning to new ways to represent outer space environments might offer a mechanism to change framings of outer space in science communication and public learning, and thus our relations with the cosmic.

THE ECOCRITICAL AS AN APPROACH TO OUTER SPACE COMMUNICATION

The ecocritical is a trope of cultural representation of the physical environment. As a distinctly political framework, the ecocritical bridges environmental and social concerns, and while originating in literary studies, it has been mobilized in interdisciplinary contexts.[24] These representations include pictorial images, popular science communication, film, TV, and other cultural artifacts. Ecocritical approaches specifically reject normative claims that the arena for resolving ecological problems is within the sciences; instead, such approaches suggest that cultural, political, and social forums are valid locations to develop, explore, or resolve such problems. Crucially, ecocritical theorizing tackles the representation of ecological problems through rhetoric—for example, about pollution, animals, or the wilderness—and sees these as closely related to wider social contexts.

Ecocritical themes challenge the modes of the pastoral that I have laid out in this chapter by offering new modes of engagement with the environment and nature. Lawrence Buell draws attention to four key dimensions of the ecocritical in any artifact (i.e., text, image, trope, etc.) that employs it.[25] First, when nature is presented, it is not set apart from human history or society as in the pastoral, but the two are mutually implicated—that is, human history is part of natural history. Second, human (white, western) interests are not the only legitimate interests in nature. Third, human accountability to the environment is a part of the orientation of the artifact to the interaction between society and nature. Fourth, within the artifact, nature and the environment are represented as a process rather than as an ahistorical, unchanging, constant.

I have selected two pieces of outer space science communication to demonstrate these processes. In so doing, I open different possibilities to orient learners to our cosmic relations. In the Wurundjeri land-based artist Sarah Baggs's self-published magazine (zine), *I'm Mad They Left Poop on the Moon (And You Should Be Too)* (2021, henceforth *Poop on the Moon*), the pollution and waste of human flight to outer space is fore-grounded to the reader. Unlike the cases I presented in the previous section, *Poop on the Moon* does not treat waste as a technoscientific challenge to be conquered in order to live in space, as is often the case in museum exhibits or blogs, or images about space. In fact, the zine opens with a challenge to this framing: "The astronauts having left their poop on the Moon is something that is often touted as a fun fact. But when I first heard that there are 96 bags of shit, piss and puke on the Moon my trivia bone was not tickled, I was angry."[26] The zine goes on to explore how this is contrary to the ways American society treats waste on Earth. Baggs perceives leaving defecation on the Moon as simultaneously quintessentially American ("dumping several brown trouts and pissing off home just smacks of American colonialist nonsense"), and deeply irresponsible in light of other legal treaties that govern the use of the Moon. Here, Baggs orients the zine text toward human accountability to the environment. Describing these actions throughout, using terms such as "indefensible," "downright crummy," and "dissatisfaction," she makes clear to the reader that she does not consider the action of leaving this polluting waste on the Moon an edifying achievement. Trips to the Moon are not run-of-the-mill events. As Baggs emphasizes in *Poop on the Moon*, the trip to the Moon was a historic one—not only as the first human journey to a celestial body beyond Earth but also important for the lunar environment too. In this way, Baggs clearly links human history and natural history, showing that the two are fundamentally intertwined. *Poop on the Moon* demonstrates that degradation in the environment is specifically tied to human use of the place: "If the only way to explore a new, untouched place is to leave it worse than you found it, one has to question the worth of the expedition."[27] This is in contrast to the *Earth* postcard, which sees the mountainous landscape as untouched by the presence of humans both as earthly inhabitants in the past and as visitors from outer space in their technoscientific suits; *Poop on the Moon* demonstrates that we have already embroiled our waste in a cosmic landscape that, physically, has otherwise not had any human interaction with it. Similarly, where *Space Zucchini* discusses creating space compost (made from "orange peels, garlic skins, apple cores, and other various food leftovers," not human waste), it is framed as repulsive to the zucchini. The liquid the zucchini drinks is "extracted from this mess" and "makes me [the zucchini] gag"—to be understood by the reader as disgusting. After appearing in this entry in *Space Zucchini*, it is unclear what happens to this space compost; it is clearly not something that humans have to be held accountable for (as it is in Baggs's *Poop on the Moon*), but something that is being operationalized for the benefit of humans and then dismissed from the narrative.

The instrumentalization of ecocritical narratives can also be seen in the director Andrew Stanton's film *WALL-E*.[28] This feature film charts a robot's discovery of a single plant on an otherwise postapocalyptic Earth, which causes a cascade resulting in the return to Earth of a spaceship that carries humans who had been forced to flee the planet following environmental collapse. Throughout the movie the environmental crisis is reinforced as being the result of human interaction with nature, including through early setting shots that advertise the departure from Earth and later through stored videos on the spaceship that describe the human-inflicted degradation of Earth. Here, ecocritical emphases are foregrounded through the tying together of humans and nature. *WALL-E* also uses seminal moments in human relations with outer space—for example, showing the rocket pass by Sputnik when exiting Earth through the orbital debris—to entangle human activity in outer space and natural history together for the viewer. This contrasts sharply to *Earth*, where living in space allows Earth to be returned to an Edenic environment free of destruction based on capitalist overconsumption. *WALL-E*'s ecocritical orientation implicates a changing (and degrading) quality of Earth even after the departure of humans. Whereas a return to pristine ecologies of mountains is perfectly possible within the imagined retro-future pastoral utopia of the *Earth* postcard, *WALL-E* recognizes that the environment does not remain static or unchanged over time, even without ongoing human interaction. The closing note of the film suggests that improving the environment will be possible (although it is somewhat unclear to what extent this will happen) through humans working *with* the environment. Rather than the *Space Zucchini*'s garden, which only grows due to the benevolence of its gardener, the discovery of the plant by WALL-E without humans to grow it demonstrates an ecocritical approach to environmental change without human intervention.

However, while *WALL-E* frames outer space (as well as the terrestrial environment) within an ecocritical rhetoric, the film relies on ideas of recolonizing the planet without change from the capitalism that caused the problems in the first place, which are underpinned throughout the film by heteronormative tropes. These include, most obviously, the gendering and heterosexualizing of the two leading robots in the film, WALL-E and EVE, whose romantic development drives some of the film's narrative and the development of the character arcs of the robots themselves. The gendering of EVE as maternal, caring, and hospitable to the plant that is found continues some of the gendered pastoral tropes I outlined in relation to *Food in Space*. The recapture of the heterosexual, heteronormative family unit rather than adults existing in isolation on the spaceship drives a subplot of the film about two humans who fall in love and end the film by looking after the babies on the spaceship. This reiterates, for example, *Earth*'s reliance on heterosexual ideals. These demonstrate how finding new narratives with which to relate to outer space can be a fraught and difficult process that brings with it older tropes too.

In bringing together *WALL-E* and *Poop on the Moon*—media with vastly different circulations, geographies of origin, and modes of communication—I have aimed to show that the introduction of ecocritical narratives into public discourse is possible at many different levels and can come from many different sites. While the zine contains a sustained, angry critique of the discard-enabled attitude of a historical event of leaving waste on the Moon, *WALL-E* captures a future brought into being by continued extractive capitalism. It is possible for multiple actors to reformulate and rework narratives to talk about outer space, and they bring different perspectives, instances, ideas, and modalities to the development of plural ways of talking about outer space.

GERMINATING, GROWING, GROUNDING

I have shown that culturally constructed ideas about how life beyond Earth might look continue to invoke a kind of pastoralism underpinned by ideas of American colonialism and dominance over the environment, and the gendered, racialized, sexualized, and metropolitan expressions that make these tropes legible within this rhetoric. I have also shown that employing alternative orientations to interactions of humans and nature in which space science takes place, such as ecocritical orientations, can pluralize the imaginaries that can be developed about outer space.

A challenge to the pastoralism implicit in the dominant narrative of plants as sources of food and their enlistment in space as part of the reproductive, agrarian labor of space—itself part of a larger capitalist orientation to outer space—might allow different relations of publics to outer space. Ecocritical approaches that situate the environment in a conversational relationship with social contexts is one avenue for this work. For example, communicating the growing of plants—such as sunflowers, tulips, and succulents of the *Kalanchoe* genus—for domestic pleasure on the ISS might bring plants within an affective register that subverts the current capitalist, militaristic, colonial focus within space sciences that disallows a place for emotion in narratives of outer space. Cultivating these possible narratives makes space for important questions about the purpose of growing plants and having food in space, opening us to ideas about plural astrocultures that prioritize giving and receiving care for humans and nonhumans alike rather than simply in service of capital accumulation.

PART II
EMPIRE

CHAPTER 4

"WE ARE AS STRANGERS HERE NOW"

Interplanetary Colonialism and Cadastral Landscapes

ALICE GORMAN

I want to begin by acknowledging that I wrote this chapter on the lands of the Kaurna people of the Adelaide Plains and pay my respects to their Elders, past and present. I would also like to make clear that I am not Indigenous myself, but my perspectives are informed by working with Indigenous communities in Australia as a heritage consultant over many years.

My starting point, "We are as strangers here now," is a quote from a poem by Oodgeroo Noonuccal, also known as Kath Walker—the first Aboriginal person, in 1964, to publish a volume of poetry. "We Are Going," one of her most famous poems, is about the alienation of Aboriginal people from the Australian landscape. She writes:

They came here to the place of their old bora ground[1]
Where now the many white men hurry about like ants.
Notice of the estate agent reads: "Rubbish May Be Tipped Here."
Now it half covers the traces of the old bora ring.
We are as strangers here now, but the white tribe are the strangers.
We belong here, we are of the old ways.[2]

In poems like "We Are Going," Waanyi author Alexis Wright argues that Oodgeroo made "a weapon of poetry," using "words as a shield to hold back the full effect of colonialism."[3] The poem is about landscapes, dispossession, and the passing of time, but the weapon is concealed in the assertion that the white tribe are really the strangers, walking blindly and unknowingly through a landscape they understand so little that the ritual and ceremonial geography of the bora ground is only seen as a rubbish dump.[4] It seems a fitting metaphor for the ways in which certain classes of human are continuing the "modern" colonial project, initiated in the fifteenth century, in outer space.

When critics of space colonization raise the concern that colonialism has been the source of inequality on Earth, with dire consequences for Indigenous people and the environment, they are often met with the rejoinder that there are no Indigenous people in space, or, indeed, environments, where "environment" is assumed to be a living ecology. The absence of humans who would suffer the ill effects of such colonialism has been recast as a justification rather than an obstacle. Such a response ignores the reality that it is as much the system of belief and action that constitutes colonialism: it is the process, not necessarily the target.

Later in the poem Oodgeroo says: "We are the shadow-ghosts creeping back as the camp fires burn low. / We are nature and the past." Here she describes how Indigenous people have become the background rather than the foreground in their own land. Oodgeroo alludes to the common colonial belief that Aboriginal people were dying out not because of European violence and introduction of diseases, but because extinction was the fate of those who were less evolved in a Social Darwinist paradigm.[5] Contemplating "We Are Going" raises an irreconcilable issue. How can you ask Indigenous people to get more involved in space—as space industry is currently trying to do—while they are still operating within the very system that disenfranchised them to begin with?

COLONIALISM IN SPACE

Regarding space exploration as a colonial process invites comparisons with the terrestrial precedents in which it is rooted.[6] But it also invites us to consider how colonialism in space might manifest in unique and different ways.

For example, settler colonialism, the system under which European colonizers gained control of the lands of Indigenous people, can be contrasted with sojourner colonialism, which is sometimes used to describe groups like Chinese migrants to Australia.[7] While settler colonialism involves claiming land with justifications such as "advancing" the state of Indigenous people and "improving" the land, sojourner colonialism denies the sojourners rights to land or the establishment of a home, regarding them as transitory. In the case of Chinese people in Australia, the presence of the sojourners is tolerated by the politically dominant settlers because they provide cheap labor, undertake dangerous or dirty work, or provide other services.

Sojourner colonialism has a certain resonance insofar as the Outer Space Treaty (1967) forbids territorial claims, and hence sovereignty, in space. Science fiction writers and new space entrepreneurs such as Elon Musk have speculated about unfree labor in the future space colonies on planets, moons and asteroids.[8] It is not impossible that a version of sojourner colonialism could be replicated in space. But there are more positive interpretations. For example, in the Antarctic, an often-used analogy for approaches to space governance, Peder Roberts, Adrian Howkins, and Lize-Marié van der Watt note that the Madrid Protocol (1998) frames humans as transient visitors.[9] To view human engagement with space environments as though they are just passing through, rather than seeking ownership or mastery, evokes the "campsite rule" (via Dan Savage): "You must leave them in at least as good a state (physically and emotionally) as you found them in."

As I write, the Moon is being mapped for the resources that are of interest for present-day space capitalists. The US Lunar Reconnaissance Orbiter, orbiting since 2009, sends back high-resolution color images, and temperature and albedo maps. It has been focusing on the water ice in the Permanently Shadowed Regions of the lunar South Pole. The presence of water ice was suspected from the 1950s,[10] but successive remote sensing missions have confirmed it. The next steps in lunar exploitation are all focused on this area.

The mapping process, coupled with the new concept of "safety zones" promoted by the US Artemis Accords, partakes of what I am calling, inspired by Denis Byrne, "cadastral colonialism," in which the geography of a celestial body changes from geological to administrative.[11] Safety zones are presented as an administrative boundary to delineate the area a lunar mission is operating in, to prevent "harmful interference" (Outer Space Treaty Article IX) between nations and for the purposes of notifying the UN Office of Outer Space Affairs. They are not designed to be a backdoor to a territorial claim, as many fear.[12] They are more equivalent, perhaps, to a mining tenement. However, there are unintended consequences. Byrne has outlined a similar process in the imposition of rectilinear land divisions and fences by European settlers on the Indigenous lands of Australia.[13] As he says: "What made the cadastral grid so ideal for the colonial project is that it could be applied with impartiality to previously unknown terrain, which is to say that it would take a landscape just as it found it, rolling over it as if it knew it in advance. In actuality, of course, it did not know it."[14] This evokes the conflicting geographies captured by Oodgeroo in "We Are Going." Following Byrne, the application of cadastral divisions would then put the lunar landscape in "immediate dialogue" with terrestrial landscapes, by using similar cartographic language.[15] Thus the administrative division is reified, although demarcated by geographic information system coordinates rather than fences. This is not to deny that safety zones have potential as a practical measure, only to note that their benignity is much more complex than might at first appear. The process of cadastral colonialism commodifies the lunar landscape, like the old atlases that featured colonial and capitalist products such as rubber, pineapples, and copra.

FROM THE STONE AGE TO THE SPACE AGE

Mustering ideas from Australia's colonial history to understand these processes on the Moon has more than passing relevance. The Space Age has built itself on the same ideas of technological and social progress as the European colonizers of the nineteenth and twentieth centuries.[16]

"We Are Going" relates to a belief commonly expressed in the space community that space technology is the result of an evolutionary trajectory that starts from the Stone Age and comes to fruition in the Space Age, as part of a natural development of human culture. Oodgeroo speaks of "nature and the past": Indigenous people are relegated to the past of the Stone Age rather than being participants with an equal voice in the present. It is presumed that Indigenous people will be left behind rather than being included as participants in the Space Age as a transformative experience that, perhaps, reshapes what it means to be Indigenous. The Stone Age narrative is being actively counteracted by Indigenous futurist movements.[17]

The "Stone Age to Space Age" metaphor draws on an archaeological framing of human culture that emerged in the nineteenth century, in which technological progress is the companion of physical and intellectual evolution. It is sometimes represented humorously, as in the famous Australian surveyor Len Beadell's cartoon of an Aboriginal man launching a rocket using a woomera or spear-thrower—a very visual representation of the evolutionary trajectory, worthy of General Pitt Rivers himself.[18] It appeals to the social Darwinist progression of primitive to barbarous to civilized society.[19] While space is meant to be the province or common heritage of all humanity according to the UN space treaty system, the metaphor draws a distinction between those who are seen to be spacefaring and those who are not—despite partaking of the "urge to explore" that is supposed to be an integral part of being human.[20]

THE COLONIAL LANDSCAPE OF WOOMERA

However, to attain the Space Age, the leading "spacefaring" nations relied on their colonial territories to launch rockets. Supplanting the Stone Age was seen as an inevitable consequence of the leap into space. This was a deeply political and colonial process that is nowhere better exemplified than at the launch site of Woomera in South Australia.

At the end of World War II, to develop rockets for the purposes of delivering nuclear warheads, Britain had to seek an area of land large enough to launch rockets and with low enough population numbers to minimize the risk of explosions and impacts.[21] It found this in a desert that appeared to be empty, in the outback of South Australia. However, the appearance of emptiness was an illusion. Kokatha, Pitjantjatjara, and other Aboriginal people had been dispossessed by ongoing processes of missionization, drought, and removal from their land, but they were still there—to the degree that the new range had to employ Native Patrol Officers to control their movements during launch campaigns.

The establishment of Woomera in 1947 provoked a nationwide protest about weapons development and the impacts on Aboriginal people.[22] A broad coalition of activist groups, including the Communist Party of Australia, the Aborigines' League, women's organizations, and temperance groups, organized a series of protests across the country. Some anthropologists, such as Donald Thomson, supported the protest, but others, such as A. P. Elkin, claimed that Aboriginal people were dying out and all this did was to hasten the inevitable. The protest movement was a catalyst and training ground for Aboriginal rights activists, such as Doug Nicholls, Margaret Tucker, William Onus, and many others; and the cultural significance of the Woomera launch site is as much about this as it is about the development of Australia's space.[23]

Rather than being a historical footnote, the protests about Woomera emphasize the entanglement of Indigenous land rights with sovereignty, security, and space exploration. The failure of the protests, despite the mobilization of a large sector of Australian society, cemented the nexus of space exploration with defense.[24] When Australia wound down its space activities at the end of the Apollo era, Woomera continued but with a primarily defense function. Space was forgotten in favor of weapons, but Aboriginal people were still largely excluded.

TECHNOLOGIES OF ALIENATION

Woomera is a cultural landscape that represents all these processes. The infrastructure of rocket launch and associated nuclear testing sits side by side with the material traces of deep and ongoing occupation of the land—stone tools, rock art, and campsites—and the intangible geography of cultural knowledge. Seen as a cultural landscape, this is not the chronological sequence of archaeological time periods, but a temporal and cultural mosaic.

Australian policies from the 1940s to the 1970s ensured that Indigenous people had no place in the new space industry. The policy of assimilation led to the removal of Aboriginal children from their families and the enclosure of "full bloods" on missions and reserves. "Half-caste" children had the potential to be made useful to society, and girls were trained for domestic service while boys were trained to be laborers. Many women ended up being employed in Woomera households.[25]

No disjunction was seen between the living presence of Aboriginal women in the township and two popular pastimes for the engineers at Woomera: collecting archaeological artifacts and obtaining souvenirs of Aboriginal culture. The Natural History Society made collections that formed the basis of the Woomera Heritage Centre. In the Old Heritage Centre, stone tools were displayed next to birds' nests and geological specimens, replicating the administrative lumping of Aboriginal people with flora and fauna.[26] There was a vigorous market in souvenirs, with a local bringing art and artifacts down from the Northern Territory for sale to the international rocket teams. Then, as now, "authentic" Aboriginal people were always elsewhere.

The colonial mechanism of collecting artifacts and specimens was practiced at other launch sites too. At the French launch site of Colomb-Béchar in Algeria, men plucked stone tools out of the desert sand like the dried and dead bones of the safely distant past. At the same time, Tuareg people were confined and surveilled to prevent them from accessing the range areas.[27] The distancing of Indigenous people through relegating them to an archaeological past was a method of justifying their physical exclusion from the new rocketscapes.

Rather than a peripheral thread, the Stone Age/Space Age narrative is integral to the construction of a collective white identity at early rocket ranges like Woomera. Those expelled outside the perimeter were the abject, evolutionarily unqualified for entry to the ranks of the spacefaring.[28] The expulsion from paradise and consignment to the lower circles was a global manifestation of settler colonialism in the Cold War.

In the US, the town of Triana is nestled outside the Marshall Space Flight Center in Huntsville, Alabama. Triana was established in the early 1800s by white landowners growing cotton with slave labor. As a local history states, "after the Civil War, Triana suffered the same fate as the rest of the towns in Madison County. Their slaves taken away from them and their lands devastated, they had to make a new beginning."[29] By the turn of the century, most white families had left. Now occupied by former slave families, Triana was described as a "ghost town." The local history cited above is silent on the details of the transition. In the 1960s, about 250 people lived in the town. While Wernher von Braun was realizing his rocket dreams, Triana languished under the shadow of the rockets without water, sewage, street lighting, and paved streets.[30] When the African American mathematician Clyde Foster, a NASA employee and protégé of von Braun, took up residence in the town, he was "appalled at the signs of poverty and despair within actual sight of the towering Saturn V test stand."[31] How Foster resurrected the town is a story beyond the scope of this chapter.

A similar process took place at the White Sands Proving Ground in New Mexico and at the European Space Agency launch site at Kourou in French Guiana.[32] At White Sands, the Mescalero Apache were moved to reservations on the edge of the mountain basin enclosing the range; they were deemed to have little claim to the land because they had migrated there in the ninth century CE.[33] The high security perimeters of rocket launch sites created social and economic boundaries with the elites in the interior and the abject on the outside. The division between Stone Age and Space Age was spatial, not chronological, and, in the claiming and zoning of lands for rockets, cadastral.

The metaphor continues in the way the story is told to modern audiences. Over a decade ago, the Woomera Heritage Centre was redesigned. I was peripherally involved and had discussions with various other participants about integrating the Indigenous story with the rocket story. The final design separated technology and rockets from social and political aspects. To learn about Indigenous history at Woomera, and the experience of local pastoralists, the visitor must walk across the atrium to a separate room. The layout reinforces boundaries between interior and exterior, even when the spaces overlap. Aboriginal people are still metaphorically on the outside, rooted to land, while the privileged reach for the stars.

These examples show how the "emptiness" of these landscapes, which made them suitable for rocket and missile launch, was not a state of being but a process that needed continual reinforcement to keep others out.

WILDERNESS, WASTELAND, AND EMPTY LAND

Emptiness is about a certain kind of footprint. Displaced people have often been transhumant pastoralists (in the case of USSR launch sites) or hunter-gatherers moving seasonally, lacking the hallmarks of "civilization" such as cities and agriculture. Not only does the land become empty, but it is assumed to be "natural" rather than the result of human curation over centuries or millennia.[34] "Remote" or arid locations were also assumed to be inert or formless, lying fallow for a new technological use. As the Australian writer Ivan Southall said of the Woomera region, "Here it was, one of the greatest stretches of uninhabited wasteland on earth, created by God specifically for rockets."[35]

Just as the cadastre creates metaphorical links between Earth and Moon, so too can concepts of terrestrial landscapes, defined in terms of their occupation or utility to humans, migrate to color perceptions of otherworldly places. Abiotic landscapes, particularly on airless worlds, are not easily characterized using existing terminology. Is it a wilderness, or a wasteland? Is it untouched by human hands, or as if it had already been ravaged to the point of death by the harshness of exposure to space?

Wildernesses can become wastelands through human intervention. There is a temporal element to this, as we see in the Wikipedia entry on mining: "Long-term planning is required to achieve sustainability and ensure that future generations are not faced with a barren lunar wasteland by wanton practices."[36] This statement implies that the preindustrial lunar landscape is productive and has immanent value; it does not become "barren" until the minerals are depleted.

Southall's conception of the wasteland is slightly different. This is not a landscape destroyed by mining—although in the 1960s, uranium, copper, and other minerals were the target of mining in this region—it is a wasteland more in the European sense of being beyond "improvement." Rocket ranges in Australia, the US, and Algeria were located in landscapes classed as deserts, synonymous with unproductive wastelands. When low populations were desirable, however, these lands suddenly acquired a new value. They were reclassified as "empty" rather than as a wilderness or a wasteland. The resonances run even deeper. White Sands has been called a "cratered wasteland," recalling the Moon.[37] Woomera echoes the red sands of Mars.[38] Colomb-Béchar is a lunar landscape.[39] The values applied to these space landscapes establish an equivalence with their imagined destinations.

The "myth of the empty land" is foundational to colonialism. Emptiness establishes legitimacy; it is a prelude to possession. The concept has biblical antecedents in the occupation of contested territories in the sixth and fifth centuries BCE Levant.[40] It has also been applied in numerous geographic contexts over centuries.[41] The reasons given for the apparent emptiness are frequently full of prejudices and assumptions, including desertification and degradation from non-Western forms of agriculture and governance.[42] Other rationales are the equation of Indigenous people with animals and the complete destruction of all prior inhabitants or settlements.[43] Critics of the "empty land" trope note that it is more accurately "emptied land" as colonial processes drive people from their traditional country.[44] "Emptiness" is then projected onto other celestial bodies, which *become* empty, acquiescing to exploitation by their silence.

There is a long history on Earth of perceived wastelands becoming valuable, whether because the low populations mean the areas are suitable for launching rockets, or because of mineral resources. What was "wasteland" suddenly has productive potential and is reclassified to legitimize the requisite activities. The division of the Moon's landscape into productive and unproductive places has already begun, particularly in the Permanently Shadowed Regions.

THE LUNAR PERMANENTLY SHADOWED REGIONS AS A CONTESTED LANDSCAPE

Permanently Shadowed Regions (PSRs) are cratered landscapes tilted at such an angle that sunlight never illuminates the crater depths. They are a rare feature in the solar system. The US Dawn mission (2015–2018) identified Permanently Shadowed Regions on the dwarf planet Ceres in the asteroid belt, and the Messenger mission (2012–2015) located them on Mercury. There are none on Earth.

On the Moon, there are PSRs at both lunar poles, where the deep regions of some craters have not received direct sunlight for two billion years. They are among the oldest shadows in the solar system. Balancing these Craters of Eternal Darkness are the Peaks of Eternal Light," which are almost always in full sunlight. The shadows trap volatiles like water ice, a critical resource in providing fuel for a lunar economy and for travel further into the solar system. The value of the shadows is assessed in economic terms, but I argue that they also have cultural value due to their rich associations, and intrinsic value as a rare landscape type that contributes to the geodiversity of the Moon.[45]

The ice is of intense interest as a source of rocket fuel. It is the focus of most lunar missions currently in planning or execution. For example, NASA's Lunar Flashlight probe was launched in December 2022 with a mission to shine infrared lasers into the craters and map the chemical composition of ice deposits (the mission failed, and the probe went into

4.1. Visualization of Shackleton Crater in the Permanently Shadowed Regions of the South Pole showing the Lunar Reconnaissance Orbiter (Credit: NASA Scientific Visualization Studio).

heliocentric orbit). The shadows have been mapped by other lunar orbiters such as Chandrayaan 1 (2008–2009), which also used infrared laser, and the Lunar Reconnaissance Orbiter (2009–present), which uses the LAMP instrument (Lyman-Alpha Mapping Project) to shine far ultraviolet light into the shadows (see fig. 4.1).

In the PSR, the gradations between light and shadow are being mustered as cadastral divisions. The Peaks of Eternal Light are reconfigured as solar energy sources and the Craters of Eternal Darkness as mineral sources. There are shadows within the shadows: the edges may receive sunlight reflected off the crater's rim, while the double-shadowed regions deeper in are illuminated only by starlight or zodiacal light.[46]

These geological or selenographic entities are being redefined as cadastral boundaries—zones that define the kinds of activities that take place there and the value placed upon them as real estate.

A form of contestation is emerging in dialogues around the environmental values of the PSR and what should be preserved intact from mining impacts for scientific reasons.[47] In line, however, with current heritage theory, which dissolves the division between natural and cultural, it is no longer useful to separate these values, or to dismiss either as individual concepts.[48]

As well as the Dantean landscape already created by the naming of peaks and craters in the PSR in relation to light, there is also a concrete example of Indigenous values coming into conflict with the overwritten "emptiness." In 1999 the US Lunar Prospector spacecraft, carrying a portion of the ashes of the renowned planetary scientist Eugene Shoemaker, crashed into the crater named after him in the PSR. Navajo president Dr. Albert Hale was appalled at this action, saying in an interview, "The moon is revered and it regulates life cycles, according to Navajo traditions and stories. To send something like that over there is sacrilege."[49] NASA undertook to consult more widely if a similar action were planned for the future. To send ashes to a dead world is not seen as problematic by a western space agency, but to contaminate a living body with death is not a trivial matter in Navajo sensibilities. In their public representation, such stories are used as examples of the Stone Age/Space Age trope, but they also highlight the poor fit of western landscape concepts. As Bawaka Country and coauthors say of the Australian context, "These are not trinkets or novel stories to share, but more-than-human modes of belonging: relationships of rights, responsibilities and kinship that bring with them particular modes of governance."[50]

The PSRs have been transformed from a curiosity to a resource, because some humans have identified a utility in the shadows. The processes of resource exploitation are reinscribing the colonial values rejected on Earth in a new location.

BECOMING STRANGERS TO THE MOON

Oodgeroo's poetic analysis speaks of the geographic process of becoming strangers at home. While anchored in the lived and historical reality of colonized Australia, I believe it also has a broader relevance to thinking about space.

As the Moon is mapped in increasing detail and becomes the focus of more intense studies, it is becoming more familiar to us, analyzed, chopped up, and dissected as thoroughly as a frog on a laboratory bench.

Yet, to exploit the Moon, we must make it strange and unfamiliar, instead of making it kin, as Dovey writes.[51] We have to distance ourselves from the eons of cultural associations, relegating them to the realm of superstition to allow capitalist industry to carve up the Moon.

In this distance lies the essence of the uncanny: the *unheimlich*, the home that has been made strange.[52] Jorgenson writes about the uncanny in the simultaneous representations of the infinite sublime and the domestic in the Apollo 11 mission, which landed humans on the Moon in 1969.[53] The moon landing—alien and distant—was beamed into American living rooms replete with consumerist domesticity and destabilized them: "The home quickly turns into the unhomely, as this vastness makes the familiar strange."[54] This, it seems to me, captures something about how the Moon gets expelled from its ancient cosmological place so it can be redefined as an empty land. It is a process of alienation at many levels and across many cultural contexts. For white North Americans, Jorgenson argues, "It represented their own alienation from a certain version of America. . . . Rather than an heroic victory for the American way of life, the broadcast represented the kinds of disorientation and self-doubt that this society was experiencing during the 1960s. Shots of the bleak darkness of space and an inhospitable moon were as estranging as the astronauts themselves, who spoke in an impersonal, technocratic language to describe the flight."[55] The unfamiliarity is a marker of another aspect of the uncanny—repetition or doubling, walking in circles because you fail to recognize the streets already passed.[56] You cannot return home because you cannot recognize it. Despite its controversial anthropological status, Carlos Castaneda captured the uncanniness of this perfectly in his 1972 book *Journey to Ixtlan*. Of his return journey to Ixtlan, the character Don Genaro says, "In my feelings sometimes I think I'm just one step from reaching it. Yet I never will. In my journey I don't even find the familiar landmarks I used to know. Nothing is any longer the same."[57]

It is the dilemma that Oodgeroo presents in "We Are Going." Country is overwritten with the cadastral grid, the rubbish tip on the bora ground, leaving only the interstices or the shadows free of surveillance. Unmoored from place, people become ephemeral shadow-ghosts. Tsing, speaking of how cadastral colonialism causes alienation for shifting agriculturalists in Southeast Asian forests, writes, "How do people become aware that they are strangers in their own lands? Sometimes they are forcibly removed. Sometimes they are just reclassified."[58]

In space it is a process of double disempowerment. The exclusion of Indigenous people from terrestrial space landscapes, like Woomera, Colomb-Béchar, Kourou, White Sands, and Huntsville, is causally linked to their exclusion from interplanetary governance. Denial of Indigenous land tenure on Earth translates into a lack of no-tenure in space, the same no-tenure enshrined in the Outer Space Treaty. According to received wisdom, there are no Indigenous people in space—and yet Indigenous people are still absent, made strangers in their own country.

SOJOURNERS AND STRANGERS

The theme of estrangement runs through Oodgeroo's poetry. Her stark language, like the utterances of the Apollo astronauts, serves to underscore a sense of alienation.[59] But despite its melancholy and nostalgia, "We Are Going" can also be read as an assertion of power: "We belong here, we are of the old ways." The bora ground, a system of earthworks used in male initiation ceremonies across large areas of eastern Australia, is connected to and mirrored in the Milky Way but desecrated by being turned into the wasteland of the rubbish dump.[60] The white ant-men, hurrying about, are seen in an aerial, or even orbital perspective, from above. They have prevailed, but they have lost too, oblivious to the rich landscape of interconnectedness with thunder, lightning, and the pale daybreak over the dark lagoon.

Oodgeroo was one of the best-selling poets of the era, her poetry bringing an awareness of the impacts of colonialism to a broad Australian audience. She made Aboriginal people present and agentic in the face of genocide; a poetic act that undermined the Stone Age/Space Age dichotomy, even as she highlighted what was being lost.[61] She created a "counter-history" that defied the myth of the empty land.[62]

Perhaps this is what is needed for the Moon and other locations in the solar system. As Byrne says, "The map-making process clearly needs to be seen as a selective, value-loaded one that renders some things invisible in the very act of giving legibility to other things."[63] Byrne talks of "counter-mapping" as a strategy to overwrite the cadastral divisions used to estrange Aboriginal people from country in Australia. In space, it can be a weapon against the designation of wilderness, wasteland, or empty land that does not separate nature and culture, past and present, and restores the strangers to their sky country. For sojourners in this country, there is a lesson to be taken from Oodgeroo's words: "Something is gone, something surrendered, still / We will go forward and learn."[64]

CHAPTER 5

INVENTING SYNCOM

Public, Private, Global

HARIS A. DURRANI

On July 26, 1963, National Aeronautics and Space Administration technicians launched a Delta B rocket from Cape Canaveral Air Force Station in Florida. The rocket lifted Syncom II—a stout satellite the diameter of a large wheel—to an altitude just shy of 23,000 miles. Then, in one action, the Delta B spun the satellite about its central antenna and ejected it into orbit, where the satellite fired a jet, "kicking" into a location high above Mozambique. At a NASA Minitrack station in Hartebeesthoek, South Africa, and a nearby Smithsonian observatory in Olifantsfontein, US and South African technicians spotted the satellite; it was traveling faster than Earth's rotational speed, drifting eastward toward Madagascar. They notified their superiors at NASA Goddard Space Flight Center in Greenbelt, Maryland, and US Army Strategic Command (SATCOM) Agency headquarters in Fort Monmouth, New Jersey. The following afternoon, these superiors sent instructions to the USNS *Kingsport*, a telecommunications-equipped Navy vessel anchored in Lagos Harbor, Nigeria. Technicians aboard the ship were from NASA, Department of Defense (DOD), and the Culver City, California-based Hughes Aircraft Company. They signaled Syncom II to fire small jets on its bottom, reversing the satellite's direction. Over the next few weeks, technicians at the stations in Greenbelt, Fort Monmouth, Lagos, Hartebeesthoek, Olifantsfontein, and two other SATCOM stations in Lake-hurst and Fort Dix, New Jersey, monitored and maneuvered Syncom II, shifting the orientation of the satellite's body and nudging its orbital position westward across central Africa, the Atlantic Ocean, and north-eastern South America. On August 26, they halted the satellite "on sta-tion" 23,300 miles above northern Brazil with its body oriented so the central antenna pointed south, allowing its signals to span the Atlantic. It now orbited Earth at the same rate as the planet's rotation at the equa-tor. On this path, Syncom II appeared from Earth to trace a figure eight ("analemma") over a single equatorial point in Brazil. It had become the first communications satellite to reach the monumental "24-hour orbit."[1]

Communications satellites in the 24-hour or geosynchronous orbit (GSO) offered novel and potentially significant advantages. A satellite in GSO orbited Earth at the same rate as the planet's rotation at the equa-tor. A system of only three satellites in geostationary or geosynchronous equatorial orbit (GEO)—a path in GSO running over the equator so a satellite seemed fixed above a single point—could enable communica-tions between almost any two points on the planet, bypassing political, environmental, and technological obstacles that accompanied telegraph cables and radio systems. Such a system would allow access to emerging commercial markets in developing countries, a tool for proselytizing lib-eralism during the Cold War. It was also viewed as a boon for military operations abroad.[2]

Syncom was a unique achievement. Due to technical and institution-al challenges, NASA, DOD, and industry had lost faith in the feasibility of GSO communications satellites. During the 1950s and early 1960s, engineers in government and industry had considered the idea. But GSO was far higher than the lower orbits occupied by earlier commu-nications satellites like American Telephone and Telegraph Company's (AT&T) Telstar I and Radio Corporation of America's (RCA) Relay I, while the control jets and telecommunications equipment necessary to orient a satellite in that orbit and direct its signals earthward added sig-nificantly to the mass of the spacecraft. As a result, most designs made GSO communications satellites too heavy for existing rocket boosters to reach the extended altitude. DOD was furthest along in developing such a satellite but faced continuing delays. Meanwhile, an interagency agree-ment that divided up NASA and DOD satellite projects slowed further progress. However, Hughes's "spin-stabilized" design (treating the sat-ellite as a gyroscope, thus lowering the number of control jets necessary to maneuver its position) and novel communications equipment dramat-ically reduced the satellite's mass, making it possible to reach GSO. In 1961, after months of institutional wrangling, DOD and NASA formed another interagency agreement on communications satellites, and then NASA signed a sole-source contract with Hughes to develop the Syncom satellites. DOD would allow NASA to use DOD's communications stations in the United States and abroad in order to operate and test the Syncom satellites. While Syncom I failed, II (which reached GSO but not GEO) and III (which successfully reached GEO) facilitated communications with leaders throughout the developing world, including Brazil and Nigeria, and, with the lower-orbit Relay I, broadcast the 1964 Tokyo Olympics.[3] Syncom II became the first satellite used in a war zone when it enabled communications with US troops in Saigon.[4] The satellites were the model for those contemplated in the 1962 Communications Satellite Act (COMSAT Act), which created the COMSAT Corporation and led to its international arm, Intelsat, created in 1964.[5]

Despite Syncom II's evidently extraterritorial qualities, the conventional story of its invention remains a domestic tale of the struggle between public power and private interests in the mid-twentieth century. Official and popular accounts claim that the satellite succeeded due to the persistence of the engineers who conceived the invention at Hughes—Harold A. Rosen, Donald Dickinson Williams, Tom Hudspeth, and John Mendel—and whose ideas were initially disregarded by corporate executives and NASA and DOD officials.[6] More cautiously, but along similar lines, David Whalen situates commercialization as the prime mover, arguing that Syncom's achievement was due not to politics or "government involvement" but the "government market" provided by agency contracts. He further states that, while government (NASA and DOD) provided the material infrastructure for Syncom—launch sites, rockets, and telemetry and command (T&C) stations—the underlying technology, particularly the spin-stabilized idea, was developed by private enterprise (Hughes).[7]

Syncom's conventional story tracks the standard, domestic narrative of the midcentury ascent of the US administrative state. Legal historians usually attribute the ascent of public institutions, for good or ill, to the ideology of public power.[8] Historians of spaceflight often provide a version of this story, depicting the Space Age as one in which belief in the advantages of big government either quashed or facilitated private enterprise.[9] Just as Whalen, using Syncom's story, argues that government acted as a crucible for private innovation, legal historians suggest that the administrative state generated private freedoms. The foundational legal historian J. Willard Hurst asserted, during this period, that the administrative state "releases energy" for commercial and technological development.[10] Meanwhile, legal historians and historians of spaceflight mainly relegate the extraterritorial scope of the administrative state to ideological battles with the Soviet Union.[11] In Syncom's story, as in legal histories of the administrative state more broadly, the search for extraterritorial power remains an unexamined cause of the growth of public institutions.

The standard narrative can be encapsulated in the Yale Law professor Charles Reich's classic 1964 essay "The New Property." Historians generally understand the essay as a reflection of legal thinking about the US welfare state, in which Reich responded to individuals' increasing reliance on "largess"—the vast resources and regulations of the mounting administrative state—by advocating the establishment of private property rights over that largess. Reich's thesis is considered a principal rationale for entitlements like the right to work, education, and social security.[12] Nonetheless, the essay employed the frontier as both metaphor and reality. Reich described the Homestead Act, with the "conquest" of Indigenous peoples, as the model for how private ownership in largess could facilitate the distribution of government resources. Reich also fixated on "research and development" in defense and spaceflight, and he addressed the COMSAT Act, describing the "communications satellite" as an "unusual form of government largess."[13] He might have had Syncom II in mind; at the time, it was one of only a handful of communications satellites in orbit. Reich, as in the standard narrative, attributed the growth of the midcentury administrative state to the rise of public power, eliding its extraterritorial roots.

I argue that, in Syncom's story, the administrative state's ascent was not only an outcome of the ideology of public power. Rather, it was also due to the alignment of public and private actors around what they regarded as a new concept of US extraterritoriality.[14] The new concept posited technological control as a basis for imagining the United States as the first thoroughly global empire. At the same time, the emergence of this concept coincided with the construction of varying ideas of the global.[15]

The chapter proceeds in two substantive parts. The first examines patent litigation related to Syncom. There were numerous patent cases between Hughes and NASA related to several patents. For the purposes of this chapter, I focus on two patents and four cases. The patents are the "Rosen-Williams Patent," filed in 1959 by Rosen and Williams (assigned to Hughes), and the "Williams Patent," issued in 1973 to Williams (also assigned to Hughes).[16] *Rosen and Williams and Hughes Aircraft Company v. the National Aeronautics and Space Administration* (*Rosen*) was a 1966 interference proceeding (defined below) before an administrative body within the US Patent Office. The administrative body decided that the office should issue title on the Rosen-Williams Patent to NASA, not

Hughes.[17] *Williams v. NASA Administrator* (*Williams*) was a 1972 interference case in which the Court of Customs and Patent Appeals issued
title on the Williams Patent to NASA over Hughes.[18] *Hughes v. NASA*
(*Hughes I*) was a 1980 case in which the Court of Claims decided that
Hughes's Williams Patent was not invalid.[19] *Hughes v. NASA* (*Hughes II*)
was a 1993 case in which the Court of Federal Claims (the successor to
the Court of Claims) decided that NASA and other corporations had
infringed on Hughes's Williams Patent.[20] I discuss the substance of the
patents and cases below.

This chapter focuses on the first dispute, *Rosen*, over the Rosen-
Williams Patent, due to this dispute's historical significance as one of
the first major cases on the patents behind Syncom. In *Rosen*, Hughes
and NASA lawyers and engineers debated the nature of the invention at
stake. Whether the invention behind Syncom was considered public or
private depended on how actors understood the material infrastructure
surrounding Syncom II's historic launch and orbital maneuvers. Much
of this infrastructure was extraterritorial, located at sea, on land, or in
space beyond the conventional boundaries of the United States. Thus,
the public/private debates about the nature of the invention were also
about its extraterritorial contours.

The second substantive part examines why NASA, Hughes, and
DOD personnel collaborated on the use and extension of government
infrastructure. They collaborated because they shared a particular vision
of US empire: technological control over the globe as an abstract space.
I sketch how this concept of empire was imagined and practiced, from
Hughes and NASA documents, to uses of Syncom II and III abroad, to
international law scholarship, to United Nations debates about the regulation of the first generation of commercial communications satellites,
which were modeled on the Syncom design.

PUBLIC, PRIVATE...

The claim that Syncom was a triumph of private enterprise begs the
question: What exactly was the thing that triumphed, and what was
public, or not, about it? If the question is about the entity, or common
carrier, that later operated the system of GEO communications satellites
based on Syncom, then the answer is that COMSAT was a corporation
whose stocks were privately owned. However, COMSAT's shares were,
in part, publicly traded. More significantly, its corporate governance was
partly controlled by the president and the Senate; in turn, Intelsat, while
an international consortium, was effectively controlled by COMSAT.[21]
Reich pointed toward these ambiguities: "The Communications Satellite

represents an unusual type of subsidy through service: the turning over of government research and knowhow to a quasi-private organization."[22] But Reich's statement is not wholly accurate, thus inviting other questions. What was public or private about the ownership or regulation of the Syncom satellites, the stations that controlled and monitored them, the electromagnetic spectrum in which the satellites' signals traveled, or the invention underlying these material elements? Because of the extraterritorial locations of most of the material elements, the question of the public nature of any aspect of Syncom was concurrently one of extraterritorial power.

Patent disputes between Hughes and NASA about the invention underlying the Syncom design provide a frame by which to interrogate the multifaceted question of the satellites' publicness.[23] They also point toward the extraterritorial story outside the standard narrative. Histories of Syncom often refer to the patent disputes as a single, brief, and significant event, evidence that the engineers, and Hughes, had invented the satellite and not NASA, thus affirming that public institutions had usurped or, as Whalen documents, were not involved in private innovation.[24] However, these references refer to a 1993 case, *Hughes II*, in which the Court of Federal Claims decided NASA and industry had infringed Hughes's patent on the apparatus and method for controlling the motion of a spin-stabilized "body."[25] Referring to *Hughes II* disregards the various cases, some of which Hughes lost to NASA, on distinct legal issues about multiple patents underlying the spinning satellite. Ultimately, citing the 1993 case reveals little about how Syncom's public, private, or global nature was understood during the launch and operation of the first successful satellite in the program, Syncom II, in 1963. At that time, one of the major cases on the horizon was a 1966 administrative law hearing, *Rosen*, before the Board of Patent Appeals (the Board) in the Patent Office. The question before the Board essentially asked what Syncom's conventional narrative presupposes: What exactly was the invention, and was it public or not? The question necessarily involved the satellite's extraterritorial scope.

Hughes II and *Rosen* differed in two significant ways. First, they addressed different legal questions. *Hughes II* concerned itself with whether NASA and other companies' satellites had infringed on a Hughes patent. This required asking whether NASA had used the patent for their satellites without Hughes's permission. By contrast, the earlier *Rosen* case was an interference proceeding that asked whether a patent should be issued to NASA or Hughes. An interference proceeding was mainly an artifact of first-to-invent patent systems (which the United States adhered to until 2011). Under first-to-file, the first party to file a patent with the Patent Office receives title. However, under first-to-invent, the first to invent received title. Invention could be constructive, wherein the party files the patent, or actual, wherein the party "reduces to practice" the invention (i.e., shows the invention to work in fact, rather than in theory). Most interference proceedings involved two parties that separately claim to have invented, actually or constructively, the same invention. For this reason, as the Board noted in *Rosen*, the hearing was technically not about an interference, because it involved two parties disputing title to one invention. Nonetheless, the hearing was treated as an interference proceeding because the National Aeronautics and Space Act of 1958, which created the agency, required that disputes about patents related to government contracts be adjudicated as interference proceedings.[26]

Second, the cases regarded different patents. *Hughes II* concerned the Williams Patent, which was issued to Hughes in 1973, while *Rosen* concerned the Rosen-Williams Patent—which, while filed in 1959, had not yet been issued, as that was the subject of the interference proceeding. For our purposes, the key difference between the two patents was the material scope of the inventions described in each.[27] In a 1980 case, *Hughes I*, on a separate legal issue (the validity of the Williams Patent), which led to *Hughes II*, the Court of Claims (the predecessor to the Court of Federal Claims) distinguished the two patents. The court wrote that the Rosen-Williams Patent defined a specific type of "ground control device," which allowed for control of the satellite's jet in orbit, whereas the Williams Patent did not define the controller but allowed for any such ground control device to fulfill that function. According to the court, the Rosen-Williams Patent defined the type of device due to events during the examination of the patent application in the Patent Office: when Hughes lawyers had negotiated with patent examiners to build the patent's claims, examiners had requested that the controller be defined in the claims. By contrast, in the Williams Patent, the Hughes lawyers had been able to leave the controller unspecified.[28]

The controllers were important because they were located in the NASA and DOD T&C stations in Nigeria, South Africa, Maryland, and New Jersey. *Rosen* determined that reduction to practice of the Rosen-Williams Patent occurred during Syncom II's first orbital maneuvers in 1963, which were operated via the "synchronous controller" aboard the *Kingsport* in Lagos Harbor. The location of the controllers mattered because the Rosen-Williams Patent defined them as part of the invention, and thus the fact that these devices were operated from government facilities meant that reduction was conducted by NASA, not Hughes, and thus that NASA should retain title.[29]

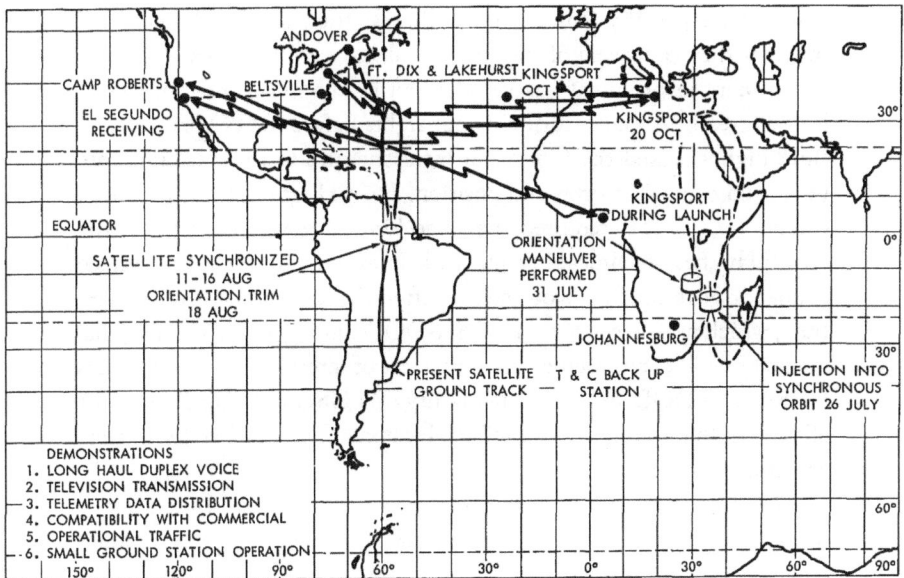

5.1. T&C and communications stations, as related to orbital maneuvers. The maneuvers between "injection into synchronous orbit" on July 26 until its position as "satellite synchronized" on August 16 are those that *Rosen* construed as reduction. *Report*, 16.

By contrast, *Hughes I* found that the Williams Patent properly reflected that the controller was not an "essential" element of the patent, as any such device would do.[30] Whalen's claim that Syncom was a product of Hughes's private innovation makes a similar assumption that these devices were not necessary to the core invention: while NASA and DOD provided the stations and launch vehicles, the idea was conceived and developed within Hughes.

However, it was less clear in the early to mid-1960s—when GSO satellites were a nascent technology—what was or was not essential to the invention. What may be essential to a court or administrative body at any particular time does not necessarily indicate what was considered essential by historical actors in the throes of the inventive process. On the one hand, as *Hughes I* retrospectively emphasized, Williams testified that the controllers were not essential and that the addition of the definition of the controllers in the Rosen-Williams Patent was solely because the patent examiners requested it.[31] However, the invention evidently could not have been tested in actuality without the use of the government's infrastructure, much of which was extraterritorial, comprising the launch vehicles and stations around the world (see fig. 5.1). Hughes engineers, especially Rosen and Williams, tried adamantly from the moment they conceived of the spin-stabilized satellite in 1959 to avoid "government involvement," as Whalen put it, and Williams was particularly aggressive about securing patent rights before the NASA contract. But Hughes lacked the funds and infrastructure to lift a satellite into GSO and control its orbit and position. Even Hughes's better-financed competitor, AT&T, required the same kind of government infrastructure.[32] Hughes's technical ingenuity likely would not have been realized if not for government involvement. After all, the Hughes-NASA contract, which was preceded by a NASA-DOD interagency agreement that made it possible, represented an alignment of interests between Hughes, NASA, and DOD to test the feasibility of a GSO communications satellite.[33] In that sense, the controllers, located at the T&C stations, were essential to the invention.

Years later, dissenting in *Hughes I*, Senior Judge Byron Skelton was so insistent that the controllers were essential that his opinion was six times longer than the majority's. Skelton's opinion is too detailed to recount, but one central claim was that the satellite's movements were "synchronized" with the controllers, meaning that these movements could not have been conducted without the spinning body (the satellite) and the external controller in separate positions while the body moved as a freely spinning object in orbit. He also showed that the engineers frequently described the controllers throughout the patent claims and testimony.[34]

Nonetheless, even if the controllers were not essential as understood in the 1960s, there remained the question of where and when the invention arose. *Rosen* partly hinged on when and where the reduction to practice of the Rosen-Williams Patent first occurred. If it occurred during the April 2, 1960, tests on Earth (which involved spinning a "dynamic wheel" that simulated the satellite's movement), then title would go to Hughes, because those tests were conducted in Hughes facilities prior to the signing of the NASA-Hughes contract that inaugurated the Syncom Program in 1961. But if it occurred during the orbital operations that put Syncom II into its position in GSO in summer 1963, then title would go to NASA, because those operations were controlled from government facilities after the contract was signed and in performance of that contract.

Hughes I distinguished the Williams Patent by highlighting that the Rosen-Williams Patent defined controllers as part of the invention. But this obscured how the invention was understood in the 1960s. In *Rosen*, the Board did not rest its finding on whether the controllers were essential, but asked whether the Hughes tests could constitute reduction.[35]

Ordinarily, it would not have been a problem for Hughes to show reduction to practice, since Hughes's 1959 filing in and of itself would constitute constructive reduction two years prior to the 1961 NASA contract. However, when the 1958 NASA Act granted the agency title to patents "made in the performance of any work under any contract," it defined "made" to include "actual reduction to practice."[36] Thus, if the invention were first *actually* reduced to practice during the satellite operations in 1963, Hughes would not be able to claim priority based on its earlier filing in 1959.

The Board found that Hughes's ground tests did not constitute reduction because reduction could only have occurred during the orbital maneuvers. It determined that reduction required showing the "utility" of the concept, which could be evidenced by the opinion of experts in the field. An aerospace expert testified that, after the tests, he "*thought*" that the model was in "excellent shape" and that the tests had "demonstrate[d] the basic *principles*" (the Board's emphasis). The Board thus characterized the tests as showing the mere likelihood of success "in theory," not the invention "in fact." While NASA signed the contract with Hughes after these tests, the Board argued that NASA's willingness to collaborate reflected not a belief that the Hughes model actually worked, but a confidence in the suitability of the Hughes model for actual testing by operation in orbit.[37] The Board alluded to the fact that when Williams and Rosen first submitted affidavits, they wrote that the patent claims pertinent to *Rosen* had been reduced during Syncom II's orbital maneuvers. A month later, they requested substitution with a second pair of affidavits stating that those claims were reduced during the Hughes tests.[38] Until a GSO communications satellite was shown to work in orbit, it remained unclear whether the control system for a spin-stabilized satellite, however tested on Earth, had been actually invented.

Meanwhile, the 1972 case, *Williams*, reasoned without even discussing whether the controllers were essential. The Court of Customs and Patent Appeals decided Hughes, not NASA, owned the Williams Patent (i.e., the equivalent legal question to that addressed in *Rosen*), on appeal from a similar dispute before the Board, which had likewise decided title should issue to NASA, not Hughes. *Williams*, reversing the Board on appeal, employed legal analysis similar to the Board's in *Rosen*, based on the same facts, but came to the opposite conclusion on the Williams Patent. The court asked whether the Hughes tests (the same as those for the Rosen-Williams Patent) constituted reduction of the Williams Patent by demonstrating the invention's "substantial utility." Likewise, the court also excerpted the same testimony from the aerospace expert and referred to the fact that NASA signed the contract after the tests.

However, unlike *Rosen*, *Williams* found that these facts were sufficient to show utility.

In *Williams*, the court did not pursue *Rosen*'s more searching analysis to demonstrate utility "in fact" or discuss NASA officials' belief that the tests merely demonstrated suitability for actual testing in orbit (not actual utility on its own).[39] Indeed, archives and oral histories show that NASA officials were consistently skeptical of the Hughes model, including during the period, lasting more than a year, between testing in April 1960 and the contract on August 11, 1961.[40]

Ultimately, whether the Williams or Rosen-Williams Patents were privately or publicly owned was historically contingent on how various actors—agency officials, judges, engineers, and lawyers at Hughes, NASA, and the Patent Office—emphasized the importance of various material elements beyond the United States: the satellite, the controllers, and the space environment. As a matter of law, the Board and the courts conceptualized these elements inconsistently. But, in historical analysis, arguing that Hughes alone developed the invention behind Syncom II erases not only the materiality of the inventive process, but also the extraterritorial contours of the Syncom satellites and their infrastructures.[41] After all, the longest portion of *Rosen* was dedicated to "the important question of whether it [the reduction] occurred in the United States," because US patent law at the time required reduction to occur within the country's territorial boundaries. The Board went to great lengths to show that the control points around the world were themselves controlled by, and thus under the jurisdiction of, the United States.[42] In the case of the spin-stabilized satellite's invention, to demarcate the intertwined relationship between public and private is to elide the extraterritorial story lurking beneath the surface. The simple fact is that Syncom II would not have been possible if not for the stations around the world, and it was out of this recognition of mutual need that Hughes, NASA, and DOD had agreed to work together.

While the patent litigation presented a tension between public and private, my point is that historians ought to be cautious about adopting those actor's categories. The way Hughes and NASA argued for and against the extraterritoriality of the system in *Rosen* differed from how they collaborated in practice on Syncom. I suggest that historians study not only conflicts between public and private but also the role of extraterritoriality in shaping what was understood as public or private in the first place.[43] Scholars ought not take cases, or even classics of legal literature like Reich's, at face value. It goes without saying that the decision of a case is a social construction. After all, as I have shown, the patent cases were filled with fissures and contradictions.

But outside of the cases, there was an alignment in practice, outside the auspices of the Board and courtrooms: public and private actors came together to share infrastructures, technologies, and ideas. In the patent disputes, public/private was a demarcation problem, where boundaries were drawn according to the significance placed on extraterritoriality when defining the inventions behind the satellite design. Instead of inquiring into the nature of the public/private split, the bigger historiographical question is why decision makers in these institutions were eager to collaborate.

. . . GLOBAL

Hughes, NASA, and DOD came to a mutual agreement on Syncom because they shared a common objective. Following the 1961 contract and interagency agreement, Syncom was explicitly imagined and practiced as a new kind of imperial project. A 1962 Hughes brochure opened:

> Since the days of the Greek City-States, population, government and civilizations have been involved in a dramatic and evolutionary process . . . their progress based on resources and the ability of men of imagination to provide access to resources. States that dominated commerce controlled communications between governments, and frequently controlled their freedoms. Today, closely allied with the emergence of new and progressive nations, there exists a fundamental need for free and adequate communications. From political and economic standpoints, present-day technology must provide new answers to the problem of world communications. This brochure portrays the historical development of many nations—and the concept of a common communications carrier between them.

The anonymous Hughes authors lauded the ability to connect "small or newly emerging nations where geography or economics have previously presented obstacles" and to "substantially increase the capacity of communication services between the world's leading government, military, cultural, and population centers that are now served by wire, land or submarine cable, microwave, and VHF [very high frequency] or HF [high frequency] radio links." Specifically, the program "represents a potentially powerful influence for improved international understanding."[44] This invoked Cold War battles over ideology in the developing world.[45] Every other page of the brochure featured an illustration of the territories of a prior empire or colony: the "Roman Empire," the "Mongol Empire," "Italy in the 15th Century," the "Conquests of the Ottoman Turks, 1418," "Europe in the 17th Century," "Colonial Latin America," the "United States during the Revolution," "Canada," and "Africa." Each

Aeolipile.

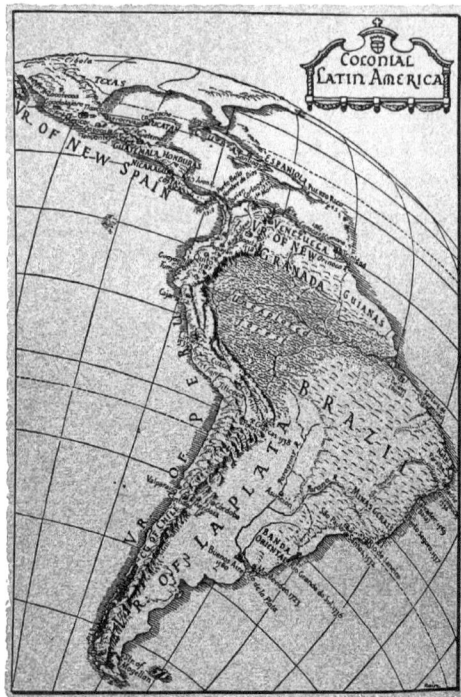

A number of overall design factors provide a solid economic basis for the Syncom system. One such factor is the five-year design life of the Mark II spacecraft. This means economy not only in the requirements for spacecraft, but also in terms of amortizing the cost of booster vehicles and in minimizing schedule demands on launching facilities.

Another key to Syncom economy is its simplicity of function. The spacecraft operates exclusively as a communications relay point in space. This not only eliminates unnecessary complexities in spacecraft design, but permits subsystem equipment requirements to be tailored to anticipated traffic loads at individual ground stations. The ground stations will handle such technical functions as switching, channel selection and grouping, and system control.

Syncom, which requires minimum radio frequency spectrum, also offers economy through the versatility of its performance. Syncom design provides for dual mode operation of the spacecraft repeater, with the choice controlled from the ground stations. In one mode, the spacecraft will receive and retransmit television or other wideband signals. In the other mode, using single sideband modulation from ground to space, it serves as a high capacity telephone exchange. Because of a design feature called "multiple access," the Syncom spacecraft can receive and retransmit simultaneously, from all ground stations, signals representative of all types of telephone communication service.

5.2. One among the many two-page spreads in the 1962 Hughes brochure. Hughes, Brochure, 12–13.

image faced a description of a technological, political, or economic benefit of Syncom on the opposite page (see fig. 5.2).[46]

The authors depicted Syncom as an evolutionary progression from prior empires' use of communications to control territories. The historical rupture was one of degree: Syncom diverged from precedent in that, to the authors, it would allow the first communications carrier to provide control over nearly instantaneous communications and thus resources, commerce, and ideology at a truly global scale. The other difference, unstated but obvious, was that the United States had not claimed territorial sovereignty over every country on the globe for which Syncom promised control. In other words, the brochure implied that technological control, via communications satellites, had replaced sovereignty as the sine qua non of imperialism.

The brochure's purpose is unclear. Perhaps it was produced to promote Hughes in its negotiations with agencies, but this is unlikely, as it is dated later than the 1961 contract. Perhaps it was meant to sponsor Hughes, NASA, and DOD as they sought support from Congress

5.3. Darcey at Goddard, "check[ing] the location in orbit of Syncom II over Africa" during the 1963 orbital maneuvers. 63-SYNCOM II-16.

in anticipation of COMSAT. Most simply, it may have been a badge of honor distributed to employees or decision makers at any of these institutions or to the public. Regardless, its ideas likely reflected an appeal to a shared, normative way of imagining Syncom. That normative imaginary was empire—specifically, a concept of empire that imagined technological control of the globe as an abstract space, a substitute for territorial annexation.

The imaginary of the satellite system as the technology of a new form of global empire was predominant in official and public discourse about Syncom. Dispersing signals throughout the world, these satellites would control communications on a global scale. A NASA-commissioned photograph depicted the Syncom project manager Robert Darcey in Goddard, studying Syncom II's location amid the maneuvers of summer 1963 (see fig. 5.3). Whether candid or staged (perhaps more so if the latter), the photograph illustrated the normative idea of empire by showing Darcey leaning against a map prominently labeled "THE WORLD," finger pressed on the center of the satellite's analemma over the Democratic Republic of Congo. The image of Syncom II's analemma on a map of the world, ubiquitous in the press and in NASA and Hughes reports, flattened the satellite's altitude to simplify planning for satellite maneuvers. Intentionally or not, the recurrent visual linked control of orbit to that of territory, a link underscored by markers for the T&C and communications stations. Images like these suggested that Hughes, DOD, and NASA could freely exercise technological control abroad, including in other nations' sovereign territories, and especially from space.

Syncom's imperial ambitions translated into the practical operations that successfully launched, placed, and used the first satellite in the 24-hour orbit. The operation of the African stations and positioning of Syncom above South America were part of a series of US efforts to exert extraterritorial power. The stations in Lagos Harbor (and perhaps South Africa) aligned with President John F. Kennedy's broader policy of cooperation in Africa, for which he imagined Nigeria as a "metropole" by which to liberalize the continent's decolonizing populations.[47] In a publicized event in August 1963, Kennedy conducted the first satellite call between heads of state when he spoke with the Nigerian prime minister Abubakar Rafawa Balewa via Syncom II. During the call, Kennedy lauded the event as an auspicious start for cooperation, recalling Balewa's recent visit to the United States and their partnership on the 1963 Partial Test Ban Treaty. They joked about the defeat of the US boxer Gene Fullmer at the hands of the Nigerian British Dick Tiger in the world middleweight title. Balewa spoke little more than sparse "Yeses" and "Thank yous" and a sentence of congratulations to Kennedy for "this very big achievement."[48] This, despite NASA and DOD officials' awareness of Nigerians' popular protests regarding the government's collaborations with the French, British, and US governments.[49] Kennedy's African policy of influence through cooperation matched Hughes, NASA, and DOD objectives to render the Syncom satellites, as the Hughes brochure put it, into "a potentially powerful influence for improved international understanding."

Likewise, Syncom's connections with Latin America and the Caribbean enabled US attempts to exert extraterritorial power over communications in the Americas. In October 1963, Syncom II facilitated the first transoceanic press conference by satellite. The *Kingsport* had departed from Lagos, conducted the first satellite communications (via Syncom II) while moving at sea amid a "squall" forty miles from the Nigerian coast, traveled north, and docked on the coast of Rota, Spain. Communications ran from NASA Headquarters in Washington, DC, north via telephone line to Lakehurst, up via radio to Syncom II, down to the *Kingsport* in Rota, and finally northeast via telephone line to an International Telecommunication Union (ITU) conference in Geneva.[50] Titled "The 1963 Extraordinary Administrative Radio Conference (EARC-63)

to Allocate Frequency Bands for Space Radiocommunication Purposes"
(Space Conference), it was the ITU's first major conference on space
radiocommunications.[51]

The Space Conference was held at the behest of numerous decision
makers in US agencies and industry. They aimed to allocate a wider portion of the electromagnetic spectrum for space services under the ITU
Radio Regulations. An expanded range would let DOD and COMSAT
operate satellites in higher orbits, particularly GSO, for the system of
satellites modeled after Syncom. DOD and COMSAT wanted to be
able to communicate with "remote areas of the world," especially for US
military operations in Latin America and the Caribbean. Cuba was of
concern in the wake of the 1962 Cuban Missile Crisis, when Kennedy's
communications with Latin American and Soviet leaders was limited.
As US agencies developed technologies that expanded the geography
of communications, interference had become a central issue. Sputnik
I's signals had already interfered with radio stations in the Netherlands,
England, and the United States.[52] Furthermore, Cubans and Americans
had been interfering in one another's radio and telegraph communications since the first US occupation (1898–1902).[53]

Cuban delegates protested their US counterparts' push for a frequency allocation regime that favored a "single global satellite communication system" (i.e., COMSAT), because this prevented other countries
from operating "several systems." They further claimed that, even if
there were a single satellite system, US promises to share space services
were not genuine. The regulations would not "permit all countries to
have access to" COMSAT, "including an equitable distribution of the
permission portion of the spectrum among all the different countries of
the world in order to enable them to establish their own systems and services."[54] This was likely true, as US delegates had concealed information
regarding their space communications capabilities (perhaps including
Syncom) during attempts to rally developing countries, particularly in
Latin America and the Caribbean, to support the US delegation.[55]

The Cubans implied that the United States, by imposing a single
space communication system, perpetuated colonial domination. When
the US delegation claimed that Cuba would attempt to interfere with
others' communications, the Cuban delegation replied: "[This] country
is rapidly shaking off the under-development imposed on it by monop-
olies and colonialism, and it would be childish to imagine that Cuba
would use its radiocommunication facilities to interfere intentionally
with other services, rather than to place them at the disposal of its people." It was precisely due to the short "distance over water," a concern
shared by the Americans, that the delegates from Cuba felt the need, as

"a small country," to "protect its [radiocommunication and, presumably, future space] service."[56] Accordingly, the Cuban delegation became the first in ITU history to make reservations in the conference's Radio Regulations. Their reservations were to the frequency regime for Region 2, which covered the Americas. This invoked the ire of the US delegation.[57]

Cuban resistance was against US attempts to use the technology developed by Syncom as an instrument of extraterritorial power. Syncom II was surely in mind. At the time, it was the only operating GSO communications satellite, and it was directly used to facilitate the press conference, as mentioned, at the Space Conference. When Cubans griped about the imposition of a "single global communications system" upon the world, Syncom II was the technological reality about which they spoke. Later ITU debates between equatorial countries in the developing world and the United States about the allocation of frequencies and orbital slots for GSO satellites would explicitly trace their origins to the twinned events of the Space Conference and the launch of Syncom II.[58]

Under the direction of Defense Secretary Robert McNamara, Syncom II and III continued to put the program's imperial imaginary into practice. Syncom III, with Relay I, broadcast the 1964 Tokyo Olympics to most of the United States and Europe, likely to demonstrate US control over the communications infrastructure in which Cold War ideologies contended.[59] Later that year, Syncom II enabled communications with troops in Saigon.[60]

FREE SPACE CONCEPTS

The story of the spin-stabilized satellite's invention is a tale of negotiation in which a variety of actors—engineers, lawyers, corporate executives, and agency officials—collaborated toward the growth of the administrative state, bringing into its institutional embrace a nascent, extraterritorial communications infrastructure. The boundaries between the invention's public and private nature were messy, but its extraterritorial materiality was undeniable. To invent Syncom, Hughes engineers, alongside their corporate bosses, joined forces with NASA and DOD officials around the production of a shared, extraterritorial infrastructure. Together, these actors articulated what they understood, correctly or otherwise, as a new concept of extraterritorial power. The concept imagined the US, via Syncom and its progeny, as the first truly global empire, where technological control supplanted territorial sovereignty as its modus operandi. Private innovation, stirred by the incentives of government contracts, may have driven Syncom's invention, but so, too, did the ideology of empire.

Scholars of the administrative state, in the histories of spaceflight and law, might do well to consider not only the perennial conflicts between public and private but also the social construction of that demarcation as a boundary drawn in response to the entanglements of extraterritoriality, infrastructure, and decolonization. This argument parallels Nicholas Parillo's claim that the shift from privateering to publicly owned naval power during the turn of the nineteenth century was conditioned by the US government's acquisition of overseas territories.[61] For Syncom, any turn toward public ownership of the patents was conditioned by an acknowledgment of the extraterritorial nature of government infrastructure. The Cold War, like the turn of the century, was a moment of the expansion of US power abroad—and the administrative state was a significant tool for this expansion.[62] Historians should seek to include these extraterritorial entanglements in their accounts of public/private conflict.

Perhaps the appeal to technological control over an abstract, borderless globe was a means to evade claims of territorial sovereignty from decolonizing peoples. Certainly, some US and European legal scholars did. Professors Philip Jessup, of Columbia Law School, and Howard Taubenfeld, of Golden Gate College School of Law, titled their influential tome *Controls for Outer Space: Outer Space and the Antarctic Analogy* (1959), defining "controls" as international "institutional or organizational device[s]." They traced such controls to a spate of precedents—including international laws governing the Scramble for Africa, the US Guano Islands Act, and the League of Nations Trusteeships—and proposed a "Cosmic Development Corporation (CODEC)" based on the British East India Company. The company, they argued, handily moderated disputes between various parties, mainly competing empires. While Wilsonian self-determination would necessitate the inclusion of Indigenous peoples in an international regulatory system like CODEC, Jessup and Taubenfeld claimed that developing nations' talk of "welfare" and "equity" consisted of untenable vagaries that ought not fix CODEC's controls.[63]

Jessup's framework even influenced Carl Schmitt's conception of US empire. Previously, Schmitt's *Nomos of the Earth* (1950) had cited Jessup on the "generalization" of the Monroe Doctrine to suggest that the United States might turn—as it did toward Hawaii at the time of Jessup's writing (1940) and to Cuba before that—to the rest of the world ("the Yangtze or the Volga or the Congo," in Jessup's words) as a "free space" (Schmitt's) wherein to exercise the "right of self-defense." Schmitt quoted sloppily from Jessup, substituting the clause "today the dimensions change rapidly" (as a reason for the new "free space" generalization) for Jessup's original line: "With the steady development of aeronautics, the range of interest [of the 'American orbit of defense'] will tend to increase."[64]

I refer to Jessup, Taubenfeld, and Schmitt not to indicate a particular reality of US empire. Rather, their expansive statements demonstrate that an idea of US empire as abstract and global was conceived by some legal scholars in response to both spaceflight and decolonization. This concept was not solely the domain of European and US legal thought. In fact, it was not far from descriptions of US empire in critiques of GSO communications satellites at the Space Conference. Cubans described the frequency region covering the Americas as "an orchestra which played to the baton of a single composer and conductor, who was always the same, and today the position is that one of the musicians in that orchestra has himself turned composer and prefers to play his own music—music with a national flavor, music of excellence, progress and complete independence."[65] The Cubans' depiction of the communications system presaged by Syncom approximated the Monroe Doctrine and extended it, as Schmitt and Jessup had, to the new dimensions wrought by spaceflight.

CHAPTER 6

BAIKONUR AS A SACRIFICE ZONE

Eco-Nationalism, Sovereignty, and Outer Space

NELLY BEKUS

On August 31, 1991, Baikonur Cosmodrome, a major spaceport of the Soviet Union, was declared the property of Kazakhstan. This sudden "gift" has not only shaped the strategies of Kazakhstani technoscientific development over the three decades of state independence but has also become one of the most conspicuous points of contention between the advocates of the country's official technopolitics and civil society groups in the wake of various issues that the operation of Baikonur has raised.[1] Described as a strategic practice of designing or using technology to enact political goals, technopolitics operates as the hybrid system of power embedded in technological artifacts, systems, and practices.[2] In 1994, having signed a lease transferring the management of the cosmodrome to the Russian state corporation Roscosmos, the Kazakhstani authorities remained committed to the idea of becoming an independent space power. Gradually, what began as a story of forced, if not reluctant, dealing with the legacy of Soviet space infrastructure has turned into an ambitious space policy that is designed to drive forward the country's technoscientific development. Various civil society groups and independent activists have, however, challenged both the agreement with Russia that allowed for the continuing presence of the former Soviet center on Kazakh land and Kazakhstan's own technopolitics.

Examining perceptions of Baikonur as Kazakhstan's sacrifice zone, where the environment and the local communities' quality of life have been compromised in the name of space development, reveals the co-existence of diverse interpretations and critical responses to the real or imagined harm caused by the cosmodrome among environmental activists and power elites. In their critique, the cosmodrome emerges as a case of space-making associated with land expropriation, its exclusion from agricultural use, and limited access for Kazakh nationals. This fully or partially expropriated zone includes the space launch port, the city of Baikonyr, and a vast "drop zone" territory for rocket debris extending over hundreds of kilometers of grasslands.[3] In this context, the positioning of the cosmodrome site highlights the connection between the space infrastructure and the Soviet and post-Soviet system of power relations, which marginalize the interests of the local population and undermine the sovereignty of the Kazakhstani state.

Being a central artifact of Soviet modernization, space infrastructure in Kazakhstan can also be placed in the context of a longer history of Russian imperial conquest of Central Asia during the eighteenth and nineteenth centuries.[4] The notion of the steppe as remote and sparsely populated—an important argument in the selection of the site for the cosmodrome—goes back to the Russian conquest of Central Asia and the early years of colonial rule. Later, during Soviet times, the area represented the other side of socialist modernity: rural, backward, Indigenous, and wasteland in multiple ways. To "amend" this backwardness, Central Asia's Russian and Soviet rulers projected onto the landscapes of the region, "civilizing" notions of modernity and progress by applying increasingly radical methods. These czarist Russian and Soviet schemes to transform the arid lands of Central Asia paralleled similar schemes undertaken in the late nineteenth and twentieth centuries across arid landscapes from North Africa to Australia, from China to the American West, driven by similar visions of modernity and what it meant to be civilized.[5] These imaginaries of emptiness continued to inform the attitude toward the territory occupied by the cosmodrome among the Russian and Kazakhstani elites after the dissolution of the USSR, a stance further complicated by the advent of the globalization of the market economy of outer space.[6]

INFRASTRUCTURAL INHERITANCE AND ECO-NATIONALISM

With the collapse of the Soviet Union, every former Union republic was left to deal with the legacy of infrastructural projects that had been meant to transform their lives and to drive forward the republics' economic and technological development. From dams, industries, railways, and roads, to nuclear power stations and test polygons—these projects were designed to benefit local society or to advance the USSR's prospects as a global power. In the last years of the Soviet Union, the rise of a nationalist agenda and the call for ethnic sovereignty precipitated the process of rethinking the rationale of these technological endeavors and their impact on the environment and native landscapes. The process of coming to terms with the legacy of socialist modernization quickly coalesced into a public outcry over the misuse of land and natural resources during Soviet rule.[7] The convergence of nationalism and environmentalism in many regions of the former USSR created the potential for the emergence of powerful mass movements that could appeal not only to environmental values but also to a sense of national identity and community.[8]

In Kazakhstan, environmental activism has had a history of successful cooperation with the authorities, when ecological campaigners forged mutually beneficial alliances with the titular elites.[9] In the late 1980s and at the beginning of the 1990s, Kazakhstan witnessed protests against the testing of nuclear weapons in Semipalatinsk. The Nevada-Semipalatinsk movement, which was formed in 1989, united prominent Kazakh cultural elites with local and international activists. Named in solidarity with Western Shoshone attempts to reclaim their ancestral homeland from the United States, and with the aspiration to end nuclear weapons testing on Newe Sogobia, the Nevada-Semipalatinsk movement became a seminal model of trans-Indigenous and transnational antinuclear activism and solidarity.[10] Their campaign for a nuclear test ban and the shutting down of the nuclear weapons testing site in Semipalatinsk proved highly effective. The test site was shut down in 1991 and turned into the International Research Centre, while the Nevada–Semipalatinsk movement has been hailed as a story of successful cooperation between domestic activists, international environmental organizations, and the national government. In Kazakhstani antinuclear activism, two separate strands of environmental engagement, identified by Gunnel Cederlöf and K. Sivaramakrishnan as the cosmopolitan and the nativist, have been combined and co-opted as a source of national pride, thus consolidating and legitimizing the nation-state.[11]

There appears to be a profound difference in the ways that environmental values have been inscribed into the history of antinuclear movements and their (in)ability to foster broader public engagement with the critique of the space program. Antinuclear activism and nationalism effectively converged in many post-Soviet states in the 1990s—Lithuania, Ukraine, Kazakhstan, and Russia—partly because they were sustained not only by local or national motivations but also by the global appeal of an antinuclear stance. Since the 1970s, popular fears and mistrust had derailed ambitious programs for the expansion of nuclear power around the world, proving that citizen mobilization can successfully force governments and industries to change their plans.[12] The public setup for the environmental activism that raises questions regarding the ecological cost of spaceflights for nations differs sharply from the one in which antinuclear campaigns operated.

In his study of Kourou spaceport in French Guiana, Peter Redfield has argued that the project of space exploration has produced a "placeless space" that can be found not only in the information landscape created by communication satellites but also at the space facilities on the ground, which obscure the ties with the local community.[13] The spaceports, indeed, are characterized by misbalanced modalities in experiencing the site of their "emplacement," as their global significance overshadows, masks, and banishes local realities with their history and culture.

Soviet space development and the emplacement of Baikonur left significant imprints on the Kazakhstani national sociotechnical imaginaries, informing the country's post-independence technopolitics. Together with the cosmodrome, however, the country's elites had inherited the idea of a "historic disposability" toward the territory it occupies that has been traced in its strategies of environmental governance. A combination of economic interests and state security has formed the core of rationality that allows the Kazakhstani state to manage the cosmodrome and city of Baikonyr while maintaining secrecy about the risks to people or places. The cosmodrome, as "a particular political and economic terrain," has been transformed into an "'inland-offshore" space . . . where the production of profit can evade or minimize contestation and public control."[14] As a result, authoritarian forms of environmental governance in Kazakhstan have normalized the pollution of lands and the suppression of any potential activism. Several ecological movements and nongovernmental organizations (NGOs) attempted to invalidate this normalization by questioning the sacrifices that Kazakhstani society and nature are made to pay for the very existence of Baikonur. This chapter draws on the notion of a "sacrifice zone" to explore major arguments developed by environmental and human rights activists in their critique of existing arrangements between the Russian Federation and Kazakhstan regarding Baikonur.

The concept of "sacrifice zone" was coined in the literature on environmental justice to represent the cost to local communities and ecologies of the destructive social and environmental impacts of development projects.[15] The concept has been deployed by ecologists to assess the harmful impact of various industries on the local ecosystem and to define the status of spaces that have been deliberately segregated and, sometimes, stigmatized for the sake of progress.[16] Applied to mega-projects of global infrastructure, the concept also enables theorists to pose the question regarding the distribution of the benefits of these projects and the responsibilities for addressing and mitigating local concerns. This chapter discusses how the perceptions of Baikonur as a sacrifice zone bring to light multiple points of contention that problematize the existing power relations in the territory of the cosmodrome.

COSMIC IDENTITY AND NATIONAL IMAGINARIES OF OUTER SPACE

Organized resistance to the outer space program emerged in Kazakhstan when the group of activists known as Anti-Heptyl was established in 2012. The group's name indicates their determination to ban a fuel used in Russian rockets that contains the toxic component Heptyl.[17] The police have broken up most of the actions organized by the activists while their organizers have faced arrest and criminal or civil charges. How can the difference in the attitude toward the activities of antinuclear and anti-space environmental activists be explained? Why did the ecological issues surrounding the nuclear test site produce a concerted societal and governmental response while the demands of the anti-Heptyl movement have remained a marginal cause treated by the government with blatant hostility? And how do those who perceive the cosmodrome Baikonur to be a Kazakh "sacrifice zone," a territory where nature and society alike have been put at risk, articulate their demand for change?

Partly, the difference can be explained by broadly conceived positive connotations associated with the meaning of the cosmos. Much of Kazakhstani society demonstrates what can be called a "legacy attitude" to the outer space culture it inherited from the Soviet past. The histories of Soviet rocketry often go back to prerevolutionary works by Konstantin Tsiolkovskii and to the nineteenth-century philosophical tradition of Russian cosmism.[18] Soviet society, from the outset, had an inherent faith in technology as the highest form of culture, the chief mode of modernization, the answer to standardization and mass production and, ultimately, the panacea for all social and economic ills.[19] Soviet achievements in space were widely covered in the Soviet press and generated what could be called a "Soviet space culture" that sustained public enthusiasm and created a sense of shared belonging and involvement in cosmic ventures across Soviet society. The Soviet space program has often been seen as a tool of political propaganda and a vehicle for cultural diplomacy that was used by the Soviet Union in promoting its self-representation not only in the international arena but also in its own multinational peripheries.

Histories of Soviet space development traditionally focus on the Russian component as its core, which it undeniably was, thereby often ignoring the broader Soviet context, in which those belonging to both bigger and smaller nationalities were offered a sense of involvement and participation in a grand story of space endeavor. More broadly, the Soviet project with its policies of what Terry Martin has called the "affirmative action" empire, was designed to bring about the rapid modernization of multiple traditional societies across the Soviet peripheries.[20] Kazakhstan, of all the former Soviet nations, had formed perhaps the strongest ties with the reality of space launches and flights, given that all Soviet cosmonauts were launched from there.

Traces of cosmic activities and elements deriving from their symbolic and material culture have been scattered across Kazakhstan's local communities and regions. They include encounters with cosmonauts who stayed at hotels during their transit to or from Baikonur, underwent rehabilitation at health resorts after landing, attended meetings with local residents, donated their cosmic possessions to enthusiasts or local museums, and so on.[21] These visits were firmly in line with the long tradition of cosmonauts who became global celebrities traveling around the world as Soviet icons who were called upon to personify the achievements of the socialist system in the context of the Cold War.[22] German Titov, the second person to orbit the Earth, traveled to the GDR (East Germany) following his spaceflight just weeks after the Berlin Wall had been built, and his visit would prove instrumental in cultivating a positive perspective on the socialist world in East Germany. Likewise, cosmonauts had an important role in the domestic Soviet context, helping to bring Soviet space successes closer to the local public. In 1968, German Titov visited the Kazakh city of Semipalatinsk (now Semey) to attend the city's 250th anniversary celebrations and receive the city's honorary citizen title. The city of Zhezkazgan, a center of the region where Baikonur is located, is proud of its "cosmic" identity manifested in multiple symbolic markers. Along with the monuments, street names, and decorative mosaics on buildings, the city also has an alley of trees planted by cosmonauts, a tradition that began in 1973 with cosmonauts from the Soyuz 12 space mission. Furthermore, the idea of developing "cosmic tourism" along the routes associated with various space activities on Earth has been advocated by Kazakhstani tourist professionals and cosmic enthusiasts, who see space themes as a compelling instrument for regions' branding.

This legacy attitude to Baikonur—one that builds upon the mythologized perception of space as an emblem of technoscientific progress— has been transferred wholesale to the perception of Baikonur's operation in Kazakhstan since independence. Cosmic enthusiasm and the dreams concerning humanity's celestial future that accompanied the history of outer space exploration appealed to the universal rationale. These claims, at the same time, remained deeply embedded in national contexts, bringing forward the ideas of space technoscience as evidence for and a symbol of the nation's modernity and progressive trajectory.[23]

Particular attention to the formation of the national imaginary of outer space in Kazakhstan had begun to be paid even before the country became independent. On March 13, 1991, seven months before the declaration of independence, the government of the Kazakh Soviet Socialist Republic adopted "The Comprehensive Program of Scientific, Technical and Economic Cooperation 'Kazakhstan-Cosmos.'" The title of the program indicated the ambition of the Kazakh republican elites to become directly involved in developing space technoscience, which was until then considered a prerogative of the central government in Moscow. The program's major objective was to ensure the "wide and effective use of the achievements of the Soviet space programme in the national economic development of Kazakhstan," strengthening the technoscientific potential of the republic's industries and developing the new infrastructure required for the integration of space services into various sectors of the national economy, including energy, telecommunications, mechanical engineering, new materials, agriculture, healthcare, and others.[24] The program also included the plan for the first flight of a Kazakh cosmonaut on the orbital complex Mir, to conduct a series of scientific experiments. As a result, in October 1991, the first Kazakh cosmonaut, Toktar Aubakirov, was sent into space, although, ironically enough, he was destined to be the very last cosmonaut ever to travel into space as a Soviet citizen. On the part of the Russian authorities, it was also a political gesture to help secure the continued presence of Russia at Baikonur after it had become the property of Kazakhstan.

On his return from space, Aubakirov was hailed as a national hero, and his image-making became a matter of national policy. A state program for honoring Aubakirov was approved, according to which the cosmonaut was obliged to go on tour around the country as part of a delegation headed by one of the deputy prime ministers. He has often been referred to as the "Kazakh Gagarin," who not only conquered the cosmos but also restored by his flight historical justice in relation to the Kazakh people, who had surrendered a part of their territory for a cosmodrome. In his book *Kazakhstan's Way*, Nursultan Nazarbayev recalled the symbolic importance of sending a Kazakh cosmonaut into space, an act that was "intended to be a form of recognition of the Kazakh people as partners in space exploration and of their signal contribution to Soviet technological achievements."[25]

Indeed, unlike other pieces of grand infrastructure built in the Soviet peripheries that served to accelerate the development of the various nations, cosmic infrastructure remained essentially unembedded within the Kazakhstani technological landscape due to the extreme sensitivity of space technology in the context of the Cold War.[26] Not only was it governed and controlled from the center in Moscow, but the political economy of the republic had no connections with the space industries. The program launched in 1991 had been the first step in supplying this missing link and bridging the gap between the Soviet space projects and the republic. With the declaration of independence, an inherited space infrastructure and the need to deal with Baikonur as a form of "residual asset" became incentives for further engaging with space exploration, providing a foundation for the newly independent state's technological future.[27] Soon, a set of unconventional governance solutions and bilateral agreements between Russia and Kazakhstan led to the establishment of hybrid governance structures. Indeed, the 1994 agreement regulated the status of technological objects and land as well as that of Leninsk town, renamed Baikonyr in 1995, with its resident population of over seventy-five thousand, which consisted for the most part of Kazakhstani nationals.[28] The agreement was criticized by the government's opponents for having been arrived at under the coercive pressure of economic and geopolitical circumstances, and is described by scholars as a manifestation of Kazakhstan's "sporadic sovereignty."[29] President Nazarbayev perceived the concession as a sacrifice made for the sake of the "preservation of the scientific-technological and intellectual resources of the space complex and a global heritage of [the] human space mission."[30] From his perspective, it was also a strategic decision with a bearing upon the future technoscientific development of independent Kazakhstan.

Over the following decades, collaboration with Russia was gradually transformed into just one element in a wider transnational dynamic, whereby Kazakhstani elites have sought and still seek to engage with multiple actors across the globe, including France, the UK, and the US. Baikonur Cosmodrome, however, continues to play a key role in the strategic imagining of the country's future space development. The possibility of terminating the Baikonur lease has regularly been discussed in the media.[31] The general direction of the changes in attitude toward Baikonur is visible in the various titles given to the lease agreement over the years. The agreement, signed in 1994, spoke of the "basic principles and conditions for the use of the Baikonur spaceport,"[32] while the 2004 document that extended the lease until 2050 mentions "the development of cooperation for the effective use of the Baikonur Complex." Furthermore, the government regularly updates the agreement, its aim being to gradually diminish the presence of Roscosmos and reappropriate the Baikonur complex for its own use. In 2013, Russia and Kazakhstan agreed on a "road map" relating to the joint use and development of the Baikonur complex and the city that ensured new gains for the Kazakhstani space program. That same year, the launch pad for Zenit launch vehicles was partially withdrawn from the lease so as to be modernized and transferred to Kazakhstan.[33]

THE COSMODROME, FROM A "SACRIFICE ZONE" TO A SPACE OF CONTENTION

Protesting against the presence of the Russian Space Agency at Baikonur and its right to exploit the territory of Kazakhstan for space launches is the focus of several NGOs, such as Baikonur for Civil Rights (established in 2009), Anti-Heptyl (active between 2012 and 2015), Baikonur Eco-Monitoring (since 2018), and the Ecological Museum (established in 1995 and registered with the authorities in 1997). Each of these actors deploys a specific set of arguments and strategies of action to contest official policy and to communicate its message to the public. In contrast to the state policy, which approaches Baikonur as a valuable asset contributing to the country's ambitions of becoming an independent space power, these organizations see the cosmodrome through the lens of its social and environmental impact.

The movement Baikonur for Civil Rights, led by Marat Dauletbayev, was established to provide support and legal protection for the Kazakhstani citizens living in Baikonyr. In its multiple protest actions, the activists raise concerns about the lack of transparency and the corruption involved in the management of Baikonur. In his media interviews and social media publications, Dauletbayev draws attention to the fact that the Kazakh people constitute the most vulnerable group living in Baikonyr, experiencing lower levels of pay and unfair treatment in the workplace, while often being denied access to social services and other rights and privileges enjoyed by Russian citizens residing in the city.[34] He compares the hybrid governance at Baikonyr to a harsh colonial administration, which "makes Kazakh people feel like foreigners in their own country," and he accuses Roscosmos of "violating environmental legislation, polluting the territory of Kazakhstan with toxic and radioactive waste and discriminat[ing against] Kazakh people."[35]

The city administration has repeatedly banned events organized by the movement on the grounds that it has not been registered with the Russian Ministry of Justice, which is the only way to operate lawfully in Baikonyr. This regulation derives from the city's status, corresponding to that of a city of federal significance of the Russian Federation.[36] Marat Dauletbayev, however, considers this requirement to be unlawful due to its violation of Article 10 of the 1994 Lease Agreement for the Baikonur complex between the government of the Russian Federation and the government of the Republic of Kazakhstan, which guarantees that Kazakhstani citizens residing in Baikonur retain their constitutional rights. Furthermore, in 2017, Dauletbayev faced two criminal charges of defamation, which were leveled against him based on complaints filed by the mayor of Baikonyr, Andrey Petrenko, and by a representative of the Roscosmos corporation. In both instances, he had accused the mayor and members of the Roscosmos administration of corruption and abuse of power in the company's dealings with the Kazakh people, whose interests had thereby been damaged. In 2017, Dauletbayev posted a text titled "Laborers of the Earth—Preparing Ships for Space," which alludes ironically to the lyrics of the Soviet song "Motovoz," composed by the bard Aleksandr Kalistratov (1981), which glorified the daily routine of the Baikonur Cosmodrome workers. In contrast to the romantic image of the Baikonur working experience portrayed in the song, Dauletbayev highlights various forms of discrimination faced by Kazakh workers, whose rights have not been protected under the current lease agreement between Roscosmos and Kazakhstan.

In June 2020, the movement launched a new campaign "for termination of the lease agreement for the Baikonur Complex," which claims that this treaty is antinational and anticonstitutional.[37] In the discourse of the Baikonur for Civil Rights movement, Baikonur features as a place where the constitutional rights of Kazakhstani citizens are not observed despite their being guaranteed by the constitution of the country and the lease agreement. Thus, according to the Lease Agreement, criminal, civil, and administrative cases should be handled by Kazakhstani law enforcement agencies (the Prosecutor's Office, the Court of Baikonyr), and the Russian side should by rights transfer all claims and cases against citizens of the Republic of Kazakhstan to the Kazakhstani courts. In practice, the Russian authorities often ignore this rule, for example, when the 26th Russian garrison military court located in Baikonur accepts for consideration civil cases against citizens of Kazakhstan, thereby violating their constitutional rights.[38] The discrimination faced by Kazakh people working and living in Baikonur, however, has been viewed as a symptom of a broader systemic problem—the government's willingness to suspend the state's sovereignty over the territory of the cosmodrome and the city.[39]

Such perceptions regarding Baikonur serve to uncover the actual reality, in which the quality of life of Kazakhstani citizens and their legal protection and social security have been compromised, with the consent of the Kazakhstani state. It emerged as a "sacrifice zone" in which the tensions involved in the post-Soviet power relations of sovereign Kazakhstan with the former Soviet center in Russia have been reinscribed within the hierarchies of rights and privileges that came to be associated with Russian or Kazakhstani citizenship. Divergence from the legal agreements existing between the Russian Federation and the Republic of Kazakhstan and the failure of Kazakhstani law enforcement agencies to address the problems their citizens face, have been interpreted by activists as a demonstration of the state's inability to protect its sovereignty projected onto individual citizens' rights.

The Anti-Heptyl group was formed specifically to respond to the series of accidents at the space launches when Proton rockets exploded over the steppe, spilling deadly heptyl rocket fuel over a broad swath of territory. The first two such accidents occurred in 1999, followed by further explosions in September 2007 and June 2013. The group deploys its protests against the harmful impact of the rocket launches on the natural environment and those residing in the regions to articulate wider environmental and political ideas. Rather than voicing their demands in terms of the categories of liberal environmental justice, with its inherent stress on the global aims of environmentalism, they equate the protection of the environment with the fight to establish the "complete" national sovereignty of Kazakhstan. In their agenda, the latter is linked with the struggle against both the "exploitation" of the Kazakhstani steppe by a foreign state, as a form of enduring, if camouflaged "colonization" and against the ruling authoritarian regime that disregards the interests of the Kazakh people, their land and the natural environment. Such openly oppositional political demands established the reputation of the movement as "a school of Kazakh democracy."[40]

Most of the group's street actions combine environmental claims with strong antigovernment slogans, and they often culminate in arrests and criminal or civil penalties for the organizers. It is worth noting that the founding of Anti-Heptyl indicated not so much the appearance of new attitudes to the space program in Kazakhstani society, but a shift from what James C. Scott designates as a "hidden transcript" of dissatisfaction shared by society and elites alike to its open articulation and the consolidation of a new agency of active resistance.[41] The Kazakhstani authorities were indeed alarmed by the explosion of two Proton rockets in 1999. They introduced a temporary ban on Proton launches from Baikonur, which was lifted after the soil pollutant at the accident site had been detoxified and compensation for environmental damage paid.[42] A series of similar crashes and launch accidents at the spaceport in 2007, 2010, and 2013 generated serious tensions between Roscosmos and the government of Kazakhstan. They led the Kazakhstani elites to question the further use of Proton rockets at Baikonur and to discuss new forms of compensation. Furthermore, they intensified the process of regaining control over the space complex as a strategic aspect of Kazakhstan's own technopolitics.

The avowed openness to space cooperation with Russia and yet at the same time apparent resistance to it has come to constitute a form of Kazakhstani governmental infra-politics that avoids the open declaration of real intentions. This veiled discontent on the part of state elites was transformed by Anti-Heptyl into a fully fledged political opposition to the government itself. The protests in Almaty (formerly Alma-Ata) and Astana have been organized under an antigovernmental banner as well as under anti-Russian slogans such as "No to Eurasian Union" or "Long Live Crimea."[43] Activists accuse Russia of perpetrating a "Heptyl-genocide" of the Kazakh nation and demand that all Russian launch sites on the territory of Kazakhstan be closed forthwith.[44] To stress that the policy conducted by the Kazakhstani government serves, in reality, the interests of the Russian state, during the performance organized at the Anti-Heptyl protest actions in Nur-Sultan on October 19 2013, an activist first appeared in Kazakh national costume only to reveal at the end of the action his "undercover" Russian folk dress.[45]

In 2013 the Anti-Heptyl activists transformed the movement into a broad coalition that included over eighty members of the Nevada–Semipalatinsk group, human rights campaigners, independent trade unionists, and Islamic religious activists. The declared goal of the coalition became the closure of Baikonur spaceport as well as the military polygons operated by the Russian Federation on the territory of Kazakhstan, and the arrangement of compensation for the damage incurred by Kazakhstani residents due to space and missile launches.

The message sent by the coalition to the wider society concerns the failure of the government to act in the interests of the people, which implies its fundamental inefficiency. In this context, the space agency Kazcosmos and its leader Talgat Musabayev, one of three Kazakhstani cosmonauts, have repeatedly been targeted by activists as representing a "weak link in the chain" of the country's ecological security and, by extension, in that of its national sovereignty.[46] The promised transformation of environmental action into political protest emerges in this context as a natural response on the part of activists to the government's failures. In 2014 the activist Ulan Shamshet picketed the office of Kazcosmos with a poster that invoked the events of 2014 in Ukraine, "Today Anti-Heptyl—tomorrow Maidan."[47] This reference signals the group's political ambitions, which go beyond their opposition to state policy on Baikonur and space programs. The revolution of Maidan not only resulted in the toppling of the ruling regime in the country but also led to a radical reorientation toward the West.

In this context, the ownership of Baikonur has also been reinterpreted as a potential foundation for an alternative astropolitical strategy, one that would allow Kazakhstan to forge alliances with partners around the world for the realization of its own space program without or with only a limited presence of Russia. The idea of such an alternative was first considered in 1994, when a NASA delegation visited Baikonur and inquired about the possibility of direct cooperation with Kazakhstan and access to Baikonur without Russian interference.[48] In 2009 the analyst A. Berkimbayev returned with an ambitious project to transform Baikonur into a "global space hub," which would consist of four autonomous segments, each providing space launching services to different regions: (1) Russia; (2) Arab states; (3) European states; and (4) East Asian states.[49] Being publicly expressed by an independent expert, the project did not represent an official stance, but it nonetheless articulated the ultimate aspirations of Kazakhstani elites to acquire more independence from Russia in its space development.[50] Being voiced by environmental and human rights activists, the argument acquires a new dimension, as it brings to the fore an argument about the particular damage associated with the all too negligent Russian approach to the territory of Kazakhstan and its people.

In 2015 the mass deaths of saiga antelope occurring in Kazakhstan were linked by Anti-Heptyl to the harmful effect of Russian rocket launches.[51] The official governmental investigation, however, backed by independent international enquiries, failed to find any evidence for such a connection.[52] A second key player in the sphere of environmental activism emerged in the guise of an environmental organization known by the name of Baikonur Eco-Monitoring (2015), led by Marat Dauletbayev. In 2016, activists from this NGO filed a lawsuit with legal support from Ecological Rights Bellona, their demands being (1) restoration of the environment, with the bill to be footed by State Corporation Roscosmos and the Russian Federation, and (2) payment of moral damages. The courts, however (both the 26th garrison military court in Baikonyr and the Supreme Court of the Russian Federation) refused to consider the demands and dismissed all these claims, on the grounds that such issues would have to be resolved at an intergovernmental level, and that the NGO was not authorized to file such an application.[53]

Furthermore, in the context of the 2016 events in Kazakhstan, when mass protests took place in Kazakhstan against the land code changes that would allow foreigners to rent land for twenty-five years, Baikonur for Human Rights and Baikonur Eco-Monitoring launched a court case to challenge the legal validity of the agreement between Russia and Kazakhstan regarding the lease of Baikonur. Essentially, while framing their demands in national terms, organizations like Baikonur for Human Rights, Baikonur Eco-Monitoring, and Anti-Heptyl have been active predominantly in the capital city Nur-Sultan, the largest city (and former capital) Almaty, and in the Kyzylorda region.[54] Their appeal has never become a matter of broader public discussion beyond the activist groups in the capital city and communities directly affected by launches.

A different strategy of dealing with the environmental impact of Baikonur can be identified at the Ecological Museum in Karaganda, which was established by a group of ecologists in 1996. Founded as a private initiative, the museum has gradually developed into a hub of environmental research, consulting, and activism in the region. In their work, the ecologists deal with a variety of issues ranging from air pollution and chemical safety to atomic energy, biodiversity, and many others.[55] The environmental work of the museum has been organized according to several strategic objectives: dissemination of information (local and national awareness-raising campaigns); environmental education (educational programs for students, creative classes, training, and consulting educational institutions), environmental research and design (monitoring and research for industrial enterprises, preparation for public environmental hearings, and creation of ecological passports for enterprises), and the development of ecotourism.

The impact of Baikonur on the environment appears on the list of the group's concerns, but its primary focus appears to be the ecological costs of the transition to a market economy. The Karaganda province is one of the most industrially developed in the country, with a high concentration of coal, nonferrous and ferrous metallurgical, chemical, and mechanical engineering enterprises that generate significant environmental disturbances requiring urgent attention from the authorities and from ecological activists. By contrast with the case of the environmental impact of Baikonur, which has been used by a foreign country, most of the ecological issues addressed by the Ecological Museum are the result of the transition to a market economy and the increase in energy use and resource intensive production that serves to boost the industrial potential of Kazakhstan.[56] This, in turn, predetermines the sorts of activities the group undertakes, which combine the promotion of education, research, and collaboration with local and national policymakers, who are much more approachable than are their counterparts in charge of Baikonur.

Due to the hybrid system of governance, in which representatives of Kazakhstan have been constrained in their conduct by the legal arrangements of the Baikonur lease agreement, the public engagement and communication regarding the environmental impact of Baikonur has proved highly problematic. For example, the work of Garysh Ecologiya, the contractor behind Kazakhstan's Space Agency, which was established in 2001 with the purpose of "implementing applied scientific research for environmental safety related to rocket and space activities in Kazakhstan," tends to confine itself to minimizing the dangers posed by the Russian-controlled space industry. Kazakhstan, in fact, operates as a subcontractor to the Russian Federation legally obliged to protect a fifty-year agreement at all costs.[57] There are multiple cases of foreign enterprises operating as "industrial colonies" in Kazakhstan, such as the Tengiz Oil enclave, where the state can be seen as pursuing its commercial interests.[58] Baikonur, however, represents a more complex venture, one in which the noncommercial benefits of future projects and the development of Kazakhstan's own space technopolitics play a significant role. It is noteworthy that a somewhat reluctant assessment of hazard associated with outer space activities and a high degree of uncertainty in the evaluations presented to the public underpin the global pattern of managing the affective politics of human engagement with outer space.[59] The environmental conduct of Roscosmos and the Kazakhstani state in this context is not dissimilar to the practices of NASA, whose Environmental Impact Assessments "are known to be fabrications but are still preferred to uncertainty," even though they are engineered and selected to function in the interests of those in power.[60]

Raising awareness about the negative side of spaceflights among the public and promoting the idea of more responsible environmental governance at Baikonur in this context appears as the only strategy available to the Ecological Museum. Baikonur and the space program have been presented at the exhibition through the lens of space junk, twisted fragments of rockets, and other space infrastructure. The photographs taken at the sites of rocket accidents display the scale of the hazards and risks accompanying space launches, which often remain hidden from enthusiastic media reportage. Rather than directly confronting state policies and decision making, experts at the museum seek to cooperate with the governmental agencies, even if with limited effect, and in this way, contribute to improving the ecological situations in the territories adjacent to Baikonur.

THE REALITY OF SPACEPORT

Baikonur, as a spectacular exemplar of "ambient infrastructure," has played an important role in shaping the way outer space has been produced, organized, and experienced.[61] Being one of the longest-serving launch ports for human spaceflight, the cosmodrome has also produced a particular place on the ground where the entangled histories of the Soviet space program, the Cold War, and the space development of the post-Soviet era became intertwined with the nationalizing aspirations of the Kazakhstani state and society. Place-making practices come with inherent hierarchies and exclusions.[62] The mode of existence of Baikonur since 1991 reveals how much these hierarchies and exclusions resist transformation and repel or exclude social challenges.

Though similar to antinuclear activism under late socialism, described by Jane Dawson as a political effort that channeled and distributed anti-Soviet sentiment and resentment about Moscow's domination across the different nations, the anti-Baikonur campaigns are also unique in many respects.[63] Unlike the Ukrainian Chornobyl and Kazakhstani Semipalatinsk environmental struggles, Baikonur represents not only the legacy of a collapsed system but also a still operative technological and commercial asset. Furthermore, the movements Baikonur for Human Rights, Anti-Heptyl, and Baikonur Eco-Monitoring effectively combine geopolitical demands directed against the Russian presence at the cosmodrome with resistance to the space and environmental policies pursued by their own state. The use of traditional folk costumes and images of wild nature in Anti-Heptyl street actions translates the protection of the environment into an ethnic identity discourse, which is combined with and mediated by nature and landscape. This nativist approach to national identity shapes their nationalism as a form of reaction against elites, with their ideological concept of a science-and-technology-centered nation inscribed within the discourse of postcolonial modernity.[64]

Notably, Kazakh activists link their discourse of ecological nationalism to concerns for the environment in times of rising global ecological insecurity, thus acquiring an image as a sort of "Kazakh Greenpeace." At the 2015 meeting of Anti-Heptyl activists with the United Nations Special Rapporteur on hazardous substances, Baskut Tuncak, they accused the Kazakhstani leadership of posing a major ecological threat not only to the Kazakh people but also to the global community.[65] The way in which Kazakh activists globalize their local environmental struggles by outlining a new space of globalization within their pro-native ecological campaign validates Bruno Latour's observation that a clear distinction between the global and the local is increasingly problematic since almost everything is at least a little of both.[66]

The conflicts and tensions generated by the existence of Baikonur in Kazakhstan reveal the entanglement between multiple modalities of experiencing the reality of a spaceport. As a major facility operated by the Russian Space Agency, it serves the technological advancement of space exploration necessitated by global conditions. This perception has also been increasingly appropriated by the Kazakhstani government as it deploys a sociopolitics that seeks to center the idea of nation on science and technology and to benefit from its technological collaboration with Russia for these ends. The ecological nationalism thus amalgamates the discourse of "native identity" with wild nature, landscape, and ecosystem, and depicts the territory used for the cosmodrome as an unjustified sacrifice made by the state for the benefit of a foreign state. Human rights activists, in turn, consider the cases of discrimination against Kazakh nationals living and working at Baikonur as compelling proof of the detrimental impact of the cosmodrome on Kazakhstani society. They bring to light the fact of Kazakhstan's suspended sovereignty, which has been sacrificed to the specific power relations underpinning the operation of the cosmodrome.

TERRESTRIAL CHAUVINISM

The Search for Extraterrestrial Intelligence
and the Inevitability of Expansion

REBECCA CHARBONNEAU

The Soviet astrophysicist Nikolai Kardashev opened his 1984 article on the search for extraterrestrial intelligence (SETI) with an observation: "At present the most important aspect of the problem of searching for extraterrestrial civilizations seems to be the need for a logically consistent agreement on what it is that we are searching for."[1] By 1984 the search for extraterrestrial intelligence had been underway in the US and Soviet Union for nearly twenty-five years, having begun with the American radio astronomer Frank Drake's search at the US National Radio Astronomy Observatory in 1960. In his summary paper, Kardashev would go on to argue that there was no consensus in the US or Soviet Union on what to look for when searching for evidence of extraterrestrials (ET). Yet this was not entirely true. While there were various theories and approaches in SETI throughout the latter half of the twentieth century, a common thread ran throughout both Soviet and American SETI—a belief in the inevitability of galactic imperial supercivilizations.

This is not entirely unsurprising considering that SETI grew up alongside the Space Race, a project that can arguably be described as a pseudo-imperial conflict between two "supercivilizations" themselves. In 1961 the US president John F. Kennedy famously drew parallels between the American frontier myth and the burgeoning Space Race during a speech in Houston, Texas, claiming, "What was once the furthest outpost on the old frontier of the West will be the furthest outpost on the new frontier of science and space."[2] In 1959 a Soviet propaganda poster displayed a rocket being launched toward an alien planet, emblazoned with the phrase "Glory to the conquerors of the universe!" What united these myths was a belief (or a suggestion) that space was a new opportunity for the perpetuation of state expansion and imperialism. In this fragment, I argue that the imperial character of the Space Race influenced both the theories and activities of SETI, ultimately shaping our expectations of what types of civilizations exist in the universe.

Much has been written about the role of the frontier myth in US space rhetoric. At the start of each episode of the American science fiction television show *Star Trek*, Captain Kirk's voice rings out, proclaiming the mission of the starship *Enterprise* "to seek out new life and new civilizations . . . to boldly go where no man has gone before!"[3] *Star Trek*, which first aired in 1966, neatly embodied the exploratory fervor of the US Space Age, a fervor that often manifested in the use of frontier rhetoric and manifest destiny metaphors to garner public interest and support. After all, as Kirk pointed out, space was "the final frontier." The aforementioned Kennedy speech was rife with references to exploration, conquering unknown lands and obstacles, and the Pilgrims' voyage to settle the Americas. At the start of the Cold War, the director of the US government's Office of Scientific Research and Development, Vannevar Bush, published the now-famous *Science, the Endless Frontier* (1945) report, which heavily relied on frontier rhetoric to promote US investment in science. The report claimed, "It has been basic United States policy that Government should foster the opening of new frontiers. It opened the seas to clipper ships and furnished land for pioneers."[4] The myth of space as an extension of the American frontier led to a framing of US space activities as a continuation of the myth of manifest destiny—the driving force moving the US toward its expansionist destiny.[5]

Soviet imperialist rhetoric has drawn considerably less attention from science, technology, and society scholars than the US frontier myth. Yet the Soviets also saw a mythic future for themselves in the cosmos. While Americans created narratives about a final frontier, an extension of the driving force of manifest destiny, Soviets admired what they saw as the destiny of communism in the cosmos, which would be the province of humankind.[6] Many early Soviet space theorists, such as Konstantin Tsiolkovsky, fantasized about a human destiny among the stars, a utopian future of human harmony and cooperation. The Soviet vision of their future in space emphasized the predictive powers of communist ideology, positing that intelligent civilizations (even alien ones) would naturally proceed along the unilinear progression of societies toward communism, eventually achieving an interstellar utopia. In an article titled "Soviet Attitudes concerning the Existence of Life in Space," published in *The Handbook of Soviet Space-Science Research* (1968), the Soviet astronomer Nicholas Bobrovnikoff claimed: "[Soviet scientists] are emphatic that their materialistic philosophy is in complete agreement with the idea of extraterrestrial civilizations. According to this philosophy life is a normal and inevitable consequence of the development of matter, and intelligence is a normal consequence of the existence of life."[7] In other words, the Soviet societal goals and ideologies were quite literally *universal*, so much so that they even applied to ET.

The belief in Soviet expansion and the myth of its ultimate destiny in space permeated the space sciences in the USSR, to the point that it even shaped the message and scientific theories themselves. Earlier in his career, Kardashev had become famous for developing what is now known as the Kardashev scale of civilizations, a theory positing that intelligent technological civilizations in the galaxy would inevitably expand their control and resource acquisition from their local stars to the entire galaxy (just as the Soviet Union aimed to do).[8] In his 1964 paper describing the scale, Kardashev proposed a three-tiered system for categorizing potential extraterrestrial civilizations, designated by their resource consumption. In this scale, a Type 1 civilization makes use of the energy resources of its own planet, while a Type 2 civilization would have use of energy equivalent to the energy output of its home star, using technologies such as a Dyson sphere to harvest solar energy. A Type 3 civilization, on the other hand, would have progressed onward to colonize and harness the energy of its entire galaxy. In line with the Soviet view of societal evolution, Kardashev did not simply imagine that there "might" be imperial supercivilizations; rather, he believed them to be inevitable.[9] Later in his career he published a paper titled "On the Inevitability and the Possible Structures of Supercivilizations," in which he argued that "civilizations always face problems that require greater activity," ergo necessitating growth and expansion.[10] He maintained that the most probable evolutionary track of extraterrestrial civilizations would be its expansion throughout the galaxy, incorporating other civilizations as it expanded and made contact; these civilizations would "join the higher civilization"—just as Earth civilizations would eventually join the Soviets in their global revolution.

The word "inevitable" in Kardashev's theories is telling; as stated earlier, communist ideology was in many ways deterministic, purporting an inevitable progression of culture (toward communism). In this interpretation, the Kardashev scale is an excellent example of how many ideas in

Soviet SETI were perfectly "of the moment"—reflecting a mythic Soviet perspective on life and the cosmos.[11] The Soviet Union was a technological, expansionist, spacefaring state, and therefore, so were the civilizations on Kardashev's scale. By creating the scale, he demonstrated how scientists used myths to inform their understanding of technological and cultural evolution of both Earth and alien civilizations. Kardashev simply extended Soviet imperialism to the entire galaxy.

Critically, imperial myth was not limited to abstract theories but also played out in US and Soviet space activities. To continue examining the case of Soviet SETI, the expansionist rhetoric of the Kardashev scale was complemented by nationalist and military involvement in their SETI activities. For example, the earliest attempt to use radio facilities to contact extraterrestrial intelligence took place in the USSR in 1962. The Soviet Union's first message used the military Yevpatoria planetary radar to send a Morse-code message, "MIR LENIN SSSR" [Peace/World, Lenin, USSR], to Venus.[12] While this message is often heralded in SETI circles as the first attempt to use radio signals to contact ET (in this case, Venusians), the design of the message shows that it was nationalist in nature, probably designed more to demonstrate the abilities of the Soviets' new radar facilities and create a show of technological power at the height of the Space Race.[13] The MIR message shows how the imperialist rhetorics of the Soviet SETI project were not simply rhetorics alone but also reflected Soviet imperialist ambitions on Earth.

This marriage of imperial rhetorics and action also played out in US SETI.[14] A prime illustration of US SETI's imperialist determinism can be found in the influential 1975 paper by the astronomer Michael Hart on the Fermi paradox titled "An Explanation for the Absence of Extraterrestrials on Earth." The Fermi paradox, named after the physicist Enrico Fermi, originated from his curiosity during a 1950 lunch at Caltech when he questioned the apparent absence of intelligent extraterrestrial civilizations, musing, "Don't you ever wonder where everybody is?" In asking this now infamous question, Fermi suggested that if the universe is teeming with extraterrestrial civilizations, humanity should have discovered evidence of them by now, perhaps by way of their expected galactic expansion. In his paper, Hart, an active but skeptical member of the SETI community, argued that SETI scientists should draw conclusions from the "only evidence concerning the behavior of technologically advanced civilizations . . . the human species."[15] Hart contended that, given the exploration and colonization of the Earth (which he frames as inevitably "human" rather than specific to particular nations and cultures), it was reasonable to assume that at least one alien civilization, if it existed, would have colonized the galaxy by 1975. Since no such evidence of galactic colonialism existed, Hart argued that humans might be the only intelligent civilization in the galaxy, and it would be we who would become the first galactic empire. The rhetoric of imperialism and expansion were so normalized to SETI scientists in both the US and USSR that they struggled to envision civilizations who behaved otherwise.

As in the Soviet Union, US SETI's relationship to imperialism extended beyond theory and discussion. In the early 1980s, a report titled "The Longest Search: The Story of the Twenty-One Year Pursuit of the Soviet Deep Space Data Link, and How It Was Helped by the Search for Extraterrestrial Intelligence," was published in the *Cryptologic Almanac*, a classified academic journal published internally by the US National Security Agency (NSA). The report began by noting the long history of searches for particular signals known to the intelligence community, but not yet found and identified. One of these "white whales" was the Soviet Deep Space Data Link, which was used, among other things, to send images from Mars and Venus to the Earth by Soviet space probes, such as those from the Venera missions. This link was the product of the Soviet Deep Space Network, of which the Yevpatoria array that sent the MIR message was a part.

Because of geopolitical instability, most notably civil war in Ethiopia, the United States had lost control of signals-intelligence intercept sites in Turkey and Ethiopia, which meant that the US "could intercept [Soviet] transmissions only during [a] short window."[16] During this same period, US scientists involved in SETI had created a spectrum analyzer designed to seek out artificial signals in space, in the hopes of identifying evidence of an ET message. But as the NSA report noted, the instrument had other uses. The NSA described the SETI analyzer as "a system designed specifically for the collection of signals from deep space" and, with the support of the designers, signals-intelligence agents borrowed the equipment to successfully identify this long-sought Soviet signal. The document optimistically noted the usefulness of SETI technology for signals intelligence, arguing that it "pointed the way to the advanced collection and signal analysis systems."[17] Ironically, the same network that was responsible for sending the first Earth message to ET, the Soviet Yevpatoria site, was found by a hostile foreign intelligence— not aliens, but the NSA. As we can see from these examples, even a seemingly benign and internationalist area of the space sciences such as SETI was still subject to national and military activities that were part of the larger struggle for control and expansion in the Cold War era.

The connection between imperialism and space exploration is perhaps stronger now than ever before. As the other authors in this book note, the use of settled land is a key part of space infrastructure in the West, and the US rhetorics surrounding space activity mimic those earlier references to the frontier, with calls to "colonize Mars" and pioneer frontiers.[18] The Soviet Union may no longer exist, but Russia has pivoted to unapologetic imperialist action in its invasion of Ukraine. In this fragment I showed how SETI scientists in both the US and USSR reflected the imperialist agendas of their respective countries, both in the rhetorics of the field and in the nationalist activities they implicitly or explicitly participated in. But just as SETI was entangled within the war and imperialism of the time in which it developed, it could also provide a universal perspective by situating its practitioners as Earthlings rather than national citizens.

One SETI scientist in the US, the astrophysicist Carl Sagan, was highly critical of the terrestrial chauvinism he saw in both countries and frequently spoke out against the frontier rhetoric common in US SETI. In 1988 he penned an essay published in both the US magazine *Parade* and the Soviet magazine *Ogonyok*. The essay, titled "The Common Enemy," highlighted the hypocrisy of the Cold War by detailing the similarities between Soviet and US imperial actions. He described the US as a country "founded on principles of freedom and liberty" but being "the last major nation to end chattel slavery."[19] He noted how it "systematically violated more than 300 treaties it signed guaranteeing some of the rights of the original inhabitants of the country" and "[conducted] worldwide covert wars in the name of democracy."[20] He similarly described the Soviet Union as a conquering nation that had forced collectivization and famine on countries such as Ukraine. He criticized its "endemic anti-Semitism" and highlighted its semi-covert imperialist agenda, noting: "Your coins display your national symbol emblazoned over the whole world . . . you can understand that citizens of other nations, even those with peaceful or credulous dispositions, may be skeptical of your present good intentions, however sincere and genuine they might be."[21] The hypocrisy Sagan highlighted in "The Common Enemy" was the impulse of the US and USSR to frame their own history of imperialism as benevolent and the other's as tyrannical. Both nations recognized the harm of imperialism and colonialism but had difficulty recognizing those actions at home as easily as they saw them apparent in their enemy—so much so that they projected these values onto extraterrestrial life. The imperial behavior of the US and USSR cannot be conflated—their histories are too different and complex to be simplified in such a manner. But it is undeniable that various forms of imperialism imbued the space sciences during the Space Age, affecting both its rhetorics and its activities. The search for extraterrestrial intelligence perhaps demonstrates this best, not only showing how the space sciences were deeply entangled in the imperialist actions of the US and USSR but also providing a window to critique the inevitability or universality of these qualities.

PART III
WASTE

CHAPTER 7

WASTE MAKES THE FRONTIER

Rethinking the Spaces and Places of the Contemporary Space Race

JULIE MICHELLE KLINGER

In 2021, Jeff Bezos and Richard Branson blasted off on suborbital flights amid global pandemics, anthropogenic climate disasters, and resurgent authoritarianism. Interrogating these phenomena as co-constituted rather than coincidental provides greater clarity on the material relations and social conditions through which contemporary human engagements with outer space are being produced. Unlike most space-related activities taking place globally, these flights served no scientific purpose but to impress upon the public—and potential markets—that the ultrarich might soon be able to turn "the final frontier" into a hyper-exclusive vacation destination, and eventually, something akin to a gated community constructed out of materials circulating in the heavens as an escape from earthly wastelands.[1] This did not come from nowhere. It expanded upon tendencies evident elsewhere, notably in two types of thanatourism, which traffic in appetites for apocalypse: disaster tourism in which people visit sites of previous social and environmental destruction, and "last-chance-to-see" luxury tourism, in which people visit sites that are disappearing under the onslaught of climate change: melting icebergs and glaciers, vanishing tropical rainforests, or soon-to-be inundated archaeological sites in coastal areas. Thanatourism describes "travel to a location wholly, or partly motivated by the desire for actual or symbolic encounters with death."[2] This may, in the context of billionaire spaceflight, provide a clearer optic than the celebrated "overview effect" for understanding the combined motivations for spaceflight, labor exploitation, and apparent support for resurgent authoritarianism among billionaire astronauts and their boosters.[3] Last-chance-to-see tourism is premised on the unfolding or predicted demise of a destination. Disaster tourism targets places that have been damaged in human or natural disasters, fixing the place in question to an apocalyptic point in the past. Both forms of tourism are driven by a curiosity or concern with the dramatic waste-making of landscapes and lives, and are enabled by a complex entanglement of affluence, curiosity, and even potentially sincere beliefs in helping humankind: an entanglement that proved effective for advancing settler colonialism, imperial science, and colonial acquisitiveness in previous eras.

The prospect of space tourism offers a different twist. Instead of the last-chance-to-see being some iconic site on Earth, the view expressed by space tourism proponents is the expectation that it is the first step to leaving behind the Earth in its entirety.[4] In a view that intersects with neo-Malthusianism and climate doomerism, all of the Earth is consigned to become a wasteland under anthropogenic climate change, and therefore either a future disaster tourist site for the few who manage to escape, or a nature preserve for the privileged few allowed to remain.[5] Billionaire spaceflight has thus been legitimized as a bold step into the final frontier, essential for human progress and essential to escape human-induced apocalypse on Earth.[6] Tourism caters to these fantasies and provides a proof of concept, but it also demonstrates how impossible planetary escape may be for the vast majority of Earth's inhabitants.

If compounding crises of our moment are interpreted as evidence that apocalypse is in fact nigh, then escape plans can be reimagined as prudent and responsible, and space tourism is a logical first step. The crises to which I am referring are those that unfolded during the time that the workshop on which this book is based was supposed to have taken place, and our eventual digital meeting. These crises, of course, have deeper roots. But during the eighteen months that elapsed between our originally scheduled May 2020 meeting to have been held in the idyllic Huntington Library in Pasadena, California, and the September 2021 virtual workshop, we were daily bombarded with frontier- and waste-making developments. Rocket launches coincided with wildfires, chemical spills, floods, and millions of deaths in underresourced hospitals. In the subsequent year, ultrawealthy clients paid "undisclosed sums" for brief trips to suborbital space while asylum seekers drowned at sea, died in deserts, or were imprisoned in illegal detention facilities.[7] The

COVID-19 pandemic laid bare long-standing social, environmental, and economic crises with unsparing clarity, revealing that what may recently have been considered exceptional disasters or unthinkable institutional failings were in fact features of a terrible new normal. Importantly, this catastrophic new normal did not derail plans for suborbital joyrides, nor was it an unfortunate coincidence. It provided the conditions for massively increasing the fortunes of a global minority, which included all three Anglophone space billionaires—Richard Branson, Elon Musk, and Jeff Bezos—through speculation, through deepening mass dependence on commercial home-delivery services in the absence of effective public support infrastructure, and by capturing government handouts.[8]

As ever greater space milestones direct our attention skyward, solid ground—safe homes, dignified workplaces, and functioning social infrastructure—is melting away from beneath so many of us. The tone-deaf media celebrations invisibilized the fact that space billionaires' adventures toward the final frontier were built on the waste-making of human lives in the form of underpayment and overexploitation, union busting, monopolistic practices, large-scale theft of the "social wage" via undertaxation, and on the waste-making of as yet uncalculated landscapes via the mobilization of immense supply chains to requisition materials for their enterprises.[9] Whether in the form of millions of tonnes of cardboard boxes, hectares of concrete, tankers of oil, or metals for vehicles and technologies of automation, there is no empire, no space travel, no pleasure tourism, and no global supply chain that operates independently of resource extraction.

In other words, no adventure toward the frontier occurs without some form of (un)intentional waste-making. For last-chance-to-see and prospective space tourists, a conviction of the inevitability of the transformation of the site in question into a wasteland—an iceberg, a forest, or Earth itself—runs deep. Viewing it as a tourist also confers incomparable bragging rights.[10] A belief in the inevitability of apocalypse, on the other hand, has the effect of erasing any possibility for corrective action, and therefore of individual responsibility.[11]

Regardless of motivation, the massive resource demands and carbon emissions required for thanatouristic escapades turn, as the *Guardian* journalist Jonathan Watts puts it, "last-chance-to-see" into a self-fulfilling prophecy.[12] Understood in this light, the 2021 suborbital flights, which each emitted much more CO_2 than a transoceanic flight, constitute the first steps taken by a cadre of the ultrawealthy who publicly fantasize about taking one last look at Earth before moving on to their libertarian space colonies in the "final frontier."[13] It is this material and ideological feedback loop between waste-making practices, the relentless imposition of apocalyptic conditions on diverse landscapes and lives, and escapist impulses that yokes together extractive frontiers on Earth and in space.

Below I examine the frontier and waste-making processes of the colonial capitalist space race across three physical sites: the mine, the launch site, and the asteroid. To understand our evolving relationships to outer space, we must connect space exploration to its material processes, which have earthly geographies. Rockets and fuels must be dug out of the Earth somewhere, but there is seldom a direct line discernible from this or that mine to this or that piece of space tech. Because future space travel takes off-Earth mining as a given, theorizing extractive practices on Earth can shed light on evolving engagements with outer space. Prevailing modes of production and consumption rest on discarding precious metals and materials even though the costs of digging up new sources and disposing of old ones requires the violent conscription of someone else's lands.[14] The inability to see this waste-making for what it is—a foolish practice to be corrected—causes powerful interests to project this same sensibility onto objects in outer space. Celestial bodies are devalued to fit extractive ambitions. This is not only prevalent in mining; similar processes are at work in lands slated for launch sites and other forms of infrastructure on Earth.[15]

MINES

An industrial mine marks a conquered frontier. It serves as a surefire sign that "unproductive" lands have been put to productive use, or in other words, expropriated for nonlocal enrichment. Five moments of waste-making characterize the mine. The first is when a new place is drawn into the sights of extractive ambitions. A sweeping, vaguely specified area—a mountain, a watershed, an asteroid, or a planet—is captured in the frontier imaginary and stripped of other values, meanings, and possibilities to set it up for extractivist enclosure, destruction, and abandonment.[16] The second moment occurs when a designated area within this vast imagined geography is enclosed, outlawing other uses, relations, and futures. This makes waste of cultures based on other relations and lays waste to futures tied to the land in question. Local knowledges are criminalized, pathologized, suppressed, or conscripted to nonlocal extractivist ends.[17] The third moment occurs when the earth is dug up, the forests razed, the rivers dammed or diverted in service of extracting a given resource, physically laying waste to what came before and thereby creating a localized apocalypse. This localized waste-making of the mine extends as it morphs through the sprawling geographies of infrastructures that connect the minerals to industrial processes, assembly, sale, and discard via their global supply chains. The fourth moment of waste-making is bivalent and is essential to fueling the whole cycle. The first valence occurs when mining sites that may in fact be abundant in all manner of minerals, metals, and possible other futures are abandoned by the extractive entity that in all likelihood disturbed billions of tonnes of Earth to extract much more modest volumes of a single commodity. The second valence occurs when otherwise valuable materials are discarded as waste, despite periodic scarcity panics, because reuse and resourcefulness have not yet been made into frontiers of accumulation on any large, formalized scale. The extractivist gaze, by and large, looks to greenfields rather than old mines or waste dumps in pursuit of the next mother lode.

This is where the fifth moment of waste-making occurs. In the extractivist move to the cosmos, Earth has been reimagined as a place that can and must be left behind whether because it has been "mined to exhaustion" or converted into a nature reserve enclosed for elite enjoyment where, in the vision of one space billionaire, "only a few will be allowed to stay."[18] Describing a mine as a barren wasteland—sometimes compared to the Moon or Mars or some other planet—places the site beyond remediation, which therefore places perpetrators of extractive violence beyond accountability. Comparing a mine to some other celestial body can infuse an otherworldly appeal into a wasted land, while more subtly impoverishing off-Earth landscapes as parallels to sites of extractive devastation on Earth. This renders prospective sites of enclosure, extraction, and abandonment as expendable in the endless pursuit of new extractive frontiers in the name of escaping the apocalypse at hand. This is why the wasteland trope is the common accompaniment to frontier-making processes of enclosure and abandonment, and why the mine is so often its literal and figurative geographical signature.

But rethinking the apocalyptic sites of abandoned mines can help temper apocalyptic and escapist thinking with respect to Earth. Like the abandoned mine, the Earth is not and cannot ever be entirely depleted. Plenty remains, even if it is disturbed, rendered residual, or placed under glass.[19] And although such sites may be abandoned by capital, or forcibly evacuated by apparatuses of dispossession, people remain or return. Memories are maintained and transmitted. This is evident in how places of death and devastation, and particularly places of mining disasters, hold a particular appeal in thanatotourism.[20] Nonlocal people come to learn, mourn, honor, and reflect. Local people work to preserve and maintain memories of lives, livelihoods, and landscapes lost. In these memorial practices, other possible futures can be excavated and revived.

Governments in settler colonial states have responded to demands to preserve landscapes and lives that preceded settler colonial domination by designating reservations and parks. Themselves a territorial instrument of expropriation, enclosure, and confinement, the creation of such "set-aside" places enables the wholesale transformation of the surface of the Earth to proceed everywhere else.[21] A similar ethos has been applied in conservationist interjections in plans to mine the Moon and "terraform" Mars: set aside a few conservation and heritage sites here and there to appease the sentimentalists so that business can get on with mining and colonizing the rest.[22] In this way, the establishment of a preserve authorizes its opposite while also creating spaces for elite access to and enjoyment of "wilderness."[23]

All these ventures require hardware, and hardware is made of minerals and metals. It is well substantiated that nonlocal demands for resources drive violent economies of extraction into sensitive landscapes.[24] It is a feature of our global economy from which space tech is not exempt. The constitutive geographies of mineral supply chains—industrial sites, land, air, and sea transport routes, factories run on sequestered labor, and smelters raining toxins on surrounding environs—channel raw materials through a series of commodified transformations until the material wrested from the Earth becomes a product that can be purchased and used. For those materials bound for space instead of the scrapyard, the terrestrial geography of space activities requires infrastructures of departure.

LAUNCH SITES

Like the terrestrial mine, noncontiguous areas far greater than the launch site itself are required for logistics and processing in order to produce the infrastructure and technologies of space launches. Writing about the San Marco launch site off the coast of Kenya, Asif Siddiqi conceptualizes sites not as discreetly bound, but as fragmented and dispersed, and characterized by blurred edges and multiple claims.[25] It is perhaps unsurprising, then, that the geographic patterns of launch sites are comparable to missile test ranges, nuclear waste sites, and detention centers: spaces of exception, concretized in logistical networks, where state and corporate dramas of exceptionalism can play out amid regimes of exclusion.[26]

A launch site is a gateway to elsewhere. The persistent and polyvalent regimes of exclusion around infrastructures of departure intersect with the appeals, practices, and effects of terrestrial tourism. Space launch infrastructure in particular refracts the now normalized practices of jet-setting tourism—specific performances of securitization, uneven spaces of militarized secrecy, and aspirational appeals to the hyper-wealthy—through the prisms of heightened exceptionalism. While thanatourism may be driven by moralistic inclinations, these are amplified in the context of space-related activities: boosters of the suborbital joyrides reframe these displays of excess as markers of civilizational progress.

Like other infrastructures of departure such as airports and seaports, launch sites are simultaneously superimposed upon and excised from their contexts. Surrounding communities must reckon with displacement, noise disturbance, respiratory burdens, and occasional debris falling from the sky as the hypermobile go about their combustive activities. But those who have chosen to make their homes nearby seldom readily accept the loss of their lands to pavement, the invasion of pollutants into their bodily tissues, or the potential to have their homes smashed under a crashing aerospace craft—at least without a protracted fight.[27] Social opposition to infrastructure siting is consistent enough that states and firms have employed a waste-making mentality: put the launch site elsewhere, away from here, in a sparsely populated wasteland that is good for little else. If locals are there, they can be cajoled, coerced, or expelled. This helps explain why launch sites are located on the lands of colonized, minoritized, and marginalized peoples. Consider, for example, the former Soviet Union launch site in Kazakhstan, China's launch sites in Inner Mongolia and Xinjiang, and France's former launch site and testing ranges in Algeria, the displacement of ethnic Somali in Kenya to make way for an Italian launch site.[28] All are the ancestral homelands of nomadic peoples, forced to make way or live with risk.

Like the mine, launch sites create long trails of waste. Rocket debris litters the plains, deserts, and forests surrounding launch sites and test areas. Spent equipment is abandoned to the ocean floor. Rocket launches entail waste-making at greater heights and depths, drawing together the mine shaft and the ocean floor with the upper reaches of the atmosphere. Aside from the colossal quantity of greenhouse gases that are emitted, rocket fuels contain a host of other materials that, when combusted, transform into highly mobile and absorbable toxins. Each test, each launch, releases a pollutant plume of combusted metals and chemicals that move in (un)predictable ways through Earth's circulatory systems. Rocket launches are the only anthropogenic source of ozone-depleting substances released into the upper stratosphere, where they do considerable damage to the ozone layer. At varying altitudes and depths, wind and ocean currents carry pollutants to far-off places, where they are eventually taken up in the circulatory systems of plants, animals, and people, sometimes settling in living tissues and exacting epigenetic harm.[29]

To bring this back to terrestrial tourism, a similar point can be made about jet fuel and other vehicle emissions. Only rarely are frequent flyers or holiday travelers confronted with the localized violence of their travels: there is no record of a private jet owner ever being held to account for the deleterious effect of leaded fuel emissions on airport workers or nearby inhabitants; highways bypass the poverty of adjacent communities.[30] At best, travelers are offered an option to "offset" emissions by planting trees somewhere else.[31] This move is entirely consistent with the waste-making approach in that it outsources responsibility and, in the words of Max Liboiron, *"assumes access* to Land as a sink" for waste as well as for clean-up.[32] Let the problems and solutions unfold anywhere other than here. In the case of tourism offsets, such practices ensure the continued diversion of funds from the adjacent communities on whom emissions, particulate, and noise constantly rain down. Perhaps, because tourism practices require selectively ignoring the environmental violence of more everyday forms of combustive travel, efforts to conceptualize the pollution and waste-making dynamics of launch sites remain preliminary.

The circulation of debris and pollutants from rocket launches are a continuation of other forms of circulation that coalesce in the launch site and the rocket. Launch sites are a convergence point for global material and energetic circuits.[33] Elements pulled out of the Earth, scraped from mountains, refined in smelters and forged in factory flames assemble at these sites: some in the infrastructure that is rooted in place, destined eventually to rust and be overtaken by mosses, rising seas, crabgrass, or vines. Others pass through this site on their way beyond Earth's atmosphere, where they may orbit in perpetuity following the end of their useful life or be vaporized by the friction of Earth's atmosphere as they descend from graveyard orbit.[34]

For earthly and space tourism marketing alike, it is the destination that matters, not the unseemly details involved in the journey. Like the view from a penthouse suite, looking at the Earth from space obscures much of the pollution, social conflict, and other unsightly necessities for pleasure travel under the glittering veil of distance. This is another appeal of the frontier, its perceived distance from here. For the tourist, being far enough away, or above, that the messy responsibilities of the everyday fade into beauty and silence is precisely the point. What I am positing here is the possibility that the widely reported "overview effect"—in which individuals traveling to space report feelings of oneness with the Earth and fellow living creatures, as well as a greater distaste for war, destruction, and division—may not be a universal human experience.[35] Many astronauts return to Earth "with a desire to do something to improve life back on the surface."[36] However, traveling to space did not appear to motivate Bezos to stem the assault on worker rights in Amazon warehouses.[37]

ASTEROIDS AND ORBITAL DEBRIS

Traveling to space does not promise an escape from waste. This is evident in the growing orbital debris problem: Earth's orbits are littered with tens of millions of fragments from obsolete, abandoned, and intentionally destroyed technology.[38] It is a curious thing, if you think about it, to compare the flurry of excited activity around figuring out how to capture and mine a far-off asteroid with the broad paralysis around, say, collecting and repurposing orbital debris.[39] Rather than sweep up platinum bits that contaminate terrestrial roadways on Earth and constrain orbital space, fortunes have been amassed and lost, companies founded and bankrupted, laws rewritten and treaties violated on the premise that platinum should instead be had from asteroids, captured and towed somewhere where they can be usefully pulverized into mineral commodities potentially for sale on Earth.[40] Mining among the stars, beyond the confines of Earth's atmosphere and regulatory systems, is redolent of settler colonial glory whereas cleaning up our immediate surroundings is the historical province of convicts, captives, and the working poor.[41] The escapist impulse that drives aversions to serious cleanup, remediation, and accountability for waste-making practices on Earth extends into outer space, where asteroids are drawn into the waste-making vision.

Among the celestial bodies celebrated in popular science and protected under international law, asteroids stand apart as potential threats. Often depicted as barren hazards, asteroids represent the new "no man's land": flecks of wasteland floating in the heavens waiting to be put to productive use. Here the frontier sensibility closely resembles its earthly counterpart. The greater numbers of asteroids relative to other celestial bodies in the solar system, as well as their apparently freewheeling character, position them as a threat to fixed powers on Earth, and this banalizes their destruction as a matter of policy, security, or development.[42] This problematization of mobile objects resonates with empire's historic unease with nomads, vagabonds, and gypsies.[43] Destroying asteroids to preserve order and facilitate surplus accumulation from captured objects readily maps onto state logics to destroy nomadic cultures to consolidate territorial control to facilitate surplus extraction from sedentarized peoples.[44]

On Earth as in space, frontiers are seldom as empty as those naming them would claim. Where they are not populated with people, they are imbued with collectively held meanings. Imposing the frontier vision involves discursively rendering places devoid of people or downgrading its life forms and other properties as able to be sacrificed to extractivist ends. Definition is important here. Just as naming a place as a frontier orients political and economic power toward conquest, the choice to value an asteroid as a harbinger of life, or a dangerous lump of metals menacing Earth, or a "cold, dead rock," mobilizes different processes. For example, NASA's Office of Planetary Protection requires special protocols "if the target body has the potential to provide clues about life or prebiotic chemical evolution."[45] This precludes pulverization.

Asteroids are some of the only celestial bodies that could transfer genetic material between worlds without cataclysmic impact. One body of theory holds that chains of amino acids, or even tiny metazoans akin to tardigrades, riding asteroids through the cosmos may have been responsible for the Cambrian explosion of extraordinary genetic variety on Earth.[46] But contemporary asteroid-related research and development is overwhelmingly geared toward capturing them, grinding them up, and turning them into the machinery of extraplanetary exploration. While there is an environmental argument to be made for figuring out how to transform asteroids or other space rocks into the hardware of space travel to reduce the earthly resource demands and launch emissions of longer-term space missions, care must be taken to ensure that human activities do not crush currently unknown possibilities circulating in the heavens. It is the waste-making vision that categorically assumes that asteroids are cold, dead rocks, condemning them all to be rendered into components for spaceships. We must not—in the manner of space billionaires who, through tax evasion, wage suppression, and union-busting, discount the value of the lives of those who labor for them—discount the value of the biotic materials and abiotic precursors to life simply because the generative potential of their asteroid home is not immediately apparent.[47] Practices in one frontier space inform practices in other frontier spaces.

Earthly environmental destruction and extraterrestrial exploitation provide mutual justification. They are processes that result from and stoke further desires for infrastructures of escape. Space as a so-called final frontier is a potent cultural fetish because at best, it suggests the possibility of cosmopolitan transcendence of an apocalyptic present, while at worst it incentivizes the most rapacious profit-driven waste-making on Earth with the promise that the ultimate escape awaits those who win the lottery of contemporary capitalism. This is not a remote or fringe fantasy. Rather, it is one that builds from broader currents in our contemporary moment: the coincidence of deprivation amid the elective mobility of terrestrial tourism, which has its roots in colonial pleasure tourism; spaceflight spectacles amid crumbling physical and social infrastructure, made so vivid in the depths of the COVID-19 pandemic.

The crass billionaire rocketry of 2021 mirrored terrestrial tourism in another way. The melting icebergs, vanishing tropical rainforests, sites of catastrophic human or environmental violence are more than frontiers for the curious adventurer, more than fodder for escapist or apocalyptic fantasy, and they are more than the eventual casualty of accelerating climate change. They are located in immutable places that will persist even after the cherished landscapes, cultural sites, and ruins are razed, melted, inundated, or dismantled, the possibilities for which we do not yet know. We might look past the spectacles to wonder, creatively, at the possible futures of such sites on Earth as a way to think with greater originality about our relations to outer space, because outer space is so much more than a final frontier.

CHAPTER 8

FIXING SKYLAB

An Alternative History of Spaceflight through Breakdown and Repair

RÉKA PATRÍCIA GÁL

On behalf of the American people, I congratulate and commend you and your crew on the successful effort to repair the world's first true space station. In the two weeks since you left the Earth, you have more than fulfilled the prophecy of your parting words, "We can fix anything." All of us have a new courage now that man can work in space to control his environment, improve his circumstances and exert his will, even as he does on Earth.

— RICHARD NIXON

The first space station launched by the United States was operational for about twenty-four weeks between May 1973 and February 1974. The three sets of crew members who lived aboard the Skylab Orbital Work-shop, as it was called, during this period were charged with using the state-of-the-art telescopes and equipment it housed to conduct ground-breaking research on extraterrestrial environments. By the time the last astronauts departed from this ship, the world record for outer space journeys had been exceeded, establishing for the public that prolonged human life in space—and indeed, dominance of planets beyond our own—was not only possible, but forthcoming. The Skylab 2 Command Module, holding the members of the last Skylab crew above water as the

USS *Ticonderoga* prepared to hoist it from the Pacific Ocean, was seen by audiences around the country. This, it seemed, was the beginning of a golden age of US spaceflight, rendering the early system failures a distant memory in the face of overwhelming success.

A year prior, however, it seemed possible that the Skylab mission would fail entirely.

"We can fix anything." This is how Richard Nixon summarized the main takeaway from the actions of the first crew of the first US space station, Skylab, on June 7, 1973, in his message to Captain Charles "Pete" Conrad Jr. Launched on May 14, 1973, Skylab experienced a mission-compromising failure immediately upon departure, which led to the delay of the launch of the first crew by ten days and required the ground crew to develop and deliver a new thermal shield as quickly as possible. The astronauts themselves performed mission critical repairs on the station immediately on their arrival.

Despite such initial hardships, long-term human habitation has ubiquitously been represented through the lens of technological utopianism in both the American and Russian contexts.[1] This techno-utopian language is particularly amplified in the contemporary moment, when the human exploration and long-term habitation of space is increasingly imagined through a privatized lens, which presents the colonization of space *itself* as technofix, a one-shot solution to the crises of capitalism and climate change.[2] The language of the "fix" appears in other areas as well: as Richard Tutton points out, both SpaceX's CEO Elon Musk and SpaceX President Gwynne Shotwell have referred to Mars as a "fixer-upper of a planet."[3] The implication in these narratives is that humans are not in a mutual relationship with their environment, but rather separate from and superior to it. Fixing is imagined as a one-directional control of the natural environment, reliant on technological innovation. This fetishization of innovation has a long history that can be traced back to industrialism, through consumer technologies that flourished after World War II, research and development labs, and, as these examples show, all the way to the modus operandi of contemporary Silicon Valley.[4]

What is ironically often obscured by the utopian rhetoric of the technofix is precisely the *labor* of fixing: the continuous work on repair, maintenance, and care that must be performed to keep technologies operational. Recent scholarship within media, science and technology, and urban studies has argued for the surfacing of the invisibilized, historically degraded work of maintenance and repair in contemporary economies.[5] But foregrounding this work is not merely about filling in cracks in the history of technology. Steven Jackson argues that centering repair work is necessary to counteract the productivist innovate-and-discard cycle that has hugely contributed to the dangers posed by the Anthropo-cene.[6] In fact, it has been established that the invisibilization of mainte-nance and repair labor is dependent on the logistics of globalization that enforces the offshoring of electronic waste to various countries around sub-Saharan Africa and Southeast Asia.[7] Focusing on maintenance and repair reconstitutes the world, and the existing literature in space histo-ry, within a corrective framework. Shannon Mattern thus argues that it also requires the acknowledgment of domestic and reproductive labor as analogous acts of preservation and conservation.[8] Understanding main-tenance, care, and repair as connected also reveals that the fetishization of innovation is not solely the result of racial capitalist logics.[9] Debbie Chachra points out that it is also patriarchal in regard to making and creating as inherently superior to historically gendered reproductive and care work.[10]

As Leah Aronowsky argues, while spaceflight has been discursively framed as an emblem of technoscientific supremacy, the reality is that human survival in space has always been "a thoroughly multispecies affair," one that highlights the interdependency between humans and their environment.[11] Indeed, the long-term survival of humans in space habitats has always been dependent on the performance of regular main-tenance and repair work: filters have to be exchanged, waste produced on board has to be carefully disposed of, and failing technologies have to be repaired continuously in order to help the station manufacture an environment in which humans can exist long term. Unsurprisingly, in their autobiographies astronauts who have served on various space sta-tions repeatedly highlight maintenance work as a fundamental facet of life in space.[12] There is clearly a conflict between the innovation-focused techno-utopian rhetoric that has defined the history of space explora-tion, and the embodied experience of being dependent on maintaining life-sustaining technologies.

My aim in this chapter is to contribute to a move toward unsettling the history of space exploration from its focus on techno-utopianism and innovation, and to grapple with the ecologies of maintenance and care labor connected to off-Earth human survival. By bringing scholarship on repair studies into conversation with the history and historiography of the first crewed mission to Skylab, I intend to demonstrate that the successes of long-term human spaceflight have always been contingent on repair work performed by the astronauts on board, as well as the crew on the ground. Ultimately, I argue that Skylab's early failures and the rescue devised for it are examples of what Alexandra Crosby and Jesse Adams Stein theorize as the dual and constructive relationship of repair and innovation within the Anthropocene: first, "repair as design," which

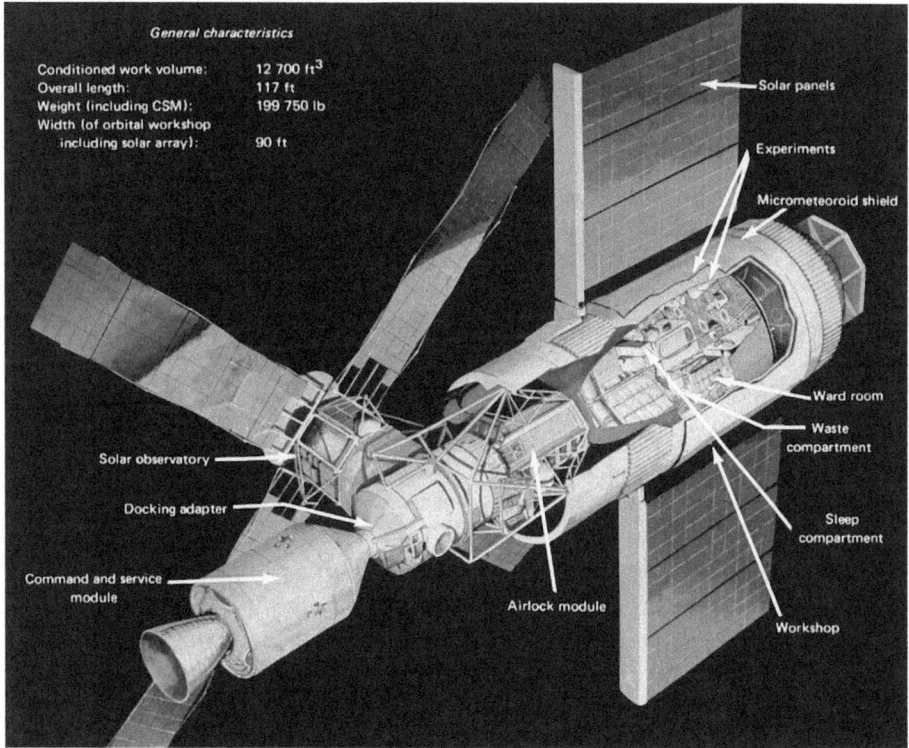

General characteristics

Conditioned work volume: 12 700 ft³
Overall length: 117 ft
Weight (including CSM): 199 750 lb
Width (of orbital workshop
  including solar array): 90 ft

Solar panels
Experiments
Micrometeoroid shield
Ward room
Waste compartment
Sleep compartment
Solar observatory
Docking adapter
Command and service module
Airlock module
Workshop

8.1. Skylab diagram. NASA.

constitutes the resourceful acts of repair that have the power to work directly against mainstream consumerism, and second, "designing for repair," or the conscious act of designing objects, spaces, and systems to be long-lived and repairable.[13] In focusing on the maintenance of Skylab and the innovative design solutions that were devised for the station's rescue, I intend to show that there are productive insights to be gained for sustainable spaceflight from thinking of space stations as entangled with their human crews of maintainers.

THE STORY OF SKYLAB

Skylab (see fig. 8.1) was the first US spacecraft explicitly intended to function as a space habitat and support a human crew for an extended period. The mission was largely built upon knowledge from Apollo, while also using some Apollo hardware, including its command and service module. During the Apollo missions it had already become clear that space infrastructures supporting human life have to be designed for maintainability, specifically in-flight repair.[14] However, while the Apollo crews only spent up to twelve days in orbit before returning to the ground, Skylab was designed to support crew for months-long scientific missions on board and could not be returned to the ground for repairs. From 1973 to 1974, Skylab was occupied for a total of six months by three crews whose scientific experiments were conducted in a variety of areas, including the life sciences, astronomy, materials science, and the observation of Earth's resources.

To maximize the time the astronauts spend on science experiments, Skylab was designed to be automated. However, the astronaut Russell Schweickart recalls that this automation brought with it its own problems: upon detecting a malfunction, the systems would go through a long set of automatic procedures, but these would not inherently allow the astronauts to be able to keep on working. Instead, if the malfunction interrupted their work—such as when it caused the station to no longer be pointing in the same direction, making it impossible to continue their telescopic observations—they would have to wait for the automated system to complete the diagnosis and repair. On the other hand, attempting manual takeover of the automatic system could disturb its procedures, which meant that the astronauts were left with no other option but to be idle while the system remedied the issue. By the time the station was in orbit, the project shifted to a hybrid solution, where the malfunctions would be handled partly by the automated system, partly by the ground crew remotely, and partly on board by the astronauts on board.[15]

Only once the station was in orbit and the astronauts began their work did it become clear that the actual crew time necessary to complete maintenance tasks would vastly exceed the design expectations. During the first crewed Skylab mission, SL-2, for example, a total of 785 "discrepancy reports" were originated, with an average number of 20.2 per day—a decrease from 22.5 per day during Apollo 17.[16] Preplanned for the mission were 16 scheduled in-fight maintenance tasks, which were considered part of regular housekeeping activities and included in the crew checklists. In addition to the 16 scheduled activities, approximately 160 unscheduled maintenance activities, most of which involved the replacement of various parts, had to be carried out on the various Skylab cluster systems. Next to these unscheduled maintenance needs, the crew performed approximately 30 contingency maintenance tasks, many of which were required to avoid compromising mission objectives.[17] This means that merely 7.8 percent of all the required maintenance activities were part of the original calculations for the first Skylab mission. Likewise, James Russell and David Klaus found that on average for all three crewed missions, the maintenance of just the Environmental Control

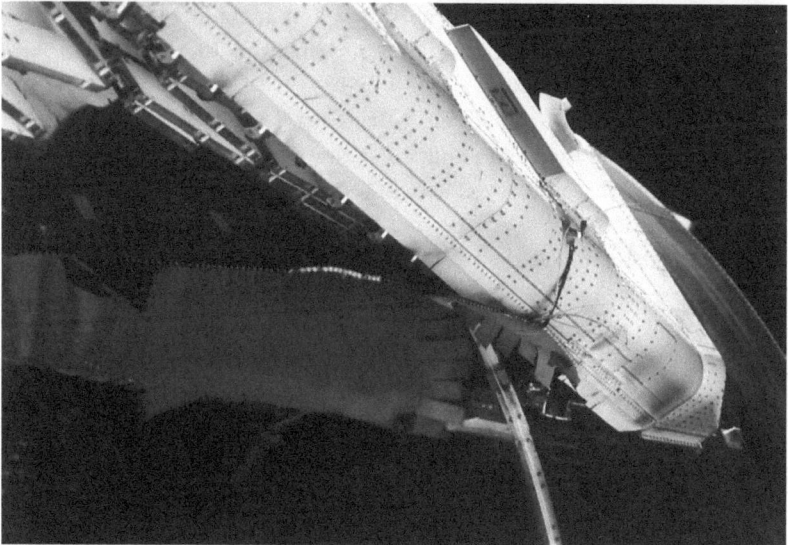

8.2. Photos taken during the first crew's fly-around, showing the remains of the missing solar array (above) and the debris of the micrometeoroid shield stuck in the second solar array (below). NASA.

and Life Support System exceeded its target by roughly half an hour each day per astronaut—adding up to about 1.5 hours per day for a crew of three.[18] Yet, as the Skylab astronaut Owen K. Garriott states, "In-flight repair of Skylab not only saved the program from near disaster but also was an essential element in the completion of most of the experimental objectives."[19] Indeed, the story of Skylab reveals that a projection of the broader cultural underestimation or undervaluation of repair work could prove fatal in space.

The station experienced a mission-compromising failure immediately upon launch on May 14, 1973, affecting two critical technologies on board: the micrometeoroid shield and the solar arrays. The premature deployment of the micrometeoroid shield sixty-three seconds into launch left the station without thermal protection, allowing the internal temperature of the station to climb as high as 60 degrees Celsius. This initial failure also caused one of the main solar arrays to be ripped off the station, while a second one was only partially deployed when an aluminum alloy strap from the broken meteoroid shield restrained it—which meant the station lost almost half of its electrical power (see fig. 8.2).[20] By themselves, neither malfunction was unsolvable. However, the solutions to each increased the problems posed by the other one: the production of electricity required the remaining solar panel to be perpendicular to the sun's rays, but if Skylab remained in this position, its full length would be exposed to the sun's rays.[21] Adding insult to injury, the overheating caused several gyroscopes to fail, making it hard for the attitude-control team to predict the movement of the station as they were conducting maneuvers.

The crew launch was delayed by ten days, from May 15 to May 25, 1973, to allow for the design and fabrication of a new thermal shield, as well as to test the best ways to release the solar arrays, which required the mobilization of all NASA centers and thousands of support contractors.[22] To tackle the problem with the heat, NASA eventually settled on the engineer Jack A. Kinzler's design of a parasol that would be mounted on telescopic spreader arms, which could spring open over the side of the workshop when mounted on and deployed from one of the scientific airlocks. The design of the parasol involved the sewing together of nylon, mylar, and aluminum to form a three-ply sunshade with a Kapton covering to protect the base nylon of the sunshade, which otherwise would have quickly deteriorated under ultraviolet light. This repair solution has widely been reported as an engineering miracle—Kinzler has come to be known within in the agency as "the man who saved Skylab," and was awarded a Distinguished Service Medal from NASA for his design of the parasol.[23] Likewise, the Smithsonian National Air and Space Museum Space Race exhibition on its website summarizes these actions by stating that "with clever engineering and improvisation, the effect of the damage was minimized, and planned Skylab operations were completed despite the reduction in power."[24]

Jackson argues that the relationship between innovation and repair is not linear, but circular, with innovation itself often being reparative, born out of the desire to solve a problem. As he explains, dominant imaginings of technology tend to situate innovation above repair in value and preceding it in time. Jackson demonstrates through the examples of Bangladeshi shipbreakers and the development of the Internet that this view simplifies the relationship between innovation and repair. He writes: "Far from being a generalized cultural tendency or a property of individual minds, innovation in the technology space, as in culture more generally, is therefore organized around problems. This makes innovation simultaneously specific and in some measure collective in nature. And its engine is breakdown and repair."[25] The example of Skylab's parasol serves as a fitting reminder that repair and innovation are always already entangled. Rather than one superseding the other in importance or preceding it in time, innovation and repair work depend on each other for their continuous functions.

But the repair of the parasol also demonstrates something else—namely, that despite gendered assumptions surrounding technological innovation and individual expertise, the creation of new technologies is no individual feat. As feminist media theorists and scholars within repair studies have pointed out, artifacts and technologies often depend on what Debbie Chachra has called "an invisible infrastructure of labor."[26] This includes both care work performed in the background by mostly women, and maintenance and repair work performed most often by poorly paid immigrants and people of color.[27] It is only through the obscuring of these gendered and racialized labors that individual inventors and engineers can continue to be cast as sole heroes of technological culture.[28]

In the case of Skylab, while only Kinzler received an award for the development of the parasol and individual credit for its success, its actual creation depended both on the expertise of hundreds of NASA employees working overtime, and on the specialized skills of New Jersey seamstresses for the prompt sewing of the three layers of nylon, mylar, and aluminum fabrics into a reliable sunshade.[29] Despite this, only one NASA website recounting these events names one of the seamstresses, and one Gizmondo article by the science communicator Mika McKinnon actually states the name of Alyene Baker, featured in a NASA archival image in which she is assisted by Dale Gentry, Elizabeth Gauldin,

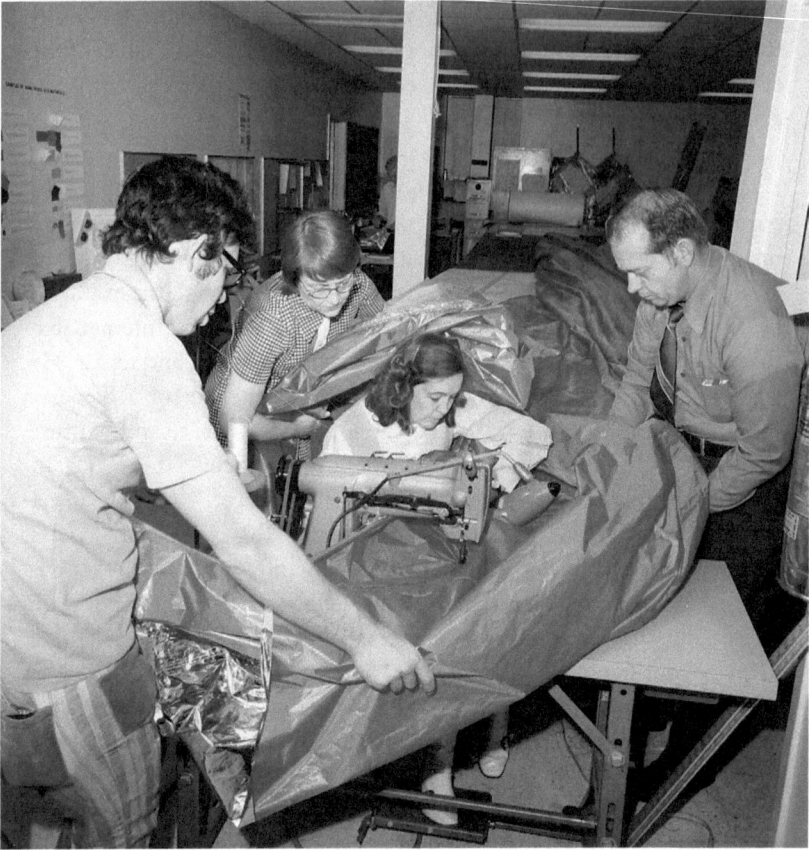

8.3. Alyene Baker sewing the replacement parasol for Skylab, assisted by Dale Gentry, Elizabeth Gauldin, and James H. Barnett Jr. 1973. NASA.

and James H. Barnett Jr. in feeding the material of the parasol through the sewing machine (see fig. 8.3 and fig. 8.4).[30] The majority of the historical sources merely state that seamstresses were hired for the process, but their expertise is not explicitly stated as necessary for the design process, even though Kinzler's design is itself very clearly inspired by the traditionally feminized labor of sewing he had seen his mother do.[31] As his *New York Times* obituary states, "He ordered a large American flag and, recalling how his mother used to hang curtains by sewing a hidden sleeve at the top and inserting a rod through it, did likewise."[32] In these narratives, the rescue of Skylab becomes dependent on the sole genius of one male engineer's design, while the expertise and the feminized labor that inspired and facilitated the creation of the technology is largely undervalued.

8 . 4 . Seamstresses working on the parasol for Skylab. NASA.

Upon arrival, the first Skylab crew, made up of Charles "Pete" Conrad Jr., Joseph P. Kerwin, and Paul J. Weitz flew around the workshop to determine and capture the exact state of the stuck solar panel via video footage to send back to the ground crew, who got started on devising a plan for the removal of the aluminum strap via extravehicular activity (EVA), more colloquially known as a spacewalk, which was to take place two weeks later. The crew decided that it was worth trying an impromptu EVA to pry the solar panel open, by exiting from the side hatch of the Apollo command module: Kerwin held Weitz by the legs, and Weitz held one of the tools on board, a shepherd's crook on a five-foot pole, with which he attempted to dislocate the aluminum shrapnel stuck in the panel, while Conrad maneuvered the spacecraft. Ultimately, they found that the piece was stuck too firmly, which meant that pulling on it sent the two spacecrafts coming toward each other, requiring Conrad and the altitude-control team to maneuver them apart, setting jet fires off around the astronauts. When they determined that this approach would not work, they encountered another malfunction: the docking latches that worked just before this EVA were malfunctioning. After trying all possible repair activities, they were left with a last option, which required another EVA, the depressurization of the spacecraft, and the entering and removal of the tunnel hatch to cut wires inside the docking probe itself.[33]

The next day, the crew shifted their attention to the deployment of the parasol, which did not pose too many problems. The parasol proved to be a temporary solution: it worked well enough for the first crew, but as the sunshade deteriorated over time, the temperatures inside the workshop crept up again at the end of the first mission. Fixing it required one last maintenance activity by the Skylab 3 astronauts Owen Garriott and Jack Lousma, who on August 6, 1973, performed another spacewalk to install a twin pole sail over the station to cover the first parasol.[34]

With the parasol deployed by the first crew, the temperature inside the station started to drop, allowing the astronauts to begin their planned science experiments. However, the problem of limited electricity on the station loomed large during the first weeks of their stay and, as Kerwin recalls, their first weeks inside the station were spent paying careful attention to the conservation of energy: turning off all lights when leaving modules, using only limited equipment.[35] Meanwhile, on the ground, the backup astronauts Russell Schweickart and the future Skylab astronaut Edward Gibson were practicing possible solutions for freeing the solar array inside the Marshall Space Flight Center's Neutral Buoyancy Simulator, and relaying their findings to the on board crew.

On June 7, Conrad and Kerwin finally departed to perform the longest spacewalk to date that would free the solar array. No extravehicular maintenance was originally planned for Skylab, Kerwin recalls, and the design of the station reflected this: there were no footholds or handholds on the outside of the station, and the outside of the workshop had sharp edges capable of damaging spacesuits. All this made the maneuvering not only difficult but also dangerous.[36] Extending a twenty-five-foot-long cable cutter toward the solar array, Conrad closed it over the aluminum strap; then using the cable cutter as a pole, he climbed toward the strap and fastened a rope around the solar array hinge. Once Conrad exerted force onto the cutter, it did cut through the jammed metal—but it did not fully free the solar array, as the coldness of space had frozen one of its hinges closed. When the astronauts pulled on the rope together, they were able to free the frozen panel, but the swing of the opening panel sent them floating away from the station. Thankfully, their space suits tethered them to Skylab.

With both solar arrays operational, the station finally generated enough power to allow the astronauts to return to their scientific experiments, while keeping up with routine maintenance tasks. Following the 28-day stay of the first crew, two more crewed missions were conducted on the station, with stays of 59 and 84 days, respectively. The incoming astronauts continued the maintenance of the station, including the installation of a second sunshade over the decaying parasol and the repair of malfunctioning gyroscopes. By the third crewed mission, the astronauts were able to perform more experiments than expected, and none of the 1,448 malfunctions on board jeopardized mission objectives.[37]

However, the station was not designed with resupplying and refueling in mind and the missions were forced to end sooner than anticipated. While experts had devised various plans to repair the station, with the conclusion of the Apollo program, no more operational Saturn V rockets were available, and without access to a reusable orbital spacecraft system for its repair and boosting into higher orbit, the station's orbit deteriorated. Taken down by a drag from the Earth's outer atmosphere caused by that year's solar maximum heating of the exosphere, Skylab eventually descended and disintegrated in the atmosphere on July 11, 1979, its debris scattered across the Indian Ocean and the towns of Esperance and Balladonia in Western Australia.[38] As repair and maintenance saved the station, the inability to sustain it led to its eventual demise.

As is evidenced by Skylab astronaut testimonies, humans whose lives have been dependent on technologies have come to recognize that innovation alone cannot make missions successful. Jack Lousma of the second mission reflects on the main takeaways from the missions for the future of long-term human spaceflight and on what can be indicated as an incisive critique of techno-utopian imaginaries of automated space colonies. He writes, "As we consider a space station we need to think about what jobs a robot can be designed to do, but we should also plan for a person to be there to program the robot, to fix it, to back it up—in other words, to generally *tend it* [emphasis added]."[39] He further points out that the historical experience of spaceflight shows that technologies *always* break.[40] Long-term human spaceflight is thus an excellent example of Jackson's argument that a world reliant on technological infrastructure is "an almost-always-falling-apart world" that requires a valuation of repair work.[41] But it also points to an understanding that coexisting with space technologies demands a recognition of the ethical obligations toward *caring* for our technologies. As Lousma puts it, we have to be aware that space technologies need to be tended to—taken care of. The centrality of repair, maintenance, and care work becomes particularly visible on space stations, where the lives of astronauts depend on performing these labors.

In his analysis of the Space Shuttle's promise, in a 1974 piece, Garriott contends that the promise of innovation can easily be curtailed by designs that make repair hard or impossible.[42] If these design choices are not integrated into these technologies—or if technologies are intentionally designed to make repair difficult or impossible—their life cycle becomes one-use, and contributes to the capitalist consumption-and-throwaway culture that Jackson criticizes. Such an approach is not feasible for space stations: if a malfunction is not repairable by astronauts on board, the habitat could lose functionalities that might require deorbiting, which is impractical due to the astronomical costs associated with both launches and deorbiting; or it could also put the lives of the astronauts on board in danger. Therefore, for Garriott, technology for spaceflight must be *designed* to be repairable. Filters ought to be easily exchangeable, panels have to be built in a way that allows the astronauts to remove them and access internal parts that may need to be repaired—if these are on the outside, designers' calculations must consider the repair being made by an astronaut wearing a spacesuit, and so on.

Similarly, Lousma pleads for the subsequent missions to consider repair work central to design decisions as well: "We need to make sure that when we build the space station, we make it so that it can be taken apart and that the proper tools are on board to do the job. We must make a space station that's fixable, not one that is put together in the factory with machines that don't permit us to remove panels to get at equipment. The station must be made in-flight repairable whether we plan to fix it or not, and it must have the capability for on-board maintenance and replacement of failed components."[43] The story of Skylab serves as a pointed reminder that, despite the technocultural undervaluation and obscuring of repair and maintenance work, it is these labors that make the world go round.

SPACEFLIGHT OTHERWISE

The historiography about Skylab engages at length with the maintenance and repair work done on the stations. Yet Skylab does not occupy the same space in the historical imagination on space travel that the Apollo or International Space Station (ISS) missions do, and the public and scholarly discourses surrounding space exploration continue to focus on stories of linear innovation and technological prowess. Emphasizing the development of new technologies hides the very real labor of maintenance and repair that is *fundamental* to how human space exploration works. Obfuscating this work serves the commodification of space exploration by keeping these difficult labors behind a curtain of mystery, and spaceflight turns into a myth of exploration, heroism, and technological valor that can be sold to investors, prospective customers, and the public. Rather than consider only the stories of new launches, technologies, and innovations, we should work to demystify space exploration by foregrounding the work of maintenance and repair on space stations. The story of Skylab's breakdown and repair is important because it opens an alternate history of spaceflight that serves as an antidote to the hegemonic historiographic valorization of technologies and their creators by instead directing our attention to the infrastructure of work and the workers that ultimately keep new technologies and space stations operational.

It is crucial to keep this in mind given the NewSpace industry's efforts to commercialize and eventually "colonize" space, especially when these plans are conveyed by a techno-utopian rhetoric that proposes going to space as a "technofix" for problems on Earth.[44] Moreover, repair and maintenance will also be particularly important for NASA's Artemis program, which aims to establish a long-term human presence on the Moon by 2030, and eventually launch crewed missions to Mars—where a round-trip would take roughly three years to accomplish. In the case of the Mars missions, this work will have to be done without possible resupply. Furthermore, due to the communication delay caused by the enormous distance, work will have to be accomplished without assistance from the ground crew, thus making sustainable systems even more important. Taking for granted or, worse, dismissing the necessary labor of space habitat maintenance would put the success of these missions and the lives of the humans on board in great jeopardy.

While Skylab was only the first US space station in orbit, its story is not an outlier. Between 2002 and 2019, the International Space Station experienced 67 high priority malfunctions requiring repair—about 4 per year. In fact, in 2007, the ISS experienced a similar issue with the station's solar array deployment that also required an EVA for its repair. Just as technologies on Earth tend to break, space technology also has its fair share of failures. In fact, as Kaitlin McTigue and her colleagues argue, there is no way to completely prevent unexpected critical malfunctions in complex engineered machinery, and despite the tremendous expertise and relentless efforts that go into making NASA's designs more reliable, spacecraft are not immune to such defects.[45] No matter how modern and robust the technologies developed for future crewed missions prove to be, ultimately, it is the labor of maintenance and repair performed by humans that will enable the success of these ventures, by ensuring that the on-board technologies remain operational for the duration of the trip.

CHAPTER 9

EARTHLY DREAMS AND COSMIC AFTERLIVES

The Failure and Reutilization of Japanese Space Facilities

SUBODHANA WIJEYERATNE

Space exploration is impossible without earthly foundations. Central to both the practice and public impact of extraterrestrial exploration are the vast facilities and physical networks the enterprise generates—launch sites, testing ranges, factories, and laboratories, among myriad others. It is no coincidence that locations such as Baikonur Cosmodrome and Houston's Johnson Space Center (of "Houston, we have a problem" fame) loom as large in the public imagination as the Hubble Space Telescope or Sputnik. In terms of employment, environmental impact, infrastructure, and economic consequences, these sites can have a greater effect on a larger segment of the population than any single rocket, probe, or space station. Furthermore, given their colossal size, expense, and the tendency to be located in remote areas, space facilities are generally understood to be enduring presences in their locales—they are too big, too hard to build, and too important to be easily relocated. Hence, Baikonur Cosmodrome has been continuously in service since 1955; China's Jiuquan Satellite Launch Center was built in 1958; and France's Guiana Space Centre was opened in 1968. All three remain major elements of their respective countries' space programs. For thousands of people in locations such as Kazakhstan, Inner Mongolia, and French Guiana, they represent massive sources of income, investment, and employment that they otherwise had no reason to expect. Furthermore, for both space programs and states, facilities can be sources of pride—an attitude informed by the particular urge of governments to physically manifest their success and importance through the generation of large networks of infrastructure.[1]

However, the presence and permanence of these flagship sites represents only one strain within the diverse experiences of space facilities. As Peter Redfield has pointed out, the construction of space facilities in remote areas has strong commonalities with the colonial experience, insofar as the process can be understood as one in which powerful authorities extend their control over not only the lifestyles of people in distant locations but also their physical environment. The creation of space facilities can thus be a contested and complex process. This is particularly the case since the 1960s, as increasing numbers of countries have joined the space race. Such places lacked the political control, geographical size, and economic might of the great powers, all factors crucial to the development and acceptance of space facilities in localities. The result is that in such places, local resistance, geographical unsuitability, and bad planning can, and did, render space facilities far more transient and ambiguous presences. Furthermore, even those sites that endured sometimes failed to provide any meaningful economic benefit to those who lived near them. The impact of space facilities in these places was thus a deeply fractured, and fracturing, experience for their neighbors.

This is particularly evident in the arrival, departure, and afterlife of several space exploration sites in Japan. Over the course of its development, the Japanese space program developed and abandoned a number of facilities across the archipelago. The experiences and behavior of the people who lived in the vicinity of these sites provide two significant insights into the development of space facilities and their associated programs in smaller spacefaring nations. First, it reveals that on smaller scales and in certain conditions, space facilities are no different in terms of their susceptibility to resistance than are other types of large-scale national infrastructure, such as power stations, garbage dumps, and airports.[2] Furthermore, space facilities also suffered from an overestimation of how much economic benefit they would bring. Second, deriving benefit from space facilities can sometimes be the result of work not by the facilities themselves, but by the people who lived around them seeking to capitalize on the presence of the facilities—both active and inactive—in novel ways. Where the earthly reality of these space facilities disappointed, locals in these areas proactively created their own modes of engagement, and devised ways in which to keep their cosmic dreams alive.

THE LOCAL REJECTION OF EARLY SPACE FACILITIES

One of the key experiences of the Japanese space program was the relative transience of its earliest space facilities. The earliest iteration of the enterprise developed under the aegis of the Avionics and Supersonic Aerodynamics (AVSA) group organized by Itokawa Hideo at Tokyo University in 1954. Itokawa, an aeronautical engineer who had worked on warplanes during World War II, had become convinced of the importance of rocketry in the early 1950s. AVSA began work on April 16, 1954, almost as soon as the United States had withdrawn from control over the country. Within a year, it had tested Japan's first postwar rocket, the six-inch-long Pencil, in Kokubunji, Tokyo. In 1955 it received a ¥17,425 grant from the powerful Ministry of Education to complement the ¥400,000 it had received the previous year—a payment constituting the first direct state investment in rocketry since World War II.[3]

After conducting a series of horizontal tests of the Pencil at various sites in Tokyo, including Ogikubo on the outskirts, the group began looking for an area to safely conduct vertical launches in 1955. Tokyo was far too built up—on one occasion, for example, a cluster of nickel-chrome-molybdenum alloy bolts from an experimental rocket ended up smashing through the roof of a hospital.[4] However, finding a suitable location was no easy feat. Inhabiting a country the size of France, strung out along three thousand kilometers of mountainous islands, the Japanese faced three major geographical challenges to this project. The first was that the archipelago, being volcanic and uneven, had few areas that are both flat and remote, and hence suitable for the siting of large testing and research and development (R&D) facilities. Second, Japan had, and has, very little recourse to inland testing—any rocket of more than a fifty-kilometer range or so would have to be launched into the ocean. Here, the Japanese had to be careful not to lob what were in effect missiles at either North Korea, or communist China—both countries with which the postimperial Japanese state had testy relations. Furthermore, Japan's coasts teemed with fishers, commercial shipping, and US military vessels, all of which needed to avoid being hit by falling debris. Lastly, even where they could find areas remote enough and flat enough, it nearly always featured a local village or two. Wherever Japan's space program went, it was going to be next door to *somebody*.

Nor did the challenges faced by Japan's earliest rocketeers end there. In addition to these unavoidable geographical circumstances, Japan's space program in the 1950s and 1960s confronted two circumstantial challenges. The first was that much of Japan's coastline was still under the influence of the US military, which made clear from the onset that it had serious reservations about a Japanese rocketry program.[5] Second, research proceeded at a blistering pace—the initial firings of the 30-centimeter-long Pencil from April 17 to April 23, 1955, for example, saw the group carrying out 29 test launches with a rocket that had a range of some 30 meters. By 1957, the group was constructing 5.5-meter-long Kappa-8s with a range of 60 kilometers.[6] By 1966, the three-stage Lambda 3-H was capable of traveling 3,000 kilometers, putting Japan in an excellent position to consider launching its first satellites (something it would achieve four years later). This rapid development meant that AVSA quickly outgrew the confines of its extant testing facilities. Failure to find better testing sites could threaten to stifle the entire enterprise.[7]

As a result, though the organizers of the space program knew from the outset that they would have to eventually find southern locations more suitable for testing, given the speed with which they proceeded, they were forced to compromise and build them in the north. They eventually settled on a strip of land provided free of charge by the government-owned Japan National Railways on Michikawa Beach in Iwaki, Akita.[8] In addition to being one of the few appropriate locations not occupied by US military forces, Michikawa Beach was also appealing in that it was not home to many fishing boats or near any major shipping lanes.[9] In fact, despite good fishing grounds (which experimenters gleefully partook of), the local fishing industry was relatively small.[10] This was important, as they planned to launch rockets westward, into the sea. Itokawa Hideo also appreciated that it was on the mainland—the man eventually dubbed 'Dr. Rocket' suffered from terrible seasickness.[11] So keen was the AVSA that it began experiments on August 9, 1955, well before completion of the rest of the facility in 1956.[12]

Despite the group's eagerness, however, AVSA was aware that the site had limitations. It understood that the use of Michikawa Beach was only feasible as long as rockets did not have the range to reach North Korea or China. The AVSA group was also keenly aware of the delicacy of public perception. The Japan in which AVSA activities took place was a society acutely aware of the dangers of technological nationalism, overweening central authority, and the risks posed by large-scale industrial complexes. Many of the signal scandals of the postwar Japanese industrial complex occurred at precisely the time Itokawa and his team were building their complexes and testing their vehicles. In 1954, for example, the fishing ship *Fukuryū Maru* was irradiated by nuclear testing in the Pacific, prompting "Japanese citizens to seek a more direct say in the politics of siting research facilities" by evoking the specters of Hiroshima and Nagasaki.[13] The 1956 discovery of mercury poisoning caused by a Chisso chemical factory in Minamata, Kyushu, and the company's subsequent attempts to deny responsibility, generated a profound suspicion of powerful corporate and institutional interests.[14] The construction of facilities that carried any sense of danger could be greeted with intense resistance particularly given the ability of locals to question central authority had been increased by the liberalization of politics under the American occupation. By the 1960s and 1970s, local groups had learned to organize in order to resist the siting of what were seen as dangerous or disruptive facilities. In places such as Hōhoku, Yamaguchi, thousands of locals organized protests against a potential nuclear power plant and voted out the pro-nuclear mayor in 1977—despite being promised special government subsidies associated with construction. Reactions could even boil over into violence, as the ten casualties of anti-Narita Airport agitation show.[15] So widespread and expected were such reactions that many institutions seeking to build facilities proactively deployed campaigns of "soft power" to convince local political bosses, businesses, and citizens' groups of the benefits of their presence.[16]

It was in this context that the AVSA group arrived in Akita, one of a collection of provinces in the north of Japan's main island, Honshu. The group initially worked to ensure that it did not alienate influential local fishers' unions, which, though small in number, accommodated the activities of AVSA by voluntarily halting their activities during launches.[17] Beyond this, AVSA deployed extensive media and social outreach in the local community. Newspapers were provided with detailed updates during construction, explaining how the completed site would include a concrete vertical testing facility, radar-tracking apparatuses, and an observation room.[18] Itokawa Hideo regularly informed locals of the group's activities, and on more than one occasion invited journalists into the facility to provide them with details on AVSA's future plans. Such invitations then became an almost yearly event, occurring in December 1955, January 1956, September and November 1957, and February 1958.[19] He also made it a point to discuss rocket failures, and explain that even a project that did not perform as intended, such as December 1958's incomplete Kappa rocket atmospheric tests, had scientific value.[20] "Dr. Rocket" was very much the project's public face; newspapers regularly referred to AVSA as "Itokawa et al" (*Itokawa Hideo kyōjura*). "Even kids know the name Itokawa Hideo," observed the *Asahi Shimbun* in December 1955.[21] In return, Dr. Rocket made it a point to pander to local pride; in an interview in September 1955, he stated that though his team had "wandered all over the country" but found that "there was no place with as favorable conditions" as Michikawa for rocket testing.[22]

Their efforts initially paid off. On an official level, various local agencies, educational institutions, government offices, and even the electric company banded together with experimenters in July 1957 to form the Rocketry Observation Cooperation Association (Roketto kansoku kyōryoku kai) to supply the project with accommodation and resources. Their largesse was justified partly on the basis that as a purely civilian project, the new space center would rely on support from the community at large—a peculiar inversion of the idea that facilities *provided* economic support, as evident in the later, larger complexes discussed below.[23] The local press was equally intrigued—*Asahi Shimbun*'s Akita supplement alone featured no fewer than 47 pieces between 1955 and 1961 discussing the organization and timing of various rocket launches, and had a regular question-and-answer section called "Roketto jikken ni yosete" (roughly, "Inside the Rocket Tests") explaining nomenclature and jargon, providing maps of launch trajectories, and describing the different types of rockets. For those with a more thoroughgoing interest, it added technical discussions on, for example, the use of solid fuel over liquid fuel, and comparisons between Japanese rocket capabilities and those of other countries.[24] The newspaper even engaged in a healthy bit of self-critique at the end of 1955, observing that "since the first four national rocket tests, newspapers have been screaming 'rockets! rockets!' almost daily."[25] At the same time, spectating launches became an affair that attracted viewers from all over northern Japan. Around 500 people were present for the launch of the earliest Pencil rockets in August 1955; the following month, similar numbers—including a contingent of some 160 middle schoolers—turned out in the rain to witness the launch of a Baby-T. Even primary schoolers were brought to spectate.[26] By 1957, it had become routine for each launch to be accompanied by crowds setting themselves up with cameras to observe and cheer.[27] Nighttime launches, such as the June 1957 Kappa experiments, were particularly popular, often attracting between 500 and 1,000 observers. Interest became so intense that when preliminary testing of Kappa engines began in September 1956, potential observers were warned that only those who had registered in advance would be allowed to watch.[28] The arrival of a five-man BBC crew to film a Kappa 3 launch in May 1957 caused much local excitement as well.[29] The activities of AVSA also stimulated and fed off interest in astronomy and astronomical observations, particularly at schools. In September 1956, for example, middle-school astronomy clubs across Akita arranged viewing events at which students made observations of an unusually close Mars.[30] In November 1956, seventeen students and teachers at the No. 2 Middle School in Yokote created a star map "about eight tatamis" (roughly 13 square meters) in size, featuring illuminated constellations, at a cost of around ¥6,000.[31] The following year, a cake shop in Akita City produced an elaborate Christmas cake on the theme of artificial satellites, featuring a depiction of the various Sputniks launched up to that point.[32]

Despite this early intimacy, however, the presence of the launch site in the long term garnered a more ambivalent response from the locality. One problem that AVSA faced was the proximity of residential housing and transport infrastructure. Locals with homes on Michikawa Beach were so close to the action that they could view the whole process of a rocket launch from their windows.[33] This was not a problem when the exclusion zone around the facility was only 30 meters, but it became one when this figure increased along with the range and power of the rockets, eventually topping half a kilometer. It was one thing to ensure that spectators remained outside this half-kilometer exclusion zone during tests, but entirely another to empty all the nearby residential buildings of their occupants for hours at a time every time a Lambda rocket launch occurred.[34] Worse, it was not just actual launches that activated the exclusion zone—newspapers regularly reminded locals that even mere engine tests necessitated its usage.[35] Compounding the danger was the fact that many people took to observing the rocket tests from boats.[36] Additionally, the facility was a scant 1 kilometer north of Michikawa Beach station.[37] During the construction stages, at least one engineer had expressed concern that trains would be arriving and departing the station 200 meters or so downrange of the facility, roughly in the direction in which rockets were being launched.[38] As early as 1955, Itokawa Hideo had to reassure local populations that the presence of a station and tracks was "alright" (*daijōbu*).[39] In this, he perhaps recalled that at Kokubunji in Tokyo, early tests of the Pencil Rocket had been carried out despite similar proximity to a railway—every time one of the "jam-packed" carriages passed, a member of staff would wave a sign by the fence, halting the entire process.[40] Rocket launches were also beginning to exert an infrastructural burden—the locality was responsible for ensuring that the local fire department was present during all tests, and for signaling local residents about imminent launches.[41] Lastly, Michikawa's location began to concern even some officials, particularly after the development of the Kappa 9L, which had a range of some 350 kilometers—which would have taken it a third of the distance to North Korea.

Of equal concern to parents across Akita Prefecture was the rise of "rocket play" (*roketto asobi*), where children would make and launch their own rockets. It was a hobby that all too often ended in tragedy. In

September 1957, for example, a group of middle-school students stole a rocket from their school in Akita City. When trying to ignite the device with a match on a tennis court, it exploded, resulting in the death of one student from shrapnel wounds. This was, newspapers reported, only the latest in a string of incidents that had occurred "since the beginning of rocket testing in Michikawa," in which students tried to make their own rockets via methods such as filling emptied pens with gunpowder.[42] A year later, a ten-year-old primary-school student in Yuzawa was injured playing with rockets in the back garden of his home.[43] In January 1959, a fifteen-year-old was injured playing with rockets in Fujisato; later that year, a twenty-year-old blew off some of his fingers, and injured a seventeen- and a six-year-old, while trying to make a rocket out of gunpowder and a beer bottle.[44] Indeed, this would prove to be an enduring problem for Japan's space problem—so many children were being injured by rocket play in Kyushu after the building of permanent launch facilities there that it prompted an official statement in 1968 from Kuroda Yasuhiro of the Space Development Promotion Agency suggesting that the activity be banned.[45]

The combined result of these concerns was an escalation in local disquiet about the activities on Michikawa Beach. Complaints became intense enough that AVSA held a local consultation conference on July 29, 1957.[46] A variety of new locations had been scouted, including in Hokkaido and Shizuoka, but many were rejected on account of high levels of fishing activity or strong local resistance. The Tokyo University group seems to have felt no sense of urgency to relocate, however; as late as September 1960, it had made no practical provisions for moving to a different location.[47] By the end of the year, the decision had been made to move to Uchinoura in Kyushu, but the move was envisioned as happening on a longer time scale than what eventually came to pass. What caused a speeded-up timeline was increasing local resistance, combined with an unfortunate mishap. On the night of May 26, 1962, a two-stage Kappa 8–10 rocket misfired and, instead of heading out to sea, turned and crashed on the testing range. Parts of the device landed in a nearby village, causing a fire that affected as many as a hundred households. In the aftermath, locals turned decisively against the Michikawa Beach facility.[48] The AVSA group now needed to find a new location as a matter of urgency. It settled first on Noshiro, farther south in Akita Prefecture—though only after initiating a series of consultations with the local population *there* to discuss the consequences of the move.[49] The location, though appropriate, was incomplete, and hastily added infrastructure was not completed until 1964.[50] Within two years, the rocketry program had left Akita altogether.[51]

SPACE FACILITIES AND LOCAL ECONOMIC DECLINE

While in Akita geography and local disquiet conspired to end the presence of space facilities, different travails were encountered in other locations. In many of these sites, though space facilities proved to be longer-term presences, their economic impact was considerably less than what had been expected when they first arrived, resulting in equally ambiguous relations with locals. The arrival of large industrial and technological facilities is often accompanied by expectations of investment and economic benefits for those who live nearby. New roads improve access; construction booms bring capital; the arrival of workers stimulates consumption. The arrival of the space facilities in Brevard County, Florida—home to both the Kennedy and Cape Canaveral space complexes—caused a 371 percent increase in population between 1950 and 1960, with growth rates of up to 20 percent well into the 1980s. Indeed the area was, briefly, America's fastest growing county.[52] Similarly, the arrival of Baikonur Cosmodrome transformed the nearby village of Tyura-Tam in Kazakhstan from an "isolated railway-stop" with "a couple of two-story houses for railwaymen, a couple of dozen small mud-plastered house, and the tents of geologists prospecting for oil" in 1955, to a town of over 30,000 today.[53] Kourou, in the vicinity of France's Guiana, similarly went from "a quiet Creole village of 650 into a town of 14,000" during the building of the space center, and then developed into a small town of around 40,000 in the 1960s.[54]

However, as Bruno Latour has observed, infrastructure is often part of a network that has stronger connections with distant locations than with the areas in which it is located. In his words: "I can be one metre away from someone in the next telephone booth, but be nevertheless more connected to my mother 6000 miles away . . . an Alaskan reindeer might be ten metres away from another one and they might be nevertheless cut off by a pipeline of 800 miles that make their mating for ever [sic] impossible."[55] In some cases, the locals were the reindeer—living cheek by jowl but unable to interact with the complex that had been constructed in their midst. This failure to have a meaningful economic impact on its hinterlands is something even the largest space facilities have experienced. The French/ESA space center in Kourou, French Guiana, for example, is a location where the "guarded official prognosis for . . . direct economic benefits has proven accurately modest"; the area as a whole remains economically "hollow" and reliant on support from mainland France. Indeed, as late as 1994, locals living nearby lacked reliable TV and phone service.[56]

Some of Japan's space facilities produced similar outcomes. Important for understanding this is the fact that both the appeal and ability of sites to provide economic stimulation is limited by a variety of exogenous factors. For example, in the aftermath of the Apollo missions and Japan's own launch of its satellite Ohsumi in 1970, interest in space exploration waned, with concurrent declines in tourism associated with space facilities. It is even more significant that the ability of such sites to integrate into local economic structures is inherently constrained. A space launch requires a swath of specialists with advanced technical training. Short of acquiring this sort of training, the number of jobs available for locals from the area is therefore limited. This is particularly noticeable in the cases of Japan's two large launch facilities in southern Japan. Both the Oosumi Peninsula, where Uchinoura Space Center is located, and Tanegashima, where Tanegashima Space Center is located, have experienced long-term economic decline. In the immediate aftermath of the construction of both facilities, nearby villages such as Minamitanechō and Uchinouracho did enjoy a "rocket boom" of construction, increased tourism, and investment. None of this, however, was able to arrest the subsidence of these areas in terms of both economy and population.

In Uchinoura, officials looked forward to the economic bounty that the new facilities promised, and initially at least, provided.[57] Tourism in particular benefited, with the arrival of large numbers of people to watch rocket launches. The arrival of specialists associated with the launch site also contributed to the local economy, and infrastructure was significantly upgraded in both areas with the construction of roads and ports.[58] Tanegashima's population similarly greeted the arrival of the space facility with celebrations and banners.[59] Here again, locals not only expected "great benefit" in terms of economic stimulation, but also enjoyed the expansion of the road network, railways, and ports in what was dubbed a "rocket boom."[60] Yet neither benefited in the long term from the arrival of the space facility. Uchinourachō's registered population actually declined after their arrival, from a peak of 11,792 in 1955 to 8,328 by 1970. By 1980, Uchinourachō's population was 6,863; by 2015, it was a mere 3,215—just 27 percent of what it had been on the eve of the space center's arrival. This happened in the context of a general population decrease within the prefecture as a whole, which lost some 200,000 people during the same period—while at the same time, its largest city, Kagoshima City, grew by some 180,000. Hence, the trend of people moving away from rural areas to large cities continued regardless of the launch center's arrival.[61] Tanegashima suffered similar declines; in fact, from the 1950s onward, the island lost roughly 0.1

percent of its population yearly. While it had been home to some 65,000 people in 1960, by 2015, it was home to fewer than half—29,847 people. Particularly striking is the fact that in the five years after the space facility's arrival—1965–1970—the area suffered its greatest five-year drop in population, a staggering 9,120—or roughly 20 percent of the population. The decline consisted overwhelmingly of young people—whereas in 1965, people over sixty-five constituted some 6.5 percent of Minamitanecho's inhabitants, by 2015, they made up over 33.1 percent. In Tanegashima as a whole, these figures were 5.7 percent and 34.4 percent, respectively.[62]

For many of those who lived on the island—particularly the elderly who were left behind—this decline represented an existential threat to the culture of a distinctive location within Japan. The island of Tanegashima had long been located on the crossroads of Japan's interactions with the wider world, beginning with trade relations with the Ryukyu Islands (modern Okinawa) farther south.[63] It was the site of the first arrival of gunpowder firearms on the Japanese archipelago; in the Meiji period, immigration had been encouraged for the purposes of securing the area for the expanding imperial Japanese state under the same impetus that eventually integrated the Ryukyus into Japan proper, as the prefecture of Okinawa.[64] The space center not only did the area's unique culture a disservice by failing to stem the flight of youth, but the arrival of newcomers—usually on a temporary basis, and only to return to their homes in other parts of Japan once their work was done—also threatened to swamp the islands' cultural distinctiveness. As early as 1969, a group of elderly locals, worried that "old Tanegashima will disappear," banded together to create a museum of traditional household and agricultural implements.[65]

Among the wider populations near the space centers, disillusion set in early on. The constant expansion of the facility necessitated the movement of many households, which caused some local resistance.[66] Despite the enthusiasm of local businesspeople, many were disappointed by the failure of visitors to stay and spend money at local businesses, or to go sightseeing on the island. A survey in 1969 indicated that 67 percent of respondents felt that the presence of the facility "hadn't improved" their lives; a similar figure said that they had not benefited from the construction boom on the island.[67] As late as 2012, people in Kimotsuki (the village closest to the Uchinoura space facility) complained about the low quality of roads as well as demographic and economic decline. Some complained that the village lacked even an accessible mall or a bookshop. "There are few places for young people to work," another citizen complained, "and they are leaving for the cities."[68]

THE LOCAL RECONFIGURATION OF NATIONAL SPACE FACILITIES

As described above, the practical consequences of association with space sites were often far less positive than locals may have hoped. Anxieties about safety bedeviled their daily lives. Booms in investment failed to persist. Demographic, economic, and cultural decline continued unabated. Despite this, however, it would be a mistake to assume that locals in these areas passively accepted the status quo without actively seeking ways to derive some benefit from the giant complexes they now lived next to. For them, keeping the space dream alive has required some improvisation, but through this very process the citizens of areas like Akita have generated new modes of engagement in which they have more power, and derive more benefit.

In the years since AVSA's departure, the Michikawa Beach facility has undergone precisely such a process. The citizens of Iwaki—the closest village—now preserve and promote the site as a tourist attraction. A plaque erected in 1986 honors the memory of Japan's first space launch facility.[69] Meanwhile, the second facility at Noshiro has had an even more energetic afterlife. When the Institute of Space and Astronautical Science (ISAS) decamped to Kagoshima, the location was handed over to Tohoku University, to use as they saw fit.[70] As late as 1985, the site was used by academics and engineers to test various experimental designs such as a 20:1 scale model of a mooted Japanese space shuttle.[71] The facility was then acquired by Akita University, which developed it into a "Space Park" where engineering students build and test their own rockets (some of these have a higher altitude capacity than the Pencil).[72] Facilities were expanded, partly to accommodate atmospheric observations via sounding rockets.[73] Noshiro eventually became home to an annual Space Festival, the "widest-scope student and amateur rocket festival in Japan." Organized by locals in conjunction with Akita University, the event has been held yearly since 2002.[74] It is presented as a key location for students who wish to pursue careers as rocket engineers to test their mettle.[75] Innovation remains one of the core priorities of the event—the latest iteration promoted the development of hybrid fuel rockets and robotic probes, foregrounding the fact that current trends in rocket development involve the "dissemination and commercialization" of rocket technology into the public sector. In fact, space has become a strong part of the locale's identity beyond just the festival and the site. Noshiro's Children's Hall (*kodomo kan*) features an entire second floor dedicated to rocketry and satellite launches, unveiled on the fiftieth anniversary of the building of Noshiro in September 1997.[76] The organizers—a consortium consisting primarily of local groups—continue to place outreach to children at the heart of their activities.[77] They are also particularly keen for Noshiro to continue to play a part in the broader Japanese space industry. One of their initiatives is to support the low-level development of novel technologies at universities such as Tokyo Institute of Technology.[78] The group is even lobbying for the return of high-altitude satellite testing to the facility.[79] In this way, Noshiro continues to contribute to Japan's broader rocketry and space industries as a whole, but as a locus of local rather than central institutional initiative.[80]

Nor are the activities of local groups limited to their own areas. Though Latour's aforementioned observations about infrastructure are primarily about the physical, this can also be applied to discourse: locations can be brought together by their *relationships* to particular kinds of infrastructure, and can create links that transcend physical distance. This was not lost on Japan's space agencies and their bureaucratic allies, who took the sharp lessons of the 1960s to heart. By the 1980s, they had begun working to ensure a new sort of relationship between localities and their space facilities. Links *between* locations that hosted these large complexes were encouraged by the Ministry of Education, Science, Sports and Culture (Monbushō, predecessor of MEXT).[81]

An outstanding example of this is the creation of the delightfully named Galactic Federation of areas hosting ISAS facilities on November 8, 1987.[82] The constituent "republics" are Taiki (home to the Taiki Aerospace Research Field), Noshiro (Noshiro rocket testing ground), Kimotsuki (Uchinoura Space Center), Saku (Usuda Deep Space Center), Sagamihara (the Sagamihara Campus), and Kakuda (Kakuda Space Center).[83] The "federation" part of a "micronation boom" (*mini-koku būmu*) of the 1980s and 1990s, which produced a variety of similar local associations, such as the Commonwealth of Cassiopeia (consisting of villages located under the constellation).[84]

The objective of the Galactic Federation, according to its founding document, is to create a "utopia of people's smiles" by promoting cultural and educational exchanges.[85] The primary activities of the group focused overwhelmingly on cultural and exchange activities between the seven participating members. The scale and frequency of these events is remarkable, ranging as they did from participation between members in each other's festivals and sports events, to the promotion of local products through schemes such as a Federation Orchard and Orchard Manager.[86] The federation also frequently staged events intended to highlight the global space industry, such as exhibitions of space-related art, lectures from figures in the space program as part of their annual "Space School" series, and their "national flag" orbiting the Earth aboard the Space Shuttle Endeavour in 1996.[87] Localities even engaged in their own form of "panda diplomacy"—or, in this case, "deer diplomacy"—such as when Iwate's "Sanriku Republic" donated some cervidae to its fellow federation member Sagamihara.[88] Perhaps its largest activity was the organization of annual "summits," events featuring up to two hundred participants, mostly children.[89] These are often timed to coincide with other localities' cultural events, such as the 1991 meeting held in Sagamihara, which shared a building with a local children's art festival.[90] Students also engage in homestays as part of a series of three-day-long visits by middle-school sixth-graders to host "states" over the course of the year.[91] This is not, however, to say that the federation focused *only* on children—it was, in fact, more than happy to encourage collaborative activities among adults, such as the hosting of a joint art festival featuring the work of senior citizens from Akita, Nagano, Iwate, and Kagoshima in Sagamihara in 1991.[92]

For all its nominal focus on culture, however, the bonds between these "republics" swiftly expanded into other areas as well. As mentioned above, most locations playing host to space facilities tend to be minor settlements with little economic power. It should come as no surprise, therefore, that many desired to cash in on their associations as much as possible. What is surprising is the extent to which members of the federation were willing to help one another out. In 1995, for example, the federation supported Noshiro, by staging its annual symposium there and organizing a bottle-rocket festival as part of the town's plans for revitalization of the town center.[93] In the aftermath of the Great Hanshin (Kobe) earthquake in 1996, the five members of the federation extended their cooperation to include assisting each other in the event of natural disasters, via an emergency mutual assistance agreement. The agreement covered the provision of materials, volunteers, and professionals in the case of a major calamity, as well as plans for taking in students rendered homeless by tragedy.[94] The extent to which members were willing to keep to the word of this agreement became apparent in 2011, when the Kimostuki rallied to the aid of fellow federation member Ōfunato in Miyazaki, which had suffered some 370 dead and over ¥107.7 billion in damage during the Great Tōhoku earthquake.[95] The aid effort featured the transfer of large amounts of food (including nearly half a tonne each of pork, cabbage, daikon radishes, and chicken), large numbers of raincoats for use by fishers (Ōfunato being a major fishing town), a volunteer magic show for displaced survivors, an exchange between Ōfunato and Kimotsuki middle-schoolers (which included a water rocket contest), and a remarkable campaign of fundraisers.[96] Total funds raised for the purpose totaled around ¥14.5 million—a substantial contribution, considering both Kimostuki's limited economic means and the small size of Ōfunato's population.[97]

This largesse is all the more remarkable when one considers the fact that, despite the presence of these various space facilities, the most rural areas within this group have continued to suffer from the population and economic decline endemic in rural Japan. Taiki, for example, peaked in terms of population in 1955, at 11,296; by 2020, this number had dwindled to 5,446.[98] Kakuda's population peaked at 37,376 in 1947, and by 2021 had gone down to 28,672—a lesser decline in overall figures, but one within which the population of the elderly has outnumbered the young since the late 1990s, and is projected to outnumber even working-age people by the late 2030s.[99] Similar population declines are ongoing in Kimotsuki, Usuda (which includes Saku).[100] In contrast, Sagamihara—a city of nearly a million people—has enjoyed a slow increase in its populace in recent decades.[101] This reality is reflective of the complex legacy of their association with Japan's space program, which in this case has worked to imbue localities with a great deal of pride—even as it fails to provide its citizens with a compelling reason to remain there.

As shown above, the relationships between Japan's space facilities and their neighbors are complex and inconsistent. Due to geographical and geopolitical constraints, authorities promoting space exploration found themselves seeking locations to host testing and R&D sites as a matter of urgency throughout the 1960s and 1970s. While these sites were more often than not initially welcomed, in many cases their arrival did not live up to expectations. In some instances, anxieties about safety and geographical unsuitability eventually conspired to force space facilities to move on. In others, the facilities remained but failed to have the sort of economic impact that their neighbors had hoped for. In these cases, locals were forced to pick up where the facilities had left off, by creating their own modes of engagement with these sites and their legacies. In some cases, such "afterlives" saw localities capitalizing on their brief association with the space program in order to generate novel narratives about both their past and their future. Certain locales came to emphasize the national importance of their previous experience as host to elements of the space program—glossing over the controversy that period of time may have brought—while also using this experience to advocate for a return to the technological center stage. Other localities used the common experience of hosting space facilities to generate intranational links and cultural initiatives. These then expanded into economic, social, and political cooperation outside of the preexisting structure of government. In nearly all cases, areas that had been economically marginal before the arrival of the space facilities largely remained so after their departure. Nevertheless, they made use of these facilities' fleeting presence to situate themselves as key nodes of accomplishment within narratives of national technological progress—while also foregrounding their cultural and social distinctiveness.

Such experiences help historians to qualify our understanding of precisely what the space age is. The focus on the big-budget, cutting-edge stories of the Cold War superpowers and post–Cold War space powers are entirely justified from the perspective of world historical importance. However, the older population living in the vicinity of Uchinoura Space Center is as much a part of the story as anybody else. The space age consists of the lived experiences of those who bore witness to it. Even those who might at first appear marginal in fact had distinct experiences during the early days of humanity's effort to reach the stars. These experiences remain a crucial but somewhat understudied part of the tale—but no less fascinating for it.

RUPTURE AND RUINATION IN THE EMPYREAN EMPIRE

LISA RUTH RAND

The Space Age began in a moment of rapture and rupture.

The first human-made satellite started its historic journey as a single object when a modified R-7 intercontinental ballistic missile lifted off from the Kazakh Steppe on October 4, 1957. Within minutes its four booster rockets detached and fell to Earth, extinguished—an initial fragmentation followed by ignition of the rocket's core stage that lifted the remainder of the assembly to orbital altitude.[1] A short time later, a polished metal sphere trailing four whips of radio antennae separated from the core stage and emerged from the nose cone that sheltered it for the ride. The trio—sphere, rocket stage, and nose cone—embarked on an unprecedented path around the planet, separate but together (see fig. Fragment III.1).

Only the sphere's formal name would endure in worldwide collective memory: Sputnik, meaning "fellow traveler" in Russian.[2] Perhaps unintentionally, the name serves as a reminder that the first human-made moon was not one object but individual pieces of a formerly unified whole. Each of these pieces played a role in shaping the "Sputnik moment."[3] Television viewers and radio listeners around the world tuned in to hear the now iconic "beep-beep" transmission emanating from the sphere. Interested onlookers consulted with local newspapers to find out when

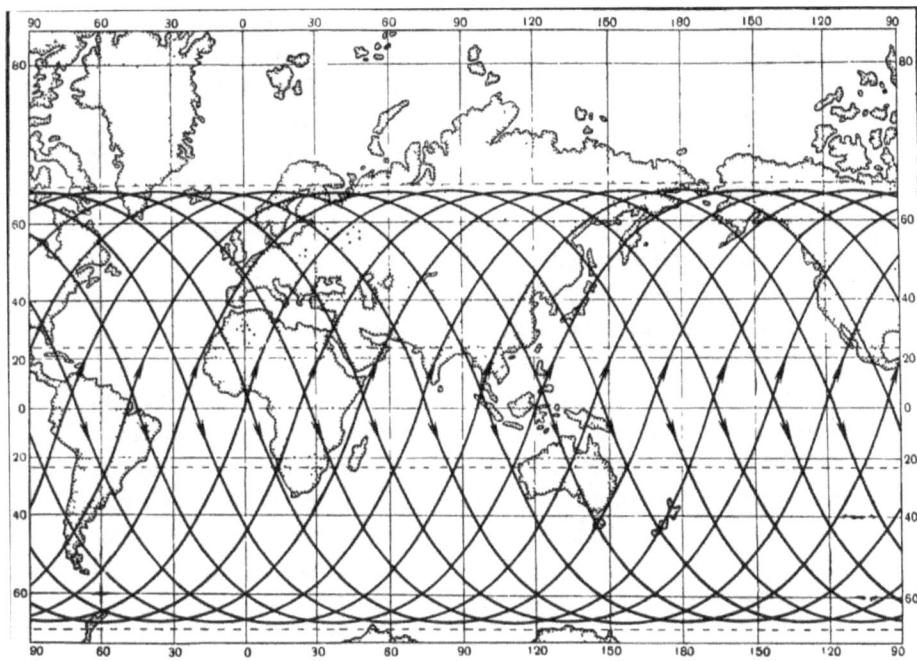

Fragment III.1. The ground track of the first Sputnik satellite as depicted in the October 9, 1957, issue of the Soviet newspaper *Pravda*. Credit: Public domain.

the bright core stage would pass overhead, gathering outside to catch a glimpse of the tiny moving star that confirmed the Soviet Union's triumph. Teams of amateur astronomers used specially designed telescopes to track the nose cone as it faintly reflected sunlight at dawn and dusk. The rocket stage and the sphere would be listed first—as 1957α1 and 1957α2, respectively—in an ever-growing catalog of artificial objects in orbit around Earth.[4] Within a few months all three pieces shattered once again when they succumbed to friction and gravity and tumbled back to their planet of origin.

From this initial set of breaking points followed decades of material and political fragmentation played out on a truly planetary scale. In over half a century since the first fellow travelers ascended, decayed, and fell through the atmosphere, thousands of rockets, named satellites, and less acclaimed artifacts would follow their path. Some remained aloft and some reentered the atmosphere, creating a more-than-planetary sweep of spatially and temporally disjointed rubble. Celebrated moments of technological achievement—world-shrinking satellites and bigger, faster rockets and first steps on the Moon—continue to occupy a hallowed place in popular and scholarly historical memory. However, the tangible traces of concurrent colonial wasting and ruination practices inaugurated by the Sputnik fragmentation endure, like all ruins, as a haunting monument to the civilizations that left them behind.

Orbital debris, both in space and returning to Earth, has become an increasingly fraught locus of space policy debate in the first decades of the twenty-first century. As the private space industry has revved up the launch tempo and so-called megaconstellations of satellites compete for space and spectrum alike, the specter of crowded orbits looms ever larger. However, neither the existence of a waste system in space nor broad attention to its multivalent risks are new. From the launch of Sputnik 1 onward, understanding the power of the outer space environment to drive orbital decay—physical, material, and discursive—has influenced the design, use, and disuse of space technologies. If the Space Age has always been and continues to be shaped by messy entanglements of convergent innovation and decay, what underexamined dimensions of this era might come into sharper focus by privileging said decay as a coequal force of change over time—or even as a starting point rather than an afterthought? What might it mean to shift the historical gaze from innovation and use to equally generative narratives of waste and ruin, and to do so from the ground up?

WASTED SPACE

The Space Age heralded by Sputnik 1 continues to be popularly celebrated for its challenges, triumphs, and innovations. Even as mainstream support for it has ebbed and flowed, the international space industry—increasingly privatized—retains a reputation for peddling the newest in high technology and accomplishments of the exploratory spirit. Much as the first satellite could be more accurately understood as a collection of formerly unified fragments, however, the Space Age can also be understood as emerging in moments of breaking—artifacts, infrastructures, geopolitical orders, alliances, and cosmologies. Even the first steps on the Moon, taken ostensibly "for all mankind," signaled a rhetorical performance of unity amid accelerating fracture back on Earth.[5]

Viewed as a single, innovative, revolutionary object, the ascendance of Sputnik 1 arguably signified a jarring geopolitical shift and accelerated what would become the Cold War proxy front known as the Space Race.[6] As an assemblage of parts, all of them designed to be disposable, the Sputnik trio represents an uneasy confluence of rupture and continuity, extending existing industrial waste practices to a planetary and beyond planetary scale. They reflect initial steps in the neocolonial construction of an orbital regime in which environmental risk, extraction, and discard would be enduringly acceptable outcomes of Space Age technosocial progress.

Neither the Sputnik trio nor its satellite descendants neatly followed a linear transition from spacecraft in use to unused debris. Like other unwanted things, waste artifacts in space are bound less by a fixed definition of utility and more by their relationships to individuals or communities of users, viewers, discarders, and cohabitators.[7] Understanding space artifacts within this dynamic set of relational categories provides a lens through which to trace the shifting priorities and practices of the groups that launch, use, and dispose of them. It also prompts a closer look at how outer space has been shaped by what Marco Armiero calls "wasting relationships" into a place of deep power imbalance undergirded by principles of disposability.[8]

When it comes to production and acts of waste, the space industry does not diverge much from its terrestrial predecessors and contemporaries. As in any industrial enterprise, technologies designed for use in outer space yield by-products—spent rocket casings, errant fuel, radionuclides, derelict satellites, and smaller detritus shed from larger objects over time. Some of these accumulate in orbit. Some fall back to Earth, disintegrating into the global atmosphere and the oceans, and sometimes falling onto land. The high-technology mystique of the rocket stages, nose cone, and any number of bits and fragments produced in the launch, use, and disposal of Sputnik 1 does not set it apart morally or materially from a factory that releases chemical discharge into a nearby river, or emissions from a refinery smokestack. Nor does the assumption of outer space as an ultimate sink for orbital effluent differentiate its use as a natural resource from prevailing industrial reliance on downstream environments—water, air, land, ice, and the bodies of humans and nonhuman animals—to carry away and contain the unwanted refuse of production.[9] In the absence of enforceable international rules for what to do with this refuse, the space industry assumes the right to many of these same terrestrial waste sinks while also instituting new ones. Disposable Sputnik laid the foundation for a perpetually porous division between Earth and Elsewhere, between inner and outer space, between old and new industrial custom—through a vertical, circumplanetary continuum of waste and wasting practices.[10]

Yet, even as these divisions materially fall apart in a wasted orbital ecosystem, the very act of wasting—the creation of a place that is wasted—reinforces centers and peripheries of power.[11] Though commonly understood as nothingness, the wasting of outer space expresses a "set of socio-ecological relationships aiming to (re)produce exclusion and inequalities."[12] Outer space itself as a natural environment has been enrolled by industrial actors in this wasting relationship, a central participant in the physical and material decay of disposable artifacts that shapes where these objects go when no longer used or controlled by human operators. In constructing objects in outer space to be, at their core, disposable, space industry has from the outset engaged in producing outer space as a wasted place. And in creating an orbital regime crowded with discarded, disposable things—with things no longer wanted by the entity that created them in the first place—wasting practices in space produce not only a targeted natural environment but also targeted communities.[13]

Wasting practices and wasting relationships in outer space provide a point from which to trace the full colonial dimensions of the Space Age including and beyond the terrestrial appropriation of land and labor to support construction and operation of the satellite infrastructure. Given the absence of land, Indigenous or otherwise, as typically defined, it can be tempting to think of space itself as a literal terra nullius, as something that cannot be colonized in the same violent manner as prior imperial projects. However, as Deondre Smiles has observed, the settler colonial imperatives of control, possession, and remaking of space undergird the frontier logics of space industry past and present.[14] The configuration of outer space from the outset as a dump site for discarded waste—as a place to be polluted—situates Space Age wasting practices within deep-rooted colonial ethics of land appropriation. As Max Liboiron notes, pollution is predicated upon the assumption of colonial access to land as a disposal site, and "assumed access to land is foundational to so many settler relations."[15] The extension of these practices into orbit and beyond confirm that colonial land relations endure even in places with no physical land to speak of.[16] And the reordering of the nature of outer space into a commodifiable resource and pollution sink surely constitutes, in Liboiron's terms, "bad relations."[17]

The upshot of these bad relations affects the "Land" of outer space, including the people enrolled and devalued in the wasting relationships that have shaped outer space into a wasted place.[18] For decades, participation in the space industry has been limited to nations and private entities with the capital and power to leave the planet's surface. The exclusion of large swaths of humanity from sharing space triggered geopolitical resistance to unequal allocation of space and electromagnetic spectrum, the extension of spheres of influence expressed through shared use of orbital information, and civil unrest in response to the perceived financial prioritization of space industry over other social and political causes.[19] It has also endured over time as the industrial logic of space as a resource to be extracted and commodified has driven the crowding of orbits and the historically unequal distribution of reentered debris primarily into regions whose inhabitants benefit the least from the use of space infrastructure.[20] These are dangerously decayed relations produced through ethics of disposability implemented at an extraplanetary scale. To paraphrase Rob Nixon, we are all living in the Space Age, but we are not all living in it equally.[21]

The fragmentation, circulation, and deposition of the Sputnik 1 assembly into the atmosphere—the exact landing site of any surviving fragments remains unknown—set a durable precedent of disposability in space technology design and use. It also set in motion the spatial extension of human-driven change beyond familiar terrestrial dimensions. The Anthropocene, or taken in a more specific direction, the Wasteocene (as theorized by Marco Armiero), is not an era contained to the poorly defined limits of a global planet.[22] Rather, the biogeochemical influence of an equally ill-defined "humanity" extends into the solar system in an extraplanetary age of wasted space. This has also had the inverse effect of potentially constraining interplanetary presents and futures, should Earth become cut off from the rest of the universe by the accumulation of disposable discards in its immediate cosmic environs.[23]

A shift in historical gaze beyond innovation, use, or even maintenance to waste and wasting practices provides an opportunity to rethink the history of human encounters with outer space along a much more nuanced spectrum of experience. The history of these encounters does not sit squarely within the realm of high technology and rarified geopolitics, but within multiple metrics of waste and decay, shaping and shaped by ongoing conflict and inequity that extend beyond the powerful states and subjects commonly privileged in a majority of historical analyses of spaceflight. However, centering fragmentation generated through decades of wasting relationships as a core attribute of space history, on Earth and beyond, also requires sitting with a potentially uncomfortable notion: that the Space Age has always been in some sense an era defined by ruin and ruination.

AN UNEQUAL SPACE AGE

The durable remnants of extraterrestrial wasting practices span a wide spectrum of size, longevity, volatility, and value. Flecks of paint shed from an aging satellite might not inspire national pride or nostalgic associations with a bygone era. However, many of the satellites themselves do, such as Vanguard 1, the second successful United States satellite, which reached orbit in March 1958. Though the historic Sputnik 1 assembly reentered the atmosphere mere months after reaching orbit, Vanguard 1 and the upper stage of its launcher have silently traced an elliptical orbit around the planet ever since the end of the satellite's six-year operational lifetime.[24] The designed utility of these objects has long since faded. However, such derelicts can and have been considered cultural heritage artifacts, imbued with new worth even in disuse and decay.[25]

Though inaccessible and unmaintainable, objects like Vanguard 1 raise contradictions about the value of unused things in outer space. Such objects pose material threats—to operational satellites and the space environment, to the aspirations of would-be satellite operators and users seeking clear orbits, and increasingly to the shared heritage of dark and quiet skies.[26] Even so, they retain symbolic significance as enduring relics of the early age of national space exploration. They also serve as rem(a)inders of the inequitable conquest of outer space from the outset, and the compounding consequences of its occupation by established colonial powers extending hegemony beyond horizontal notions of center and periphery.[27]

When viewed this way, the remains of Vanguard 1 fit congruously within a coterie of ruins generated atop uneasy intersections of cultural value, mutual memory, and incremental violence. The rubble of physically decaying orbits and crumbling material objects takes on additional significance when imbued with the symbolic power of the ruin.

Ruins are expansive things. A ruin can be a fetishized monument to its absent creators, or an uneasy invocation of destruction and loss that endures beyond individual or collective memory. Ruins also do not occur passively or in a vacuum of intent—they are made through active processes of ruination.[28] Ruination can occur in the aftermath of an immediate, intentional act of destruction, but it becomes fully articulated over longer temporal scales. Per Ann Laura Stoler, ruins can represent colonial dispossessions and dislocations that take place over time, such that the world-shaping contours of ruination render themselves imperceptible by comparison to more acute acts of imperial violence.[29] Much like wasted places, ruins are things that are perpetually made and remade, through continuous, active deterioration and degradation.[30] As William Viney notes, ruins "stand between survival and termination, continuity and annihilation," a site or object whose meaning draws from the material and temporal collapse between past and present through persistent decay.[31] A ruin by its nature is a thing that lingers, that gives substance to an uneasy disjuncture between use and disuse, a thing produced by acute acts of violence and slow, crumbling decline. And the actions behind making or becoming ruins may come to be obscured behind nostalgic narratives constructed around material remnants of a treasured past.[32]

The ruins of the Space Age did not occur passively. The United States and the Soviet Union each approached the creation of ruins at the Cold War proxy front steeped in contemporary national principals of posterity in the shadow of an uncertain future. The launch of a fractured Sputnik as the first fragments of a perpetual orbital ruins echoes Soviet imperatives to construct built environments that obscure historical difference, in which the new and the ruined coalesce together contemporaneously in a "timeless present."[33] The architectural historian Grigorii Revzin claims that post-Stalinist architecture was never built to last, predicated on time scales of individual human lifetimes rather than indefinite posterity like classical ruins. In Revzin's estimation, this planned rapid ruination enabled Soviet designers to experiment and, importantly, make mistakes.[34] This presumed national right to err may have likewise inflected the Soviet approach to a disposable built environment in orbit, and aligned Soviet space technology practices with material and literary evocations of the Soviet project as having been in a "constant state of decay—in ruins—from the very beginning."[35]

The same powerful states in Cold War conflict fostered orbital ruination in synchrony with nuclear ruination. The postwar acceleration of atomic weapons testing on both sides of the Iron Curtain yielded a sharp rise in atmospheric radiation. It also drove the expansion of global networks of extraction, production, and disposal built upon long-standing colonial relationships. Being a human after 1945 also meant being nuclear.[36] The era of mutually assured destruction kindled an embrace of ruination as a fundamental, even cherished practice in American politics and mainstream culture. Joseph Masco has argued that this surge in nuclear weapons testing produced a national imaginary marked by an acceptance of widespread, comprehensive destruction that shaped governance and technoscientific enterprise—and that even the act of imagining one's home in ruins became central to democratic participation in Cold War American society. No wonder that the very nation that invested so deeply in ruination as nation-building during the nuclear age would extend these practices into another realm of geopolitical conflict.[37]

Like ancient imperial ruins and the world-enveloping nuclear ruination of the mid- to late-twentieth century, space industry ruins have changed shape over time, as the slow burn of natural forces in orbit assimilated a built environment in the absence of maintenance. The crumbling of a ground-based space industry infrastructure constructed and abandoned along well-worn colonial pathways attests to the ongoing creation of imperial voids—what Gastón Gordillo calls such "feared, uncontrollable spaces" produced by states through conquest and then left to fall apart in places too wild or violent to merit said maintenance.[38] As Asif Siddiqi has shown, in the fragments of abandoned launch complexes and tracking stations local sites of decay give rise to a truly global view of the Space Age grounded in highly specific, colonial place.[39] Ruination binds together the durable remnants of space industry infrastructures on the ground to those in orbit—a literal and figurative imperial void.

By comparison, the colonial effects of degraded vestiges of empire in orbit, on and around the Moon, or beyond, may initially seem less clear than its earthbound analogues. These are ruins in motion, after all— messy remains of a past era that refuse to stay put or emplaced. They cannot currently be visited like "ancient" sites revered by enthralled tourists, photographed like abandoned launch and tracking infrastructure, nor always easily viewed or controlled even by the state and private powers that created the objects in ruin in the first place. Ruins in orbit are more elusive and mobile than those on the ground, moving through time and space, visibility and invisibility, in a manner incongruent with the more indelible traces of bygone eras memorialized in crumbling stone.

Perhaps this very placelessness and unknowability of orbital ruins opens generative ways of thinking about the unevenness of human encounters with the outer space environment. Or put another way, thinking of outer space as ruins provides an opportunity to emplace material changes in the orbital landscape within larger histories of conquest and spatial decay.[40] Taking a vertical measure of Space Age ruination, making discernible the wasting effects of extraterrestrial ruins on people, places, and things, connects interspatial voids, drawing together the unequal experiences of the Space Age at not just a global scale but a truly planetary one. As much or perhaps more so than the occupation of a global environment-turned-resource by active satellites, the monumental extraplanetary scope of ruins and ruination reveals the outsized colonial contours of an unequal Space Age.

CURRENT AND FUTURE RUINS OF A GILDED SPACE AGE

From the removed perspective of decades between Sputnik 1's ascent and the rise of an era of space industry dominated by a growing number of influential national and commercial actors, it is also worth taking a measure of the potential cultural power of outer space as ruins. In her 1953 book *Pleasure of Ruins*, the British novelist Rose Macaulay wrote of the dizzying experience of encountering the ancient in decaying material form: "The intoxication, at once so heady and so devout, is not the romantic melancholy engendered by broken towers and moldered stones; it is the soaring of the imagination into the high empyrean where huge episodes are tangled with myths and dreams; it is the stunning impact of world history on its amazed heirs."[41] Macaulay's description of the breathless sublime inspired by the famous ruins of ancient Greece could just as aptly apply to the myths, dreams, and nightmares of the societies that built, used, and abandoned what became the accumulating relics of the Space Age. It might be tempting to encounter the orbital ruins as similarly spectacular, if evasive, conduits between the present day and a "tremendous past" in the highest of empyreans.[42]

Indeed, anthropologists of outer space identify ample historical value in space remnants past and future—in the dead satellites, expended rocket bodies, and obscure artifacts circling the planet; in abandoned lunar landing sites; and in the ruin-to-be of the International Space Station.[43] Together with the ground-based infrastructures that launch, support, and communicate with spacecraft, the ruins of the satellite age constitute a cultural landscape shaped by both natural and anthropogenic activity. This landscape provides a material record of the myriad ways that human interactions with outer space have changed over time.[44] In their ruined states, they also may tell tales beyond what their creators' archives reveal.[45]

Even so, there is danger in summoning outer space into the exalted company of the Acropolis, Macchu Picchu, Tikal, or Chaco Canyon. Popular veneration of ruins honors resilient reminders of a noble civilization that once was but has ceased to be—those displaced from ruined lands also displaced temporally into the past outside the hegemonic gaze of a settler present.[46] Engaging in such uncritical reverence for an orbital ruin would obscure the verticality of empire that followed in the wake of the Sputnik fragmentation and the violence enacted in the ruination of a literal high empyrean. Whether on the ground or aloft, in the words of Gordillo, ruins are not politically innocent. Those who encounter them may "glorify, fear, or ignore the same ruin."[47]

Perhaps ignorance of extraterrestrial ruins has obscured the reality that while the cultural and geopolitical milieu of the Cold War space race may be past, its wasting praxis endures into the present. The project of ruination is ongoing and expanding, with the neoliberalization of space exploration central to a "nostalgic" reclamation of imperial imperatives in the twenty-first century.[48] Like any colonial project, the conquest of outer space has not occurred in a void of private participation. The shift toward a particular kind of powerful, independent aerospace enterprise since 2000 has also shifted the power landscape in the space industry. This has affected the construction of orbital ruins in sometimes surprising ways.

Within two decades of the turn of the century, private space enterprise introduced reusable rockets whose first stages can take off and land multiple times, in contrast to single-use expendable launchers that had been a primary method of getting to orbit since Sputnik 1.[49] At a glance, this innovation promises to stem the ruination of space and Earth alike by reducing the number of orbiting derelicts like the Sputnik 1 rocket core. Supporters of private space extol the "democratization" of access to orbit that is also enabled by the lower cost of launching a reusable rocket. Indeed, financial barriers to entry have plummeted, with more nations, companies, and communities achieving pathways to orbit aboard privately owned and operated rockets.[50]

Yet the sharp upward spike in launch tempo that the shift to reusable rockets has enabled through economies of scale represents a continuation, or even escalation, of previous ruination practices. Parts of these rockets return to be spruced up and flown again, but these are only the most visible components of a vast space infrastructure still built upon an ethos of disposability.[51] The subsequent proliferation of single-use satellites, upper rocket stages, and other spacecraft, and the emissions produced in launches and operations, have arguably intensified the wasting relationships of a prior era under a rhetorical veneer of economic and environmental virtue—on multiple levels a Gilded Space Age. What is past is also present in the displacement of Indigenous communities by launch infrastructure construction and in the exclusionary technopolitical colonization of rapidly crowding orbits.[52] These innovations so far do not mark a distinct rupture with a glorious if flawed past; instead, they signify a perpetual, even unremarkable acceleration of long-standing practices of extraplanetary ruination.

Fantasies of progress and modernization carried aloft on these new technologies also endure, consistent with modernist visions of eras past. The utopian future in the cosmos lauded in the current moment conceals the dust and decay created in its ongoing production. As Anna Tsing reminds us, dreams of modernization, historical forgetting, and ruination go hand in hand.[53] The construction of private space stations and orbiting office parks promised by private space in a near future rhyme discordantly with the nineteenth-century expositions and Parisian arcades whose decaying, dusty ruins stood out to Walter Benjamin as testament to crumbling novelties of imperialism and capitalism—and a haunting preamble to the accelerated production and consumption of the post-45 era to come.[54] What might the orbital ruins reveal to a future wanderer, an "amazed heir" to the Space Age contemplating what remains of the glorified consumption of the post-Sputnik world, an era both revolutionary and always already in ruins?

PART IV

RUPTURE

CHAPTER 10

FROM SPACE RACE DRAMA TO HAPPY HANDSHAKE

The Role of Audience in the Apollo–Soyuz Test Project

DARINA VOLF

"Three decades of race between East and West"—this description of early space efforts in a report on the occasion of the 2013 annual congress of the Association of Space Explorers (ASE) is hardly surprising.[1] As a rule, the popular narrative of the dawn of space exploration tends to emphasize the competitive aspect, and the advance into outer space is usually told as a story of Cold War rivalry, culminating in a space race between the United States and the Soviet Union. Space cooperation, in contrast, is presented as a relatively new development characteristic of the post–Cold War era, with the International Space Station as its most powerful symbol. However, international space cooperation did not start with the collapse of the Soviet Union; it accompanied space endeavors from the very beginning. Moreover, space cooperation during the Cold War was not limited to partnerships within the blocs, but sometimes transcended the bloc divide. On top of that, there was even cooperation between the two major space rivals—the United States and the Soviet Union. For many years, this superpower space cooperation had been limited to the less publicly visible fields of space medicine and biology, but in the 1970s the two rivals cooperated in the prestigious and highly visible field of human spaceflight. Today, the joint US–Soviet mission, known as the Apollo–Soyuz Test Project (ASTP) and conducted at the height of the

10.1. The famous handshake in space between the Apollo commander Thomas Stafford and the Soyuz commander Alexei Leonov. NASA.

détente period, is almost forgotten. After all, it does not fit the dominant narrative that explains the breakthroughs in space technology during the Cold War in terms of system competition and the race into space. This is also the reason that, even in research literature, the joint US–Soviet mission usually appears at best as a marginal episode with no more than political significance as a symbol of détente, but not necessarily as an integral part of the history of pre-1990 space exploration. But at that time shaped by Cold War anxiety, the whole world watched as the Soviet Soyuz and the American Apollo spacecraft docked in space on July 17, 1975, and their commanders, the Soviet cosmonaut Alexei Leonov and the American astronaut Thomas Stafford exchanged their memorable handshake (see fig. 10.1).

The ASTP was the first-ever cooperation in human spaceflight and the first large-scale space project between the United States and the Soviet Union. Such cooperation was unimaginable just a few years earlier, when the rivals were mobilizing enormous resources to beat the other in the race to land on the Moon first. Although there had been some limited space cooperation such as the exchange of certain scientific data, the ASTP required a more substantial flow of information and several years of joint work by the American and the Soviet teams of engineers, managers, and other space industry personnel. The uniqueness of this event explains why the ASTP attracted a great deal of attention in the 1970s and became one of the most powerful symbols of the détente period.

The joint flight was covered by media all over the world, and the United States Information Agency discovered that the ASTP preparations were not only one of the top stories in US-related newspaper coverage around the world at the time, but also "by far the top story in terms of favorable bias."[2] On the one hand, coverage of the ASTP emphasized the unexpectedness and novelty of the event. The *New York Times*, for example, called it "a political as well as scientific miracle."[3] The Soviet newspaper *Pravda* emphasized that for the first time in history, there was a major joint scientific experiment in outer space.[4] On the other hand, the handshake in space was seen as only the first step on the way to more substantial cooperative projects and to truly joint international space exploration. *Pravda* and other newspapers in communist countries expected the ASTP to usher in a new era of cooperative space exploration.[5] In the *New York Times*, John Noble Wilford proclaimed the end of the space race and, like *Pravda*, saw the joint flight as the beginning of a new era of space cooperation.[6] The British *Daily Telegraph* went even further and, reflecting on the political implications of the mission, raised expectations that the friendly handshake in space would be followed by a deepening of détente on Earth.[7] In short, the media on both sides of the Iron Curtain agreed that the ASTP was a mission of historical significance.

But, as this chapter shows, the role of the mass media cannot be limited to reporting on the flight and reflecting on its significance. Rather, the mass media and the publics they represent played a crucial role in the development of the joint mission, its progress, and the eventual cessation of space cooperation. Some scholars have already pointed out that the media played a key role in the space race from its inception in the late 1950s.[8] Through the media, the rival parties observed each other but also shaped the way they were perceived and observed by the other side. The space historian Asif Siddiqi instructively demonstrated this effect on the decision to launch the first satellite, the Soviet Sputnik.[9] While it is widely acknowledged that the public and the media played a crucial role in the space race, their role in space cooperation has not yet been discussed in depth.

The ASTP mission is usually explained in terms of the interest of politicians in promoting the policy of détente, or the interest of space managers in gaining access to the other side's space program to strengthen their own.[10] Only recently, newly declassified Russian sources and more thorough research have allowed for some nuanced interpretations and insights. Andrew Jenks has offered a more complex picture of the shift toward space cooperation in the 1970s, emphasizing the synergies between political and societal developments on the one hand and transnational ideas about the peacemaking potential of collaborative space exploration on the other.[11] In addition, the sources now available allow for a better understanding of the working of the preparations for the ASTP and the problems accompanying the mission's realization.[12] Even though the interpretations differ, especially with regard to secrecy in the Soviet space program, these accounts have in common that they focus on Soviet and US actors who were more or less directly involved in the mission. The role of the mass media and its specific contribution to the US–Soviet space cooperation has not been studied so far. In this chapter, I discuss the role of media and public communication processes in the development and stabilization of US–Soviet space cooperation in the 1970s. I argue that bringing the media and the public into the ASTP story contributes to a better understanding of the turn to space powers' cooperation, its course, and its end.

THE LINK BETWEEN PUBLIC, POLITICS, AND SPACE PROGRAMS

Space enthusiasts from different countries dreamed of space travel long before the first satellites were launched, but World War II and the fear of an escalation of the emerging Cold War created the conditions for the rapid development of rocket technology and big science to make these dreams come true. Advancing into space required enormous resources, which at that time could only be mobilized by powerful nation-states. Therefore, the support of political elites was essential, and the first task of the advocates of a space program was to convince political leaders to allocate the resources necessary to realize the visions of spaceflight. The mass media proved to be an important ally in this task, as they soon gave space activities a great deal of attention. Because of the Cold War rivalry, the media were particularly effective in securing political support for space ventures to point out the space efforts of the other side, to spread the impression that one was lagging behind, and thus to frame spaceflight as a competition between the superpowers.

Given the specific features of the political decision-making processes, there were, of course, some differences between the United States and the Soviet Union. The leading figures of the American space program were not prevented from publicly criticizing the positioning of the space efforts in the political and public agenda of the United States. On the contrary, they often did so to send a signal to US policymakers to increase support for the space program. In an interview in November 1957, shortly after the launch of the second Soviet Sputnik, Wernher von Braun, then director of the Development Operations Division of the army's Ballistic Missile Agency at Huntsville, claimed that there would be more Soviet space firsts in the near future.[13] The NASA administrator James Webb warned in 1963 that the Soviets could reach the Moon first.[14] In doing this, the US space program managers perpetuated the idea of a space race in the public and at the same time exerted pressure on the political authorities to fund space programs to stay in the race.

For the Soviet space program authorities, it was not possible to openly criticize the space program—in neither the national nor international mass media. However, to secure political support for their space activities, they could use international public attention to space exploration. They collected, translated, and presented articles to political leaders from various foreign media dealing with the superpowers' space programs, thus creating awareness of the Soviet Union's involvement in the space race.[15] In addition, US space plans and activities, which were widely reported in the press, found their way into memoranda requesting approval of space activities in the Soviet Union. For example, a memo from representatives of industry and science concerning flights of Voskhod spacecraft mentioned NASA's "intensive works" on the Gemini and Apollo programs.[16] Finally, Soviet scientists could spread rumors in the West by speaking vaguely about future Soviet plans in space. Such statements by scientists mistakenly believed to be in charge of the Soviet space program were closely followed and sometimes exaggerated in the international press. For instance, in August 1960, the Academy of Sciences member Leonid Sedov was quoted in Western newspapers as saying that the Soviet Union would soon launch a manned satellite. Because the newspapers referred to Sedov as the "head of the Soviet space program," his statement may have seemed to reveal actual Soviet space plans. However, Sedov was neither in charge of nor even directly involved in the space program.[17] Nevertheless, such statements in the Western media put pressure on the Soviet government to meet the expectations raised in the public abroad.

The media were not only important in the communication processes between space managers and political elites but also served as a platform for political elites and the national public to discuss and agree on the relevance of spaceflight and the direction of national space policy. The massive publicity of space activities made politicians realize that they could use space achievements to strengthen their position. Of course, performance in the space race was a double-edged sword, and failures could affect a political leader's popularity as much as spectacular space triumphs. In the United States, this became evident in the wave of criticism directed at President Dwight Eisenhower following the Soviet launch of Sputnik in 1957. The early space accomplishments of the Soviet Union and the seemingly weak position of the US in this field made spaceflight an important issue in the 1960 election campaign. At that time, catching up with the Soviet Union in space was presented and perceived not only as a matter of national pride but also as an important factor in national security. Research has already shown that, to win election, the presidential candidate John F. Kennedy was able to exploit and further inflame the American public's anxieties about Soviet preeminence in space. He presented space exploration as a strategic race that the United States could not afford to lose.[18] The result was a mobilization of enormous resources for NASA and its program aimed at a manned lunar landing. Although the American public's support for the Apollo program was by far not as strong as is sometimes assumed, it was enough to create the impression of a broad social consensus on this topic and to establish the lunar landing as a national priority.[19]

The nature of the Soviet political system makes it more difficult to assess the level of Soviet public support for the space program. Moreover, the link between political authorities, public preferences, and policy decisions is less straightforward than in political systems with competitive elections. Nevertheless, even though the Communist Party was the uncontested ruling party, it was interested in the Soviet public's opinion for several reasons. On the one hand, rule based on coercion and terror was costly and difficult to maintain in the long run, so efforts were made to pay attention to public sentiments, especially after Stalin's death. On the other hand, the Communist Party was not a unified entity, and internal power struggles led to frequent reshuffling of national priorities. Similarly, the Soviet mass media, which at first glance might appear to be little more than a mouthpiece for the Communist Party, played a certain role in political communication between the public and the authorities—albeit a different role from that of the mass media in Western societies.[20] Public opinions, interests, and criticisms were not transmitted to the political system primarily through the media, but through more hidden channels, such as the monitoring of rumors and complaints or reports from local political elites and security services.[21] In the media, rare critical comments were usually couched in hardly understandable, overly formal language or presented in rather diluted form. Of course, the trained Soviet citizen was able to read between the lines. It was not uncommon for articles to begin by describing some complaints and critical comments allegedly overheard on the street, and then go on to counter the criticism and justify the Communist Party's policies. From today's perspective, it is hardly possible to determine whether the authors of such articles were really trying to defend communist policies or were using the justification narrative "trappings" to be able to voice criticism. Nevertheless, in the Soviet system of censored press, this was a way of public communication, and today, it allows us to gain insights into the attitudes of the public at the time.

There is reason to believe that in the Soviet Union the beginnings of spaceflight were accompanied by a wave of genuine enthusiasm not only among the political elites, who considered space achievements as an opportunity to improve the image of the Soviet Union and themselves, but also among the public. Not surprisingly, cosmonauts returning from space were greeted by cheering crowds. The Soviet elites had enough experience with orchestrated ceremonies. In the case of the cosmonauts, however, their encounters with the public seem to have generated more interest and participation than the usual mass jubilations.[22] Moreover, the political elites sought to establish a close link between the Soviet communist system and the achievements in space, which were presented as a direct result of the unfolding of science and technology in the wake of the Bolshevik revolution. If space travel and its enormous costs had been perceived in a rather negative way, it would have been unreasonable to use it as a source of legitimacy for the communist regime. Finally, and perhaps most important, the Soviet triumphs in space came at a favorable moment, when Soviet society was beginning to recover from the Stalinist terror and to look to the future with confidence. The advance into space proved to be an ideal symbol of this new and promising era.[23] The Soviet mass media played an important role in disseminating these propaganda images, linking the communist past and future with space exploration. In addition, they conveyed the message that the entire Soviet society stood behind the successful venture into space with its hard work, thus contributing to the impression of national unity, mastery, and space euphoria. In this way, even in the Soviet Union, the support of the political elites for space exploration and the preferences of the public were to some extent interconnected, although not as directly as in a democratic system with opinion polls, electoral campaigns, and competitive elections. Summarizing the above, in both the United States and the Soviet Union, the public had the means to influence the space program, its position on the list of national priorities, and its budget by shaping policy decisions, whether through electoral choice, public debate, or tacit approval.

The public's preferences and expectations not only shaped government support for space activities but also had a more direct impact on the direction of the space program by providing space officials with guidance on how best to appeal to the public. Space officials considered the public, both domestic and international, to be the target audience for the space theater. The space race, as the primary driver of the development of space exploration, worked only as long as the audience was willing to watch, applaud, and, in the case of the national audience, pay for the spectacle. Because of the worldwide interest in space activities, the audience was large, heterogeneous, potentially global, and not easy to grasp by means used for communication within national public spheres. As a result, the leaders of the space programs had to think in rather abstract terms to identify some key public norms and expectations that they believed were relevant to this imagined audience they were trying to address. This was important not only to keep the interest up but also to project a favorable image of the national space program, since the space race was deeply embedded in the Cold War rivalry for supremacy.

Based on beliefs about the norms and expectations of the imagined global audience, the public presentation of space activities on both sides of the Iron Curtain followed similar patterns.[24] In each case, the national space program was portrayed as peaceful, cooperative, oriented toward the common good, and dedicated to scientific progress. In contrast, the rival program was portrayed as hazardous and threatening—a fact that justified the imperative of competition for supremacy—that is, the space race—even though both space programs generally tried to appear open to cooperation. In their coverage, the mass media pushed forward the idea of rivalry and competition, evaluating the two rivals and eagerly awaiting the outcome of the race. The dominance of the idea of a race, the worldwide attention that Soviet space firsts attracted, and the expectation of an American response to Soviet space achievements gave US and Soviet space officials a powerful impetus to captivate audiences by maintaining competition rather than seeking cooperation. For more than a decade, this seemingly broad consensus among politicians, space officials, and the public about the importance of competition and the idea of outer space as an arena for playing out the superpower Cold War rivalry remained unchanged and shaped the practice of space exploration.

THE END OF CONSENSUS AND FRAGMENTATION OF ATTITUDES TOWARD SPACE EXPLORATION

Throughout the 1960s, although space officials, politicians, and at least parts of the public seemed to prefer superpower competition in space to cooperation, there were already voices favoring the latter. The idea of US–Soviet cooperation, or at least assistance in case of technical difficulties in space, was present in several movies and novels on both sides of the Iron Curtain.[25] With President Kennedy's proposal for a joint American–Soviet lunar mission, such considerations even reached the highest levels of politics, but the realization of this idea failed.[26] For one thing, the Soviet Union's reaction to the proposal was far from enthusiastic, despite some subtle and ambiguous indications of interest from the Soviet leader Nikita Khrushchev and the scientist Anatoly Blagonravov. Second, the American public and politicians were divided over the possibility of a joint American–Soviet trip to the Moon, and skepticism seems to have prevailed. After all, Kennedy himself had declared the Moon landing a top national priority aimed at demonstrating American prowess. Commenting on Kennedy's surprising proposal, some experts suggested that it was not meant to be taken literally, while others speculated that the offer was the first step in attempting to pull out of the costly lunar landing program. In addition, space engineers pointed out the various technical problems associated with a joint mission. The fact that space cooperation with the Soviet Union was not politically feasible became clear when the House of Representatives declined the proposal for a cooperative American–Soviet lunar mission in October 1963.[27]

In the second half of the 1960s, the situation began to change gradually as the space race lost some of its public appeal. Along with declining interest in the space race, international space cooperation began to gain importance and gradually became the new norm to be followed not only in general but also in the relationship between the two space rivals—a process that culminated after the lunar landing. There were several reasons for the shift away from the dominance of the competitive paradigm. First, NASA itself fostered this development. From its inception in 1958, its public relations people considered an emphasis on data sharing and cooperation with the international scientific community as a good way to appeal to the global public they imagined.[28] Openness, peacefulness, and cooperativeness were some of the key qualities that should demonstrate US superiority over its Soviet rival in a situation in which the Soviet Union was accumulating one pioneering space achievement after another. As early as the 1960s, NASA concluded cooperative arrangements with several Western allies after they were invited to participate in NASA's program at the Committee on Space Research meeting in 1959.[29] The expansion of international cooperation in space allowed for the participation of various countries without spaceflight capabilities, and thus could positively affect the other countries' interest in space as well as the worldwide image of the American space program.

The second reason that the idea of international space cooperation took hold in the public mind was the increasing awareness of vulnerability of the Earth and the belief that global problems such as environmental degradation could hardly be solved by individual countries alone. Paradoxically, the findings and results of the space race helped demonstrate the need for international cooperation. Photographs of the Earth from space, such as *Earthrise* (1968) taken by the Apollo astronauts, contributed to the understanding of the Earth as a single entity and triggered global concern about its condition.[30] Third, with the achievement of the lunar landing in 1969, the attention to space spectaculars, the space race, and spaceflight in general waned even further. At the same time, the geopolitical atmosphere characterized by a gradual easing of international tensions, known as détente, opened space for rapprochement between the superpowers and increased public receptivity to cooperative efforts transcending the bloc divide.

The shrinking attention of the world community to the space theater already had a certain negative impact on the relevance of space programs in the list of national priorities, but the loss of interest on the part of the national public posed a vital threat to space endeavors, since it also affected the level of support by political elites. After years of steady growth of space expenditures on both sides of the Iron Curtain, the second half of the 1960s saw a decline in government support and funding. This was related to the changing attitudes of the national public. In the United States, enthusiasm for the space program, which, as already mentioned, was never as great as sometimes assumed, started to dwindle, and there was increasing criticism of the high cost of the space program at the expense of solving problems on Earth. The number of those in favor of cutting space spending grew.[31] The US print media picked up on this issue, publishing articles critical of space spending. For example, the *New York Times* columnist Howard Rusk reflected upon alternatives to the costly space program: "Man continues to be plagued by his mortal enemies—disease, hunger, ignorance and intolerance. It is impossible not to conjecture what the state of the world would be today, if the same amount of money, talent and dedication had been focused on their solution."[32]

In addition to criticism of high space spending, the American public changed its attitude about the importance of being ahead of the Soviets in space exploration. The interest in the space race declined, while support for cooperation between the superpowers increased. By 1965, more than 50 percent of respondents said it did not matter at all whether the Soviet space program was ahead or not. A report by the Space Task Group established by President Richard Nixon in February 1969 to formulate recommendations for the post-Apollo space program, pointed out: "Today, new Soviet achievements are not likely to have the effect of those in the past."[33] Since Nixon was "a careful reader of opinion polls and other indications of public sentiment," the declining support of the American public for the space program resulted in a shrinking budget for NASA.[34] The situation further deteriorated after the lunar landing. It not only led to a further decrease in funding and loss of human resources but also caused a kind of identity crisis triggered by uncertainty about the future direction of the space program, which had been unidirectionally focused on the goal of winning the race to the Moon in the previous years.

A similar reshuffling of national priorities was taking place behind the scenes in the Soviet Union. Newspaper articles attempting to defend the Soviet space program against public criticism and to justify space spending became more frequent in the second half of the 1960s. This indicated that public support for the space program was decreasing. Some of the articles vaguely mentioned that there were doubts about the usefulness of the space program and the amount of space spending. For example, an article in *Izvestiia* in 1967 by the director of the Soviet Space Research Institute, Georgy Petrov, titled "Why We Are Storming Space," pointed out that "the press in many countries and even more often people in their conversations have raised this question."[35] And in June 1970, the *Izvestiia* correspondent Boris Konovalov criticized those who grumbled that there was no need for a space program, even though they watched television "somewhere in the Extreme North or the Far East" and relied on weather forecasts obtained with the help of meteorological satellites.[36] Such statements did not go unnoticed in the United States, where the media concluded: "From speeches made by Soviet leaders . . . it seems that proponents of the Soviet space program have been pressed in the same way as their American counterparts to justify the costly appropriations that a dozen years ago were given almost without question."[37]

As in the US, the decline in public enthusiasm for space was accompanied by a downgrading of the space program in the hierarchy of national priorities. In addition, Leonid Brezhnev was less enthusiastic about the space program than his predecessor Khrushchev, unless it promised some political benefits.[38] But at the time, the potential political benefits were rather low. Apart from growing public skepticism about high space spending, the space program's standing as a flagship of Soviet mastery suffered several severe blows. This started with the sudden passing of the head of the space program, Sergey Korolev, in 1966, followed by the fatal accident of Vladimir Komarov during the Soyuz-1 mission in 1967, and the death of Yuri Gagarin in a plane crash in 1968. Eventually, the image crisis culminated with the lost Moon race in 1969.

The leading figures of the space program on both sides of the Iron Curtain had to deal with these challenges. In the United States, NASA sought to expand its international programs. In his address to the United Nations in 1969, President Nixon called for "internationalizing man's epic venture into space" and announced concrete steps toward that goal.[39] The UN General Assembly was an appropriate forum to demonstrate the cooperative nature of the US space program to a broad international audience. Moreover, in their statements to the mass media, NASA officials contrasted the cooperative US approach with Soviet noncooperation. Several articles described NASA's attempts to initiate cooperation and its disappointment with the Soviets' "tepid response." For example, in March 1970, the NASA administrator Thomas Paine mentioned to Richard Lyons of the *New York Times* that NASA had tried to initiate discussions on cooperation with the Soviets twenty-seven times in the previous decade. According to Paine, the Soviets either refused these offers or ignored NASA's letters completely.[40] In doing this, American space officials disseminated a positive image of the US space program while exerting pressure on the Soviet Union to open its space program to international cooperation.

In the late 1960s, such articles about Soviet unresponsiveness to NASA's cooperative initiatives were more damaging to Soviet prestige than in the early 1960s. First, the Soviet Union could not surmount the unfavorable coverage with impressive space achievements as it had in the early days of space exploration. Second, the idea of space cooperation resonated more with the audience worldwide, as reflected in the media coverage of the time.[41] In addition, the UN General Assembly, as a global forum, called for broadening international space cooperation during its annual session in 1969.[42] Archival evidence suggests that Soviet space managers paid great attention to the coverage of their (non)cooperativeness. A 1971 report lists several documents on the state of US–Soviet cooperation.[43] These also include some press articles dealing with the Soviet reluctance toward cooperation. For example, a July 1969 article by Paul Scott in the *New York Daily News* accused the Soviet Union of obstructing international space cooperation by sending useless data to NASA and ignoring all cooperation proposals. In February and March 1970, the Academy of Sciences translated two articles into Russian, one from a French magazine and the above-mentioned article by Lyons.[44] Only a few weeks later, on May 13, 1970, the Soviet authorities approved the start of talks with the US on a joint space mission.[45] Thus, there is reason to believe that the coverage of the Soviets' negative attitude toward American cooperative initiatives contributed to the change in Soviet willingness to cooperate, as it posed a disadvantage in the competition over a country's national image at a time when the global norm seemed to be cooperation rather than competition.

In summary, the consensus on the significance of space leadership, which had been rather fragile right from the start of space exploration, began to fall apart at the end of the 1960s. Worldwide attention to space achievements declined, many people considered solving problems on Earth more important than spaceflight, and the space race was perceived as a waste of money and unnecessary duplication of effort. At the national level, too, the number of people in favor of a strong space program declined, as did the political support behind it. The success of the Moon landing program failed to regain global attention for space exploration; on the contrary, it accelerated the general decline in space enthusiasm even further. The fragmentation of opinion about the relevance of spaceflight and the criticism of the space race forced space officials on both sides to adapt to the new conditions and expectations. Since the mass media repeatedly emphasized the need for cooperation between the superpowers, and since détente created favorable conditions for rapprochement, one possible response to this pressure on the space programs seemed to lie in a turn toward cooperation.

THE ASTP IN THE MEDIA AND THE MEDIA IN THE ASTP

"In the Columbia film *Marooned* a Soviet cosmonaut rescued an American stranded in space . . . an unlikely scenario for both political and technical reasons. Next week Life follows Art as the Americans and Russians attempt the first-ever link-up of their spacecraft in the Apollo–Soyuz Test Project."[46] With these words, the British *Daily Telegraph* introduced an article on the ASTP a few days before the start of the mission. The story of a movie that inspired the first cooperation in human spaceflight also found its way into the official NASA history of the ASTP, a book titled *The Partnership*, and a book on the ASTP published in socialist Czechoslovakia.[47] The popularity of this story about an inspiring movie encouraging space cooperation can be explained by the fact that it fit perfectly into the effort to present a convincing rationale for the joint mission to the public. Space officials of both space programs had every reason to believe that the imagined audience they were trying to target to regain worldwide attention for the space program would appreciate a space endeavor motivated by the noble goal of saving human lives. Moreover, the goal of increasing the safety of astronauts and cosmonauts could also count on the support of the national public because of the losses of lives on both sides in the previous years. Furthermore, long-term space cooperation had the potential to reduce costs and thus to win the approval of those parts of the public that were skeptical of the high space expenditures.

Consideration of public sentiments and the image that would result from the ASTP affected not only the public presentation of the project's aims but also several technical and managerial aspects of the ASTP. After the long struggle for supremacy in space, none of the space powers would have agreed to arrangements that would in any way create an impression of their inferiority. Quite the contrary, after the lost Moon race, the Soviet Union was particularly interested in demonstrating parity with the US and made every effort to organize the joint mission in accordance with this principle. In the realm of technology, an androgynous docking mechanism was developed, which enabled both spacecraft to play an active role in rendezvous and docking. Moreover, there was no single working language; instead, Russian and English were used on equal terms, at least formally.[48] On Earth, communication within American–Soviet working groups was ensured using interpreters, while in space the astronauts were supposed to speak Russian and the cosmonauts English. During the 1973 Paris Air Show, a mockup of the planned Apollo–Soyuz docking was exhibited in a special joint pavilion equidistant from the American and Soviet pavilions.[49] The great interest of the international public in the exhibition might have inspired the Soviets to propose a joint movie on the ASTP. However, the American side rejected this proposal, and one of the arguments against it, revealed in internal documents, was precisely the awareness that the film would further help the Soviets in their effort to demonstrate parity.[50] The concern for parity was also evident in the official designation of the joint flight. While the US side used the term "Apollo–Soyuz Test Project" abbreviated ASTP, the Soviet media and the media in socialist countries preferred to name the participating spacecraft in reverse order—"Soyuz–Apollo."[51] Despite the US–Soviet cooperation, the thinking and actions of space officials were still deeply embedded in the mode of competition.

The Soviet work on the components for the joint flight was accompanied by reflections regarding the impact of the performance in the joint mission on the country's international prestige. The Soviet practice of keeping all preparations and exact dates of spaceflights secret was not possible for cooperation with NASA. Unlike the Soviet missions that were not announced in advance, a delay in preparatory works for the ASTP could have had negative consequences for the public image of the Soviet space program. Thus, work on the ASTP became top priority over all other works.[52] A document related to the draft design of the rocket system stressed that it was extremely important to ensure the reliability of the systems for reasons of cosmonaut and astronaut safety as well as for reasons of prestige. The same document also pointed out the necessity to make provisions with the US side for the publication of certain materials and documents to avoid possible damage to Soviet prestige.[53]

In the end, participation in the joint spaceflight with the United States did not cause any damage to the image of the Soviet space program. On the contrary, all experts and journalists agreed that the Soviet Union had gained international prestige thanks to the successful ASTP flight.[54] Moreover, many of them believed that the Soviet space program took most of the credit for the joint endeavor. The German *Süddeutsche Zeitung* pointed out with some irony that most of the international applause for the final flight of the excellent American Apollo spacecraft went to the Soviet Union.[55] Ultimately, Soviet space officials considered the most important benefit of the ASTP to be the increase in prestige and the fact that the Soviet Union presented itself as a leading nation in science and technology, whose participation was indispensable for major international space projects.[56] Not surprisingly, the rise in the popularity of the Soviet space program in connection with the joint mission was met with displeasure in the United States. In addition, there was criticism of the American technology giveaway.[57] These concerns contributed to a rather reserved attitude toward a new joint mission on the US side and, together with the deteriorating international situation in the following years, led to an almost complete standstill of cooperative efforts between the two space powers.

THE ASTP'S LEGACY

In the history of human spaceflight, the mass media played a key role that can hardly be overestimated. By linking the national and international public, politics, and the various actors involved in the space programs, the media not only became the central site for societal negotiation processes regarding the directions of space exploration and, more generally, its relevance and position in the list of national priorities. They also had a more indirect impact, as the actors of the US and Soviet space programs observed each other's plans and actions through the various media, attempted to control the media coverage of their own program to serve their purposes, and made their decisions considering the anticipated effects of these decisions on media images and the public reactions. This was true for the space race but also for American–Soviet space cooperation. The joint ASTP mission was a cooperative effort, but it also allowed for a more direct comparison between both space programs. Any failure on the part of one cooperation partner could be measured against the performance of the other. On July 15, 1975, just before the launch of the Soyuz spacecraft for the joint mission, the Soviet space minister Sergey Afanasyev was faced with the decision whether to delay the launch because of a broken television camera on board the Soyuz. He decided to launch the Soyuz despite this problem because otherwise the Apollo launch would also have to be delayed and "the world would wonder what was wrong with the Russians."[58] The image of Soviet space technology and its reliability was at stake, as the whole world was now watching the performance of both rivals simultaneously.

Despite the "happy ending," this sensational joint US–Soviet mission, which culminated in a melodramatic handshake somewhere in the sky over Europe divided into blocs, remained an anomaly in the Cold War era. This may help explain the dominance of the competitive paradigm in the narrative of the first four decades of spaceflight. Additionally, the perception of the ASTP was shaped by interpretations that presented the mission as a mere political project. The Soviet scientist and then director of the Space Research Institute, Roald Sagdeev, is often quoted as describing the ASTP as the most expensive handshake, a sacrilege to science, and something that could only be of interest to politicians.[59] Indeed, the ASTP did not meet the expectations of those who hoped for the opening of a new era of cooperation in space. Rather, it represented a path not taken during the Cold War. Nor did it make a substantial contribution of scientific value to space research. Nevertheless, from today's perspective, it is not accurate to dismiss it as a dead end in the history of space exploration. The joint mission was the first international cooperation in human spaceflight. Space officials on both sides gained experience in conducting international manned space programs. This became important with the first flights of international spacefarers aboard Soviet and American spacecraft in the late 1970s and 1980s. When inviting cosmonauts from other socialist countries, Soviet space officials referred to their experience with the ASTP to convince the allies that they were capable of carrying out spaceflights with foreigners.[60]

What is more, the ASTP demonstrated to the global public the possibility of superpower cooperation in complex space endeavors and promoted the idea of space cooperation in general. In the 1980s, marked by heightened tensions in international politics and space, the spirit of the ASTP was revived in the arts. The novel *2010: Odyssey Two* by Arthur C. Clarke and the movie *2010: The Year We Make Contact* (1984, directed by Peter Hyams), based on the novel, not only took up the idea of American–Soviet cooperation in space, but both contained a reference to the ASTP in the form of a spacecraft named after the ASTP commander Leonov. Promoting space cooperation and environmental protection are the goals of the aforementioned ASE founded in 1985 by astronauts and cosmonauts, including Leonov. All these repercussions of the ASTP helped preserve the idea of large-scale space cooperation between the superpowers at a time when it was politically infeasible and prepared the ground for the post–Cold War shift toward international space cooperation.

CHAPTER 11

INTERNATIONAL COOPERATION, COMPETITION, AND CHANGING TRENDS IN AMERICAN SPACE CINEMA, 1997–2019

ESTHER LIBERMAN CUENCA

In the 1997 film *Contact*, Dr. Ellie Arroway (Jodie Foster), gets a second shot at a mission when it is revealed to her that the Japanese had built a spaceship in secret. After being passed over in favor of a male astronaut by an international space committee, Dr. Arroway goes on a solo mission into the outer reaches of the unknown on this ship to find the source of mysterious sound waves. Watching *Contact* in the theater as a sixteen-year-old, I was struck by this turn of events. It was given little screen time, but it presented Japan, a country that never launched a person in space, as having the capacity to build a trillion-dollar spaceship. Whether the Japanese could build a project worth a trillion in movie dollars is debatable, though it is worth noting that when *Contact* was released, Japan's Institute of Space and Astronautical Science (ISAS) had launched the M-V, a solid-fuel rocket capable of launching satellites into high orbit, representing, at the time, the culmination of an investment worth billions of yen and a decades-long initiative that began in the 1950s with the founding of Japan's space program. Some of ISAS's most notable milestones included sending a robot spacecraft to fly by Halley's Comet in 1985 and one that flew into lunar orbit in 1990.[1]

The appearance of the Japanese, however brief, in a big studio picture like *Contact* also anticipated the launch of a major initiative that occurred a year later: the International Space Station (ISS), a joint project funded by the United States, Japan, Russia, and several European countries. Though it followed previous collaborations, such as the Apollo–Soyuz Test Project (1975), and other small cooperative programs spearheaded by the US National Aeronautics and Space Administration in the 1960s and 1970s, the ISS was not only a considerable financial investment on NASA's part but also a deep commitment to international partnerships moving forward. This partnership was dedicated not only to science and diplomacy but also to the prevention of potential hostilities that eschewed the national prestige that came with complicated human spaceflight missions into deep space.[2]

In certain respects, the ISS has been a rousing success when one considers American public opinion. In a 2019 poll taken by C-SPAN/Ipsos, three-quarters of the respondents were supportive of NASA and its programs, which include the ISS. In a Pew Research Center poll taken in 2018, a little over 70 percent said it was important that NASA remain a global leader in space exploration.[3] The latter sentiment seems almost nostalgic considering that the cancellation of the Shuttle program resulted in NASA's dependence on SpaceX and Russian carriers to transfer both supplies and astronauts to the ISS. The launch of the ISS may have ushered in a new international age in space flight, but it is other programs—especially those of China, Japan, and India—that are ascendant.

Unfortunately, the national surveys of American interest in space exploration do not account for how most Americans are exposed to news of NASA's programs and initiatives. But what is clear is that most Americans do not consume "space news." Of those surveyed in the Pew Research poll, only 7 percent consumed news about space exploration with avid interest, while a fifth of those surveyed have heard nothing at all, and the rest, 70 percent, have heard only a little about the space program.[4] The American enthusiasm for NASA, difficult to quantify but clear, must come from other sources.

Popular interest in NASA's space activities is both generated and reflected in popular culture writ large—in media such as television, comic books, and films.[5] This is not by accident. In the 1960s, public relations personnel publicized NASA's achievements to justify the enormous costs of the Mercury and Apollo programs to American taxpayers. This carefully coordinated onslaught of publicity not only included glossy photospreads, human interest stories, and ticker-tape parades but also seeped into spheres of popular culture not sponsored by NASA's public relations office, including television shows like *Star Trek* (1966–1969) and feature films as varied as Stanley Kubrick's *2001: A Space Odyssey* and Robert Altman's *Countdown*, both released in 1968.[6]

Table 11.1. List of spaceflight movies released between 1997 and 2019 in chronological order, featuring fictional corporations or non–US space programs, their equipment, or personnel.

No.	Year	Film title	International/Corporate program, equipment, or personnel (both fictional and nonfictional)	Non–US country or fictional private corporation
1	1997	*Contact*	Mir, Japanese-made spacecraft	Russia, Japan
2	1998	*Deep Impact*	One cosmonaut	Russia
3	1998	*Armageddon*	Mir, one cosmonaut	Russia
4	1999	*October Sky*	Sputnik-1	Russia (Soviet era)
5	1999	*Virus*	Mir, several cosmonauts	Russia
6	2000	*Mission to Mars*	World Space Station, various astronauts	Various countries
7	2000	*Red Planet*	Kosmos (probe)	Russia (Soviet era)
8	2000	*Space Cowboys*	IKON (satellite)	Russia (Soviet era)
9	2004	*The Day After Tomorrow*	ISS, one cosmonaut	Various countries, Russia
10	2007	*Sunshine*	Icarus II (spaceship), various astronauts	Japan, China
11	2009	*Moon*	Sarang Station (lunar base)	Lunar Industries
12	2011	*Apollo 18*	LK (lunar lander)	Russia (Soviet era)
13	2013	*Europa Report*	Europa One (spaceship)	Europa Ventures
14	2013	*Gravity*	ISS, Tiangong, Soyuz, Shenzhou	China, Russia
15	2015	*The Martian*	CSNA, Taiyang Shen (a booster rocket)	China
16	2017	*Life*	ISS, Soyuz, various astronauts	UK, Japan, Russia
17	2016	*Hidden Figures*	Mentioned in the context of the space race	Russia (Soviet era)
18	2018	*First Man*	Mentioned in the context of the space race	Russia (Soviet era)
19	2018	*Geostorm*	International Climate Space Station, various astronauts	Various countries
20	2019	*Ad Astra*	A shuttle, a lunar base, and a Norwegian Biomedical Space Station	Unnamed private company, Norway

In much the same way that historians and media scholars have examined the importance of popular culture in both expressing and shaping American enthusiasm for the space program in its early years, in this chapter I draw from a similar pool of evidence—American, mainly big- or mid-budget Hollywood films released between 1997 and 2019—for insights on how American audiences perceived their space program and the presence of foreign agencies in this global space age.[7] The films released in this twenty-two-year period indicate an increasing American awareness of other countries' space agencies, though what is presented in these films lags behind global trends. Nevertheless, the release of these films is significant given that they usually represent a considerable financial investment on the part of movie studios, which aim to make broadly appealing entertainment for American audiences. Additionally, these films were released in international markets, thereby projecting abroad certain images of the US and its role in the development of spaceflight. Roger Launius contended that films like *Armageddon* (1998), *Deep Impact* (1998), *Contact* (1997), and *Space Cowboys* (2000) have perpetuated enthusiasm for NASA into the twenty-first century.[8]

In the sample of twenty films released between 1997 and 2019, this study does not include films with heavy elements of fantasy such as those in the *Star Wars* or *Star Trek* franchises, but mainly focuses on movies (even if highly fictionalized) that depict the ISS and non-NASA space agencies and equipment that are grounded in some sense of verisimilitude, at least when compared to their real-world counterparts (see table 11.1). This is not to say that there are no fantasy elements in the films chosen for this sample, but that the fictionalized or fantastical elements seen onscreen aim for a type of realism because they present their stories within a world of nonfictional possibilities. Some films that fit these criteria, such as *Interstellar* (2014), have not been included because they only depict NASA. The sample thus includes only films depicting either non-US space agencies or private corporations, even if NASA does not appear in the film.

This study begins in 1997, before Zarya, the first ISS module, was launched in November 1998, and the news of this international effort was therefore well into circulation. The year 2019 closes this distinct two-decade period in film, just before moviegoing habits and distribution models changed drastically with the COVID-19 pandemic. It is also the end of an era marking a steady decline in NASA's dominance in space, giving way to other space agencies that are dedicating far more resources to space exploration and human spaceflight missions. The latter also includes the capitalistic efforts of private enterprises, such as

SpaceX, which may link their vision to American prestige but whose long-term goal is profit and corporate glory.

This period can be divided into two distinct phases of artistic output. The first phase, from roughly 1997 to 2003, consists of spaceflight films that suggest either the friendly or antagonist presence of Russian cosmonauts. When a foreign space agency is referenced or shown in film, it is either Soviet or Russian. The second phase, from roughly 2004 to 2019, is characterized by films suggesting not only Russia's presence in space but also that of other countries, including China and Japan, and corporations that have turned to spaceflight for private profit. Crucial to understanding this second phase of spaceflight films is the sharp turn toward themes dealing with human-caused environmental catastrophes, which reflect to some degree the shift that began to occur in public consciousness about climate change and pollution around this time. Because these films are American-made and produced primarily for American audiences, the protagonists and their conflicts are firmly rooted in a world of American exceptionalism, whether in science, ingenuity, or heroism. In this sense, these two phases have reflected a kind of funhouse-mirror reality of the American space program in these past two decades, one that depicts US dominance over Russia, its main competitor, and ends with the US still a major player, but now one of many, in a fragmented and crowded international cosmos.

CINEMATIC NOSTALGIA FOR THE SPACE AGE

In the sample of spaceflight films, Russia, of course, looms large. Because most of the films take place in a contemporary setting, the glory and competition of the Soviet space age are a thing of the past, though there are spaceflight films that take a nostalgic view of this history.[9] The syrupy sweet *October Sky* (1999) is based on the memoir *Rocket Boys* (1998), written by the former aerospace engineer Homer Hickam (Jake Gyllenhaal), whose small-town dream of working at NASA was inspired by the launch of Sputnik-1 in 1957.[10] In the film, the moment of transformation for Hickam comes when he spots Sputnik-1 shooting across the sky, the camera framing his face similarly to when characters in other films fall in love for the first time. The Soviets are a force for creativity and a catalyst of change for the young Hickam, who represents the ideals of a fledging American space program that, at the time, seemed hopelessly behind the Soviets in the race for space.[11] The release of Clint Eastwood's *Space Cowboys* the following year takes a darker view of Soviet accomplishments and how they threaten American supremacy. Unlike *October Sky*, *Space Cowboys* has a contemporary setting, telling the story of a retired Mercury-era air force pilot and engineer, Frank Corvin (Eastwood), who is brought to NASA to repair a decaying Soviet-era satellite.[12] Eventually, it is revealed that the satellite, called IKON, is secretly housing nuclear missiles, a remnant of Cold War hostilities that now poses an existential threat to a world in a post–Cold War era. At least prior to 9/11, the post–Cold War era was one of comfort and idyll, with the imminent dread of warfare replaced by the security of America's unquestionable supremacy in global politics.

Despite their differences, *October Sky* and *Space Cowboys* are essentially spaceflight nostalgia films for the baby-boomer set, sepia-toned stories that imagine histories of a golden age of rugged male pilots, handsome male astronauts, and intrepid male scientists, like Philip Kaufman's *The Right Stuff* (1983). It is notable that two more recent films that traffic in nostalgia for the American space program, *Hidden Figures* (2016) and *First Man* (2018), are either far more critical or introspective about the personal costs of achieving American dominance in space. *Hidden Figures* takes place prior to the Apollo era, set against the backdrop of Yuri Gagarin's first spaceflight and the pressure to get the Mercury astronauts into space.[13] The title *Hidden Figures* has a double meaning: although it chiefly refers to the climax of the film, in which a Black mathematician, Katherine Goble Johnson (Taraji P. Henson), cracks the code of a difficult problem by uncovering its "hidden" numbers, it is also a reference to the women of color who worked in computing and as engineers for NASA and were rarely acknowledged for their work.[14] *Hidden Figures* thus puts their contributions at the center of its story as a corrective to the types of histories that tend to ignore women and people of color. The film portrays NASA first and foremost as a place of (segregated) work, in which success was possible not from genius breakthroughs or male camaraderie but from everyday labor that went into launching an American man into space to compete with the Soviets for international bragging rights. Achieving this feat comes at a real cost for NASA's Black workers, who are often hampered by petty jealousies, misogyny, and racism.[15]

The Neil Armstrong biopic *First Man*, like *Hidden Figures*, also turns nostalgia on its head. It briefly references the race to the Moon with the Soviets, but forgoes the classic nostalgic tropes found in films celebrating NASA's achievements. It focuses, instead, on the deeply personal trials of Armstrong's life, beset by the tragedy of his young daughter's death and a rocky relationship with his wife (Claire Foy). Armstrong (Ryan Gosling) cuts a lonely and distant figure, almost as if the film is asking us to ponder the chasm between Armstrong's national reputation as first man on the Moon and Armstrong the man, who happened to be first on the Moon and rather ordinary in his everyday life.[16] These nostalgia films show how NASA's history has been told according to changing tastes. *October Sky* and *Space Cowboys* look to the past as a golden age. *Hidden Figures* and *First Man* look to the psychological costs of the Mercury and Apollo programs—the business of space is not so much national glory as it is the mundane work of computing, piloting, fixing equipment, navigating workplace politics, and balancing work with having a family life.

Some of the nostalgia films, such as the horror flick *Apollo 18* (2011) and the Russian mockumentary *First on the Moon* (*Pervye na Lune*, 2005), both indulge and critique their subjects. As Brian Willems has argued, *First on the Moon*, which shows "recovered" footage of the Soviet Moon landing in the 1930s, is a meditation on the intractability of the future. Even with Soviets "winning" the lunar race, this alternate past did not exactly lead to a brighter future.[17] The critically reviled *Apollo 18* is also a "recovered," or found-footage, film in the vein of horror classics *The Blair Witch Project* (1999) and *Paranormal Activity* (2007). Its story revolves around an alternate history to explain, ultimately, the reason for NASA's cancellation of the lunar program before the planned Apollo 18 mission. It was not lack of funds or public interest, but sentient alien space rocks that killed the astronauts during an Apollo 18 mission that NASA then covered up. The astronauts in *Apollo 18* were able to use a deceased cosmonaut's technology, the Soviet LK, to try to escape the Moon aliens. As implied in the film, the Soviets had been landing on the Moon for a while and NASA had known about it, but what really stopped the Moon missions were aliens outside the control of geopolitical forces hundreds of thousands of miles away. *Apollo 18* is deeply silly and surprisingly light on scares, but its premise is contingent on a popular acknowledgment that NASA had once been cutting edge but, since the Apollo missions, has been in decline—at least when it comes to human spaceflight programs that *really* excite people. *Apollo 18* as well as the AppleTV+ show *For All Mankind* (2019–), though wildly different in their critical reception and execution, ask the same question through the lens of alternate history: Was Apollo the best we could ever do? Nostalgia, as it turns out, throws into sharp relief the deficiencies of the present.

MIR, RUSSIAN TECHNOLOGY, AND THE "WORLD SPACE STATION"

When the ISS was first launched, it was merely a shell of human possibility, unable to sustain living and breathing residents until the new millennium, though the Russian space station, Mir, had been inhabited by cosmonauts since 1986. Mir was thus already well-established in American popular consciousness by the time the ISS first launched in 1998.[18] The Hollywood films released between 1997 and 2000 are a testament to how Mir had been woven into the larger fabric of spaceflight entertainment representing the post-Soviet condition in a world in which American capitalism and military might have "won" the Cold War.[19]

Through documentary cinema at least, American audiences were likely first exposed to Mir through the always-popular IMAX films that were, and still are, largely shown in specialty theaters able to project the towering images captured by IMAX cameras, which were placed in NASA's Space Shuttles to document life aboard Mir. The forty-minute IMAX film that resulted from this cinematic experiment, *Mission to Mir* (1997), documented the tail end of the successful Shuttle-Mir program, which ended in 1998 and represented one of several international collaborations between NASA and Roscosmos since the 1970s.[20] The full-length documentary German film *Out of the Present* (1995), which marketed itself as the first film produced in outer space, was a monumental achievement of technical and political filmmaking, given that it documents the cosmonaut Sergei Krikalev's peaceful tenure on the Mir between May 1991 and March 1992, during which the tumultuous transition from Gorbachev to Yeltsin and from Leningrad to St. Petersburg occurred when the Soviet Union transformed into the Russian Federation. *Out of the Present* had a slow rollout internationally between 1995 and 1997, though American audiences, if they had seen it at all, likely caught it on home video years later.[21]

Mir's American feature film debut was in *Contact*, the film that opened this chapter, and had delighted me as a teenager when it was released in the summer of 1997, months before the November IMAX release of *Mission to Mir*. Based on Carl Sagan's 1985 novel of the same name, *Contact* is largely concerned with age-old philosophical questions: What is more important, faith or reason? What drives the universe? What, if any, are the singular qualities of our shared humanity?[22] The Mir space station makes a brief appearance. In that scene, the protagonist, Dr. Arroway, returns to her abode and sees a computer monitor, a portable satellite phone, and a widescreen television with a mounted camera that had been mysteriously set up in her absence. She then notices a string of Russian words on the monitor, clicks the keyboard, and suddenly two cosmonauts floating in zero gravity appear on the television screen. The eccentric billionaire, S. R. Hadden (John Hurt), comes onscreen, bald, and floating upside down like a space Count Dracula, telling her he has taken up residence on Mir because he believes the weightlessness of space will slow his rapidly advancing cancer. It is here that he informs Dr. Arroway that the American government has funded a secret, copycat vehicle, located off the coast of a Japanese island and built by Japanese independent contractors, whose labor and product Hadden's company has acquired. He tells her, "They still want an American to go . . . wanna go for a ride?"[23] This three-minute scene speaks to the American perception of foreign space agencies as much as it foreshadows what was to come in the decade following. Mir, it was implied, could be a refuge for rich space enthusiasts like Hadden, whose money could butter up the Russians who are hard up for cash. Their technology is sound and advanced, but the Russians can be bought by wealthy weirdos with fanciful pipe dreams about curing cancer—suggesting, perhaps, a sort of old-school Soviet corruption.

By the late 1990s, it was not so out of the ordinary to hear of space tourism, which allowed private citizens or corporations to buy seats on the Soyuz and into an experience that had previously been reserved for a select few.[24] The first millionaire to buy his ticket on the Soyuz was a Los Angeles businessman in 2001, but Russian space tourism had been up for sale a decade before that. Toyohiro Akiyama, a Japanese journalist, traveled to Mir in 1990 and stayed there for a week as a reporter for the Tokyo Broadcasting System, which paid the Soviet government $12 million for the privilege. In 1991, Helen Sharman, a British chemist who worked for the Mars chocolate company, won a contest to train at Star City and conduct scientific experiments on the Mir space station as part of Project Juno, a private enterprise that promised the Soviet government £7 million. Though the project's partners failed to raise the necessary funds, the Soviet government allowed Project Juno to proceed, and Sharman became the first British astronaut in space.[25] Roscosmos allowed for a broader international presence in space even though doing so may have painted it as a stridently capitalist, and perhaps predatory, agency.

The presence of the Japanese as deep-pocketed outsiders in *Contact* attested the perception that Japan was a technological powerhouse and a sometimes "menacing" threat to American hegemony in the 1980s.[26] In 1997 the prominent image of Japan in American culture was one of its ubiquity in the market of consumer electronics and automobiles, which was cultivated thoroughly in the 1980s during a time when Japan was perceived to have control over the American economy. What colored anti-Japanese sentiment in the 1980s was, at times, a racism rooted in envy and fear. The envy, at least, was palpable. Ernest Boyer, the former US commissioner of education, said in a congressional hearing in 1983 that Sputnik had stimulated better math and science education in

American schools between 1958 and 1975. He remarked sarcastically that "maybe what we should do is to get the Japanese to put a Toyota in orbit," effectively comparing Japanese ingenuity with the Soviet threat of technological dominance decades earlier.[27] In the late 1990s, however, Japan's economy was in decline and its threat to American economic superiority no longer a pressing concern, a trend best reflected in the scene that references, but does not really show, the Japanese as the faceless, expert technicians of Dr. Arroway's spacecraft in *Contact*. Significantly, the Japanese are neither the primary funders nor the visionaries of the project that would launch Dr. Arroway's expensive adventure.

The 1998 releases of two blockbuster films, *Deep Impact* and *Armageddon*, were heavily commented on by the entertainment press for their similar plots, though their depiction of international cooperation with Russian cosmonauts significantly differed. Disaster films were a mainstay of 1990s blockbuster filmmaking and, as space films go, *Deep Impact* and *Armageddon* represent how NASA might respond to world-ending extraterrestrial threats, although they come to different conclusions about the nature of heroism and cooperation. Both films are about humanity's attempt to prevent destruction by impeding an extraterrestrial object from hitting Earth, but in *Deep Impact* it is a comet and in *Armageddon*, the more financially successful of the two films, an asteroid. *Deep Impact* was released first, in May of that year.[28]

In *Deep Impact*, the astronauts and cosmonaut aboard the spaceship Messiah work together to obliterate the comet before it reaches Earth, ultimately sacrificing their lives after all other attempts to break apart the comet fail. In a similar plot point, the astronauts in *Armageddon* plan to blow up the asteroid before it reaches Earth, though it is the salt-of-the-earth oil rigger Harry Stamper (Bruce Willis) who ends up sacrificing his life to save his crew and the world. The cosmonaut aboard the Messiah in *Deep Impact* is a stock character: he is the lone foreigner (played by the respected Russian stage actor Aleksandr Baluev) in a crew of Americans and depicted as a serious scientist whose only role is to deliver believable scientific-sounding exposition. Ultimately, he is easily forgettable. The Russian cosmonaut in *Armageddon* demands a few more words. Played by the well-known Swedish character actor Peter Stormare, it is a supporting role, but a memorable one. His performance is buck wild, equal parts hero and comedic relief, and Stormare chews the scenery with aplomb. In the film's second act, the crew of oil riggers turned astronauts arrive at Mir to refuel their Space Shuttles en route to blow up the Earth-threatening asteroid. They encounter a lone cosmonaut, Lev Andropov (Stormare), who has been driven mad by isolation, a persistent trope in science fiction films such as *2001: A Space Odyssey*

(1968), *Marooned* (1969) and, later, *Sunshine* (2007) and *Moon* (2009). This is a workingman cosmonaut, a technician, doing the equivalent of a long, thankless nightshift, trapped on a rickety space station hundreds of miles above Earth.[29]

A peculiar element in *Armageddon* is that, although the action is contemporary to the time of its release and supposed to take place in the late 1990s, the film's Mir is still stuck in time, circling the Earth as a relic of the Cold War era. The hammer-and-sickle flag hangs in the station. Lev is wearing a CCCP (USSR) T-shirt, the logo encased in a red star. The astronauts feel sorry for Lev. Oscar (Owen Wilson) expresses regret about "the upheaval and strife in your country," saying: "Man, that must be tough up here with the loneliness, but I want you to know that up here there are no Russians and there are no Americans, we're just spacemen." It is almost as if the fall of the Soviet Union occurred without Lev being any wiser to it, echoing the real cosmonaut Sergei Krikalev's journey in *Out of the Present*. The Mir space station itself is a complicated maze of failing parts and appears to be a heap of space junk that might once have been considered sophisticated in a bygone space age. The role of Mir in the story is (improbably) to be a refueling station for the astronauts on their way to blow up the asteroid. Its role is also to break apart in a fire to add another explosive set piece to the film, but it was likely also a nod to the actual fire and collision accident that occurred on Mir in 1997 with two astronauts and two cosmonauts onboard.[30] At one point, once Lev is on the Space Shuttle and they are flying away from a nuclear explosion, he fixes a failing component by beating it with a hammer, saying: "American component, Russian component, all made in Taiwan!" In the end, Lev and the American astronauts cooperate and save the world, but they do so with (or perhaps despite) faulty Russian assistance.[31]

Dueling spaceflight movies with similar plot devices were also released in 2000, and in these films Mars and the search for (and meaning of) intelligent life were the thematic through lines.[32] Unlike *Deep Impact* and *Armageddon*, both *Mission to Mars* and *Red Planet* were box office disappointments, though they pointed toward increasing optimism about international space expeditions led by American astronauts and NASA. The first of the Mars films to be released that year, *Mission to Mars*, was directed by Brian de Palma, who is not exactly known for his science fiction fare and brought little of his distinctive style to the overly earnest film.[33] *Mission to Mars* is about an American crew on its way to Mars to recover a lone astronaut played by Don Cheadle. This lone astronaut's crew, an international group including a Russian cosmonaut and a French astronaut, had been killed in a mysterious accident years earlier. *Mission to Mars*, however, is a forward-looking film, and sees a future in 2020 in which the ISS is now the World Space Station, a cavernous mass orbiting earth with artificial gravity and numerous astronauts from countries such as Belgium, Egypt, Spain, and Russia. The NASA logo is emblazoned everywhere, indicating that, despite its international character, the future of spaceflight still belongs to the United States.

Mission to Mars shows a high-tech future in which missions to the red planet have been ongoing, launching regularly from the World Space Station. It is a similar plot point in *Red Planet*, which takes place in 2025 and tells the story of an American crew launching from a "high-orbit space station" to a Mars that has been terraforming for decades. Predictably, the crew's landing craft gets damaged, and they have no way to return home, all the while they are being hunted by an artificial intelligence military robot dog. To get off the planet, NASA provides a solution to the crew, who are to travel one hundred kilometers to the site of an abandoned thirty-year-old Russian "rock probe" that failed to launch, jumpstart the machine, and fly it back into orbit. This probe, with the generic name "Kosmos," is a reference to Soviet-era technology.[34] The mission commander Kate Bowman (Carrie-Anne Moss) mentions that the Russian designer of the probe who had generously provided the specs to NASA now works at a deli in Brooklyn. This detail is undoubtedly a reference to the fate that many Russian scientists met after the dissolution of the Soviet Union as they migrated out of their homeland in search of better economic opportunities.[35] As the IMDb trivia page for the movie explains, "The Russian Mars probe may also be a subtle dig to Russia's less than stellar record with Mars missions."[36] This might be true. The Russian technology that appears toward the end of the film when the astronaut Robby Gallagher (Val Kilmer) finally reaches the probe is some sort of combination of a 1970s Mars lander with a Lunokhod moon rover—not the sophisticated machinery of the American-made spacecraft that appears in the film. Even so, it is the half-broken, Soviet-era technology that the astronaut reconfigures to save his own life.

The Mir space station makes one last appearance in *Virus* (1999), a big-budgeted film with a well-known cast that disappointed at the box office. Mir, as in *Armageddon*, exists to be taken over and destroyed to further the action of the story. Unlike *Armageddon*, the Mir in *Virus* is still a functional spacecraft, with cosmonauts (sporting the standard of the Russian Federation not the Soviet Union) speaking Russian, playing chess, and communicating with a Russian navy vessel in the South Pacific. The film's title refers to an extraterrestrial virus that takes over Mir, kills the cosmonauts, and then beams itself down to the vessel below to infect computers and people. It is difficult not to read broader Y2K anxieties into the film, which is nevertheless based on a Dark Horse comic book first published in 1992 (in the comic, it is a Chinese, not a Russian, vessel that the virus infects).[37]

Two years after the release of *Virus*, Mir was formally decommissioned to make way for Russia's partnership with the ISS. The destruction of Mir in both *Virus* and *Armageddon* suggests the perceived waning power of the Russian space program from an American perspective. The new era of spaceflight was international and coming out of the long shadow of old Cold War tensions.

INTERNATIONAL COOPERATION, CLIMATE CHANGE, AND THE PRIVATIZATION OF SPACE

The new millennium represented a shift in the way big-budget Hollywood films produced for mass consumption depicted the role of spaceflight in an international context. Hollywood films in general were using the topic of climate change as fodder for the often-lucrative business that was the "disaster" film genre. Disaster films have been, in one form or another, a staple of blockbuster moviemaking for several decades. These are films in which natural or human-made disasters wreak havoc on the lives of the protagonists and their environments, sometimes on a global scale.[38] Disaster films were common in the 1970s, with movies such as *The Poseidon Adventure* (1972), *The Towering Inferno* (1974), and *Earthquake* (1974), and, having suffered a dip in popularity in the 1980s, became dominant once more in the 1990s with mega hits such as *Independence Day* (1996) and *Twister* (1996), and more modest successes of two exploding volcano movies released in the same year, *Dante's Peak* (1997) and *Volcano* (1997). Disaster films are flexible in that they can combine elements from different genres, as did *Deep Impact* and *Armageddon*, two films that mixed family drama and science fiction with world-ending disaster. The first decade of the 2000s, however, saw a shift in the way natural disasters were seen in Hollywood spaceflight films. The focus was increasingly on climate change and how human action (or inaction) on that front had endangered the world. In this respect, the disaster genre of movies responded to these changes, as they did in their depiction of a new player on the scene: the private space corporation.[39]

The Day After Tomorrow (2004), directed by the disaster-film aficionado Roland Emmerich (who also helmed *Independence Day*), was a worldwide hit grossing $550 million; it signaled the direction spaceflight movies would take, although the scenes in space take up mere minutes of screentime compared to the action set on Earth.[40] *The Day After Tomor-*

row imagines a world in which human-provoked climate change leads to a global freeze so dire that the president of the United States is killed, the cold weather chases down the protagonists like a serial killer, and sheets of ice cover the Earth's surface in the aftermath.

In a public opinion report published in May 2004 just before the release of *The Day After Tomorrow*, the authors recognized the importance of film as responding to trends in public sentiment about climate change (at least, for this report, in the United Kingdom), saying that: "The film will have a potential audience of 500 million people, and is set to put climate change on the 'mainstream' agenda. Environmentalists hope that the Hollywood blockbuster will change people's perceptions about climate change, while others have warned that the film is a manipulation of science to serve a political agenda."[41] As predicted, the film fueled debates about climate change.[42] The ISS appears in *The Day After Tomorrow* as itself for the first time—no more Mirs or World Space Stations from here on out—and serves as the audience's bird's-eye-view of Earth as the planet suffers a climate disaster. The astronauts onboard are spectators of the chaos from a distance. The ISS is indicated in the first shot of it orbiting around Earth, its design reflecting the early stages of its modular build in 2002. The first shot of its interior shows various flags: Canada, Belgium, Italy, Switzerland, Norway, the UK, the Netherlands, and, curiously enough, Australia and Brazil, though the latter two countries did not (and do not) have partnerships with the ISS. Since 2000, a permanent international crew has been aboard the ISS; however, it was unlikely that the average American moviegoer could name the participating countries in the program. One wonders whether the placement of the flags was meant to represent the primary worldwide markets for the film.[43]

At the film's conclusion, the image of the ice-covered Earth from the vantage point of the ISS is what the audience is left with. *The Day After Tomorrow* is not necessarily a thematically rich film, but for summer popcorn fare that forces its audience to contemplate the consequences of environmental destruction, it is heady entertainment.[44] The last shot of the film involves the camera panning out from the ISS, which circumnavigates an icy planet resembling Earth, and a cosmonaut (played by Vitali Makarov) uttering: "Look at that! I have never seen the air so clear." Millions have died, the Earth resembles a snow cone, but with this death comes regeneration, and perhaps, from now on, the film is telling us, people will learn from their mistakes and treat their environment with more care. What the film signals in a more subtle way is that there are real-world applications for the kind of science that space agencies like NASA can do. World-ending comets and fanciful adventures to

Mars seem not as urgent as dealing with the real and immediate threats of climate change.

Although the disaster is first detected by climatologists of the National Oceanic and Atmospheric Administration and a fictional British agency called the Hedland Climate Research Center (standing in for the real Met Office Hadley Centre in the UK), the experts involved in averting disaster also include a scientist from NASA (played by Tamlyn Tomita). This detail points toward an expanded role of the agency in cinema that deals with scientific matters other than spaceflight. A decade and a half later, most of the American respondents surveyed in the C-SPAN/Ipsos and Pew Research polls indicated, in the former, that it was important for NASA to monitor natural disasters (73 percent) and, in the latter, that NASA's top priority should be to monitor the Earth's climate systems (63 percent).[45] In other words, public opinion is trending in one direction concerning climate change and, ultimately, it is the climate extinction event in films such as *The Day After Tomorrow*—nasty, quick, and beyond human control—that make lasting impressions on popular awareness about the environment.[46]

Sunshine (2007) also points toward ecological disaster, but unlike *The Day After Tomorrow*, it is a slow, freezing extinction that can be prevented if an international crew on the spacecraft Icarus II can deliver a nuclear payload into the Sun (in a plot similar to the 1990 film *Solar Crisis*).[47] Although *Sunshine* is probably best categorized as horror or a psychological thriller in the vein of Ridley Scott's *Alien* (1979), it still gestures toward familiar themes in disaster films: impending doom and a cast of characters trying to escape or avert the disaster. The setting of the film, almost entirely on the spacecraft, feels claustrophobic, and director Danny Boyle and writer Alex Garland smartly avoid any images of Earth until the very end, when a frozen, snowed-over landscape illuminates with newly unleashed sunshine over the Sydney Opera House, signifying the success of the Icarus II mission.

The real tension, the film argues, lies not in superficial national boundaries that divide us, but forces, such as religious fundamentalism, that seek to undermine science.[48] The film begins deep into the second expedition, with the crew approaching the planet Mercury on their way to the Sun. The first expedition, which was mysteriously aborted around Mercury before the events of the film, was led by a crazed astronaut, Pinbacker (played by the British actor Mark Strong with what sounds like a German-tinged accent), who is revealed to be a religious zealot. The climate crisis on Earth and the attempt to "restart" the Sun is really a larger story about the seemingly irreconcilable differences between religion (represented by Pinbacker) and science (represented by the astronaut Robert Capa, played by Cillian Murphy) as approaches to solving humanity's problems. It is a neat allegory for the potential ruin that might befall humanity if climate deniers and their political enablers, at times unified in both willful ignorance and love of corporate profits, get their way.

The Icarus II mission, while named after a mythological character that exemplifies hubris, is ultimately a collaborative, international effort to save the Earth. The spaceship carries a Japanese mission commander (Hiroyuki Sanada), a Chinese botanist (Michelle Yeoh), a Chinese navigator (Benedict Wong), and several American astronauts on a dangerous mission, all of whom represent different ideologies on the faith-versus-reason spectrum. Boyle, who consulted with famed astrophysicist Brian Cox on the science presented in the film, had also asked NASA experts what the makeup of a crew sent on a mission to deep space would look like fifty years in the future. According to Boyle, he was told that the countries that could fund and organize a mission as ambitious as the one depicted in his film would be the "emerging economies" of the world. NASA experts had told him: "America will be lucky if we're still involved [in deep space missions] because the principle [*sic*] drivers will be the Asian economies. India, China, maybe Japan."[49] For this reason, Boyle decided to represent these "emerging economies" within the ensemble, casting the Japanese actor Sanada in the role of the commander.

The second decade of the twenty-first century continued to depict the ISS, international crews or space agencies, and the larger theme of environmental anxiety with the appearance of a new player driving space expeditions: the private corporation. The depiction of a corporation clashing disastrously with human interests was perhaps most memorable in the *Alien* franchise, the first of which, *Alien*, introduced the evil Weyland-Yutani Corporation. The Weyland-Yutani Corporation's soulless indifference to the working-class members of the *Nostromo* is what drives the plot forward, the greedy corporation willing to sacrifice the lives of the crew to obtain the titular alien as a biological weapon.[50] The plot of *Alien* was largely copied beat by beat in the mid-budget film *Life* (2017), which has no scary corporate bogeyman but is set on the ISS with an international cast, depicting astronauts from the US, UK, and Japan (*Sunshine*'s star Sanada makes an appearance here, this time as an ISS systems engineer), as well as Russian cosmonauts.[51] Much as in *Alien*, the crew gets hunted down one by one by an alien life-form that was inadvertently awoken by one of the astronauts. Unlike in *Alien*, the whole crew fails to prevent this mutating life-form from descending onto Earth, leading to the destruction of both the Soyuz and the ISS in the process. In *Life*, the evil corporation is swapped for human naiveté and hubris, which play a part in unleashing a destructive force beyond human control. It is this kind of carelessness that could (and does) lead to larger population disasters—the mutating life-form was an obvious metaphor for a pandemic-causing disease, a theme that has become resonant in the post–COVID-19 era.

The image of blue-collar astronauts and the corporate overlord popularized in *Alien* was influential on Duncan Jones's debut film *Moon* (2009). Though modestly budgeted at $5 million and making only $10 million worldwide for Sony Pictures Classics, it was critically beloved and nominated for a slew of international film awards. *Moon* is a story about isolation, of a working-class astronaut and Lunar Industries employee, Sam Bell (Sam Rockwell), whose multiyear shift in mining precious lunar soil from the airless and lonely desertscape is ending. Eventually, it is revealed that Lunar Industries has been cloning Sam Bell for decades—the original Sam Bell is safely on Earth—as a cost-efficient way of reproducing cheap and disposable labor.[52] Astronauts are (and can be) merely cogs in the capitalist machinery of space exploration.[53] In the end, one of the clones escapes to Earth and testifies against Lunar Industries. Its stock price drops, which is possibly the only way that unethical corporations can get their comeuppance in both fiction and the real world.

The premise of *Moon* is reminiscent not only of the song "Space Oddity" (1969) by director Jones's father David Bowie but also of the classic ballad "Rocket Man" (1972) by Elton John, the story of a workaday astronaut on his way to Mars on another mission in a line of several expeditions that keeps him away from home. In other words, the idea that astronauts in the future will be considered ordinary employees who suffer forms of alienation like that of other proletarian workers is not exactly new. The blue-collar astronaut has been around not only in music but also in films such as *Alien*, *Silent Running* (1972), and *Outland* (1981), all of which Jones cited as inspirations for *Moon*.[54] These stand in sharp contrast to the heroic, square-jawed astronauts popularized in *The Right Stuff* and *Apollo 13* (1995).[55] But contemporary context also matters. Changes were afoot in the first decade of the 2000s, a period in which the corporatization of spaceflight got underway, and which no doubt influenced *Moon*, *Europa Report* (2013), and *Ad Astra* (2019).

In that decade famous billionaires announced, with great fanfare, their intentions to transform their considerable private capital into space expeditions that only world governments were able to launch previously. Elon Musk's company, SpaceX, was founded in 2002, but it did not begin launching robot cargo ships to ISS until 2012.[56] The British billionaire

Richard Branson founded Virgin Galactic in 2004 for the purpose of providing space tourism, with Branson himself making a much-vaunted, though unimpressive, spaceflight in July 2021 that did not even manage to enter Earth's orbit. Jeff Bezos, the CEO of Amazon, rode his phallic spaceship Blue Origin into suborbital space about a week later.[57] The so-called billionaire space race—a pale, ersatz version of the US–Soviet space race—achieved less, and for more money than Yuri Gagarin and Alan Shepard's first flights six decades earlier.

Public opinion, at least in the period under study, is considerably rosier than mine on the partnership between NASA and private corporations, which are heavily subsidized by taxpayers, although it is not entirely clear whether respondents know this. Private companies such as SpaceX are underwritten by subsidies, propped up by existing infrastructure, and kept in business by generous contracts—all paid by American taxpayers. According to the 2018 Pew Research poll, a third of those surveyed believed that "private companies will ensure that enough progress is made in space exploration, even without NASA's involvement," but a majority (51 percent) have little to no confidence that the private sector will prevent the proliferation of space debris (or space "junk"). The increasing presence of private companies in spaceflight ventures, which over 70 percent of Americans are either fairly or a great deal confident will be profitable, had not been possible without government action that ensured the contraction of NASA's own programs.[58]

The Bush administration announced in 2004 that the US Space Shuttle program would end after the construction of the ISS, a policy that the Obama administration upheld when Congress passed the NASA Authorization Act in 2010. The last Shuttle, Atlantis, flew in 2011, and from 2012 until 2020, US astronauts hitched a ride on the Soyuz to reach the ISS until SpaceX had the ability to start sending astronauts of their own in 2020. The Brad Pitt vehicle *Ad Astra* shows how a partnership between corporations and NASA could look in the future: Roy McBride (Pitt) is taken on a private spacecraft to a bustling lunar base, filled with tourists, to rendezvous with US military and NASA personnel to travel outward toward Mars (and beyond) to search for his missing father (Tommy Lee Jones). The perspective in *Ad Astra*, however, is decidedly bleak. The film's sprawling lunar base might seem technologically advanced but is in fact in a highly contested warzone with pirates aboard lunar rovers ready to kill astronauts for profit. There is also a mall inside the base, for space tourists and shoppers. It has become a capitalist nightmare. It looks like hell.

Though films like *Ad Astra*, *Moon*, and *Sunshine* were artistic achievements rather than big, international moneymakers, *Gravity* (2013) and

The Martian (2015), both well-received box office hits worldwide, were disaster films that referenced environmental problems.[59] They center on NASA and American astronauts, but they also acknowledge the presence of a major competitor: the China National Space Administration (CNSA). China's government long had an interest in developing space technologies, such as satellites and rockets, since the late 1950s. Famously, China was not offered a partnership with ISS because of the US government's concerns about national security, and thus developed its own spaceflight technologies, which include the space station Tiangong, first launched in 2011, and the spacecraft Shenzhou, a souped-up copy of the Soyuz.[60]

Both Tiangong and Shenzhou, albeit without Chinese astronauts, make memorable appearances in *Gravity*. The instigating action of *Gravity*'s propulsive plotting is the threat of space debris colliding with the NASA astronauts Dr. Ryan Stone (Sandra Bullock) and Matt Kowalski (George Clooney) while they are working on the Hubble telescope. The increased media attention to the issue of space junk and its relationship to climate change undoubtedly shaped the concerns of the movie.[61] As several scholars have shown, orbiting debris makes it potentially lethal to launch missions into space. The equipment that burns up in the Earth's atmosphere upon reentrance also contributes to rising levels of carbon dioxide.[62] The space junk in *Gravity* destroys not only the Hubble but also the Shuttle *Explorer* and the astronauts onboard (it is difficult not to see this as a metaphor for the cancellation of the Shuttle program two years earlier, or a reference to the Shuttle disasters of 1986 and 2003). Eventually, Dr. Stone makes it to the ISS, which is also, in a spectacular sequence, destroyed by the debris. She escapes on a Soyuz and eventually makes it to the (evacuated) Tiangong, where she boards a spare Shenzhou to land on Earth.[63] One of the most potent images that viewers take away from watching *Gravity* is that space is *crowded*. Bullock's character might be isolated in what is essentially a survivalist story, but if anything, the film shows the presence of active, international space programs and an outer space that feels *lived in*. Space is a busy place; it is cluttered, environmentally hostile, and full of traffic.

The success of *The Martian*, released two years later and based on Andy Weir's best-selling novel of the same name, also, in its own way, deals with environmental anxiety, although *The Martian* frames the resolution to this anxiety optimistically.[64] Much like *Gravity*, *The Martian* is also a story about rugged individualism and survival in hostile environmental conditions against considerable odds.[65] Unlike *Gravity*, the characters in *The Martian* inhabit a fantastical future, in which NASA is a robust, well-funded agency responsible for the colonization of Mars.

It is a NASA that is in the middle of a series of missions that mirror those of the Apollo program. In fact, the film begins in *media res*, with a harmonious group of American astronauts (and one German) happily working side by side on the Martian terrain, mining soil, and conducting experiments. A sandstorm cuts a monthlong mission short and the crew inadvertently leaves Mark Watney (Matt Damon) behind on his own to "science the shit" out of his predicament.[66] The film and book imply a near future in which exploratory missions to Mars, including its early colonization, could lead to more human settlement. As Watney says in both the film and novel, "They say once you grow crops somewhere, you have officially 'colonized' it. So technically, I colonized Mars," a statement that echoes the destructive colonialism of global, imperialist powers.[67] But the colonization of Mars is not shown, at least narratively, as a necessary dystopian measure because Earth is in danger of becoming uninhabitable. If anything, people back on Earth are reasonable, caring, and living in a world where there seems to be boundless enthusiasm about space exploration and the future of humanity.

The Martian is utopian in its storytelling. At the end of the novel, Watney is saved because of the multiyear collaboration between NASA, Jet Propulsion Laboratories, and CNSA, which prompts him to reflect on the state of humanity: "The cost for my survival must have been hundreds of millions of dollars. All to save one dorky botanist. Why bother? Well, okay. I know the answer to that. Part of it might be what I represent: progress, science, and the interplanetary future we've dreamed of for centuries. But really, they did it because every human being has a basic instinct to help each other out. It might not seem that way sometimes, but it's true."[68] These are words not spoken in the film adaptation, but they describe the future that Watney represents in the film. A utopia of scientists and problem solvers and the possibilities of using science to break out of a problem.[69] It is a world that includes global competitors such as CNSA's scientists, who put aside politics and open their agency to NASA's personnel to save the life of one American astronaut.

The film shows NASA's chief Teddy Sanders (Jeff Daniels) and flight director Mitch Henderson (Sean Bean) traveling to China and getting a tour of CNSA's facilities. It ends with the agencies working together to send the fictional Taiyang Shen rocket to resupply a NASA spacecraft to Mars. NASA is depicted as the superior agency, with superior science and better politics, and CNSA a bit behind, with its secrecy and technology dating back to the "Apollo 9 era" (as said in the film). But ultimately, it is Chinese technology that assists the Americans and, most important, brings China from the outside and into a global community of science. These story beats certainly made the film sellable in China to audiences there and to one of its key investors, the Bona Film Group, but this Chinese connection was already present in Weir's novel, written years before the adaptation was made.[70] One of the last scenes shows a Chinese astronaut seated alongside one of the American astronauts, Rick Martinez (Michael Peña), on another mission to Mars—symbolically showing how humanity can be united in its quest to colonize beyond Earth.

The very expensive summer feature films of the late 1990s, *Armageddon* and *Deep Impact*, located the sources of destruction in extraterrestrial objects—disasters outside of our control—that American-led ingenuity could avert with superior technology. The Russians remained in the picture, although their technology was shown to be in decline. A noticeable shift appears in the concerns of space movies in the first decade of the twenty-first century, with two prominent themes: the first, a fragmentation of the specter of US dominance and, second, the simultaneous emergence of climate catastrophe as either a backdrop or principal foil in these films.

These themes—internationalism and environmental anxiety—are conveyed in the movie *Geostorm* (2017), which aimed for blockbuster box office and worldwide appeal but failed to emulate the success of previous Roland Emmerich films (the director of *Geostorm*, Dean Devlin, was Emmerich's producing partner). *Geostorm* is a rich text that sums up the major themes in space films of the past decade, framed around a climate catastrophe of never-ending natural disasters. All the world's governments have come together to transform the ISS into a network of highly advanced, interconnected space modules with artificial gravity called the International Climate Space Station (ICSS), which controls a series of satellites, nicknamed "Dutch Boy," which can change the climate in any region on Earth. When the film opens, the US is firmly in control of Dutch Boy, but on the cusp of granting its access to other countries. The plot revolves around Dutch Boy's mysterious malfunctioning, which is orchestrated by an evil secretary of state (Ed Harris), who wants to prevent it from being an internationally controlled technology. From the film's perspective, the bad guys are bad because they fear losing hegemonic control over global affairs, which includes space. A select number of astronauts from Germany, France, Nigeria, and the UK on the ICSS, under the leadership of Dutch Boy's American architect (Gerard Butler), are chosen to investigate the matter. The UK astronaut eventually betrays the crew on the ICSS, from which seemingly dozens of astronauts are evacuated in several commuter NASA Shuttles when the station needs to shut down to take away its control from the villains. In the end, the bad guys are caught, Dutch Boy becomes an international initiative, and humanity saves itself using the best space technology.

What *Geostorm* and other films since around 2005 have shown is a greater acknowledgment—through plotting and characters—of the global and multinodal possibilities of spaceflight. They hold a mirror to geopolitical conditions, including the development of the ISS, the growing presence of many other countries' space programs, the cancellation of the Shuttle program, and the corporatization of space. Notably, China's space program has been more prominent in some of the later films, reflecting its ascendance and gradual replacement of Russia's dominance in space culture. They also show an awareness of the damaging effects of climate change. In movies such as *The Day After Tomorrow*, *Sunshine*, *Gravity*, and *The Martian* in the 2000s and 2010s, the growing awareness of environmental decay, climate change, the accumulation of waste, and the potential for Earth to become uninhabitable—all marks of the Anthropocene—shaped the types of stories told.

Because these are Hollywood films made primarily by filmmakers working within American corporate systems that have vested interests in global box office receipts, they depict NASA as an organization that is at the top of its game, and American technology in its prime, but these stories also integrate astronauts from other countries and space programs as partners in the struggle against a host of environmental problems. This theme is an important one. It is a recognition that the US does not stand alone in space, and that even more harmonious partnerships with other countries, even those in competition with the US, can represent possible futures. This is certainly the message in one of the biggest space movies of the past two decades, *The Martian*, a hopeful film that illustrates how difficult it is to survive on Mars, let alone to colonize it. It is a celebration of the possibilities of publicly funded agencies and scientists working toward a common goal, and that, despite near-apocalyptic conditions, we just might navigate our way out of a sea of detritus and challenging political conditions to forge a better future.

CHAPTER 12

IRONY, SLAPSTICK, GORE

Humorous Fragmentation of the Mythology of Soviet Space Supremacy in Post-Soviet Film

NATALIJA MAJSOVA

The history of space programs has been gaining in popularity as a film topic over the past two decades, especially in spacefaring countries such as the United States and Russia. This development can partly be explained by the recent fiftieth anniversaries of the inaugural events of the Space Age, such as Yuri Gagarin's 1961 pioneering spaceflight and the Moon landing of 1969, and partly by the swift advances in visual effects (VFX) industries that make it possible to produce increasingly realistic depictions of a space mission. In the Russian context, filmmakers' heightened interest in the history of the Soviet space program has correlated with recent cultural policies (of Vladimir Putin's government) that have, since 2000 and especially since 2010, encouraged cultural production that explores especially glorious aspects of Russian history. Scholars specializing in Russian film studies have repeatedly noted the persistence of an imperialist tendency, characteristic of twenty-first-century Russian cinema on the levels of funding policies, genre development, and narratives and iconographies of specific films.[1] This tendency may be seen as a reaction to a range of reflexive, darkly interrogative, and humorous film productions of the 1990s and early 2000s, which sought to destabilize certain persistent Soviet master narratives; including the one about the advent of the Space Age.

Following Nancy Condee's argument about imperialism as an important preoccupation of recent Russian cinema, Justin Wilmes has argued that this tendency is reflected in at least three persistent formal and thematic tendencies: World War II films that glorify the USSR's victory over Nazism, the "new Orthodox blockbuster" that celebrates the distant past and the origins of Orthodox Christianity on Russian territory (and adjacent foreign territories, such as Ukraine, which are considered Russian in these productions), and films that conjoin "traditional imperialism and neoliberal globalism."[2] Wilmes has further stipulated that modern Russian film production reflects "the growing sophistication of ideological cultural products in an era of channeled discourses, release valve policies, and postmodern authoritarianism," using devices, such as irony in humor in markedly complex ways.[3] The number of space-exploration-history-themed feature films produced in Russia over the past two decades, and their mythological undertones indicate that the Soviet space program might be another aspect of the outlined trend to glorify Russian (and Soviet) identity through historicization.

In line with Wilmes's observation about the growing sophistication of the operations of state ideology in Russian film, this chapter sets out to elucidate the complexity of the uses of humor in Russian post-Soviet feature films preoccupied with the Soviet space program and released since 1995. The chapter aims to outline the structures, functions, and effects of irony in different registers of space-themed cinema and to assess the capacity of humor for destabilizing some of the key tropes and narratives of the Soviet imaginary of the Space Age and the imminent communist future. The chapter aims to show that humor, especially irony, has the capacity to function as both a decolonization strategy that fragments the dominant (Russian) master narrative about the glory and greatness of the Soviet space program and an element of neocolonial attempts to reinstate this master narrative.

THE FUNNY TURN IN POST-SOVIET ASTROCULTURE

Modern Russian space-themed film production comprises films of diverse genres and their combinations. Even the basic distinction between documentary and fiction is sometimes blurred because fact and speculation have been closely intertwined in narratives about space exploration ever since the advent of cinema and the beginnings of space programs in the early twentieth century. This chapter argues that space-themed films are ambivalent; many science fiction feature films make implicit truth claims, and many documentaries incorporate tropes developed in and for science fiction films. Bypassing the documentary/fiction dichotomy, and foregrounding the above-mentioned ambivalence of space-themed films as a feature that facilitates both mythmaking and the debunking of master narratives of cosmic triumphalism, this chapter engages with feature-length films as generally more publicly resonant than shorts, setting out to explore the artistic and entertaining dimensions of space-themed cinema, such as engaging ideas, interesting characters, and inspiring stories. Scholarship has demonstrated that this aspect of popular astroculture and its history are important because they help trace various interpretations, attitudes, beliefs, and tropes related to the Space Age, its history, and its popular imaginary.[4]

Throughout the twentieth century, popular astroculture was largely associated with the genre of science fiction, as the inaugural events of the Space Age closely resembled bold ideas conceived within this genre. In the popular imagination, links between science fiction film and the Space Age were reinforced by the film's capacity to convincingly visualize events (e.g., spaceflight or the Moon landing) long before they could materialize.[5] Finally, especially in the Soviet context, until the late 1970s, science fiction cinema developed in close consonance with the educational genres of popular science and youth film, which sometimes even overlapped.[6]

The Normative Framework of the Soviet Imaginary of Outer Space

Especially during the "dawn of the Space Age" (1957–1969), visualizing outer space and narrating space exploration (and conquest) were considered a part of the larger Soviet citizenship project, which is evident not only in popular astroculture, such as science fiction cinema, but also in reports about the first feats of the Soviet space program (reportages, hand-colored photos of cosmonauts, etc.).[7] The aesthetics of space exploration and space conquest in science fiction and documentary genres were therefore closely interlinked and bound by the same normative framework and distinct tropes, which integrated the Soviet space program into the communist modernization project.

A good case in point is the report in the Soviet popular science journal *Nauka i zhizn'* on the director Pavel Klushantsev's 1957 popular science film *Road to the Stars*, in which the launch of the Sputnik-1 satellite was narrated and visualized as the direct outcome of decades of space-oriented thought, and as the first step toward a communist, Soviet Space Age.[8] No documentary footage of the highly confidential Soviet space program was used in this film, but Klushantsev's fictional visualizations were first interpreted in the context of the factual history of the early days of the Space Age and then embedded into speculations about the future, narrated in an authoritative and predictive tone.

Completely dependent on state funding and subject to rigid content control, especially until the 1970s, Soviet space films generally depicted the Soviet Space Age as reliant on a highly traditional society with rigid hierarchies, separating the old from the young, and men from women. Society was shown as governed by reliable (international, but Soviet-dominated) communist authorities, and cosmonauts as highly modernized "(wo)men of the future" who performed especially difficult jobs but were also ordinary family men and women outside of work.[9]

This imaginary of the Space Age, consolidated in Soviet astroculture between the launch of Sputnik-1 and the Moon landing, was peaceful, but also rather solemn. The serious atmosphere of Soviet space-oriented astroculture in general and especially of pre-1980s cinema has been noted in existing scholarship, as has the gradual transformation of space-related chronotopes (constellation of time and space, which partly determines the characters' actions and plot) after the American Moon landing in 1969.[10] In contrast to the first Soviet feature films concerned with space, *Aelita: Queen of Mars* and *Cosmic Voyage*, these later films—in the genres of popular science, science fiction, or space program history— did not merely entertain.[11] Equally, if not primarily, they were intended to amaze, convince, and inspire spectators to contribute to building the bright future that the films depicted.[12] These goals reflected both the importance of the Soviet space program to the citizenship and modernization project of the Communist Party, and the genre constraints of socialist realism, which haunted Soviet science fiction film over a decade after its condemnation by Nikita Khrushchev at the Twentieth Congress of the Communist Party of the Soviet Union in 1956.[13]

Humor as Subversion in Late Soviet Space Cinema and Its Post-Soviet Legacy

By the time of the Soviet Union's collapse, the "serious" sensibility found in these films was replaced by humor and irony. One can, of course, ask whether anything about the Space Age can be laughed about, and consequently, whether humor is at all important to discuss in the context of astroculture. Modern theories of humor maintain that laughter can be elicited through manifestations of incongruency—that is, misalignments between expectations, customs, or beliefs, and actions. Such incongruency can occur on the basic level of physical activity, resulting in "slapstick" phenomena, such as a protagonist slipping on an unnoticed banana peel, or on more complex levels, in the domain of our understanding of the world, resulting in irony. Irony can occur along two axes: a horizontal axis (self-irony, self-empowerment, and peacefulness), and a vertical axis (where irony is used to highlight feelings of hurt, humiliation, and power disbalance).[14] Systematic employment of irony along the vertical axis throughout an entire cultural production, such as a film, painting, or performance, is defined as satire.

Satire is not an uncommon or new element of global astroculture, even if we limit ourselves to film production. Ever since Georges Méliès's trick film *A Trip to the Moon*, humor has played an important role in astroculture, both as a plot dynamization mechanism (slapstick elements) and as a strategy for questioning the rationale behind space programs and idealist tropes, such as the astronaut as the modern hero (through irony and self-irony).[15] In this context, the seriousness of Soviet space cinema in various genres (popular science, biopics, and science fiction), which makes use of humor only sparingly, completely avoiding irony until the late 1970s, is thought-provoking.

Many official Soviet and Russian space policy documents are still secret, so it is difficult to directly pinpoint a direct correlation between official (post)Soviet policies regarding the popularization of the Space Age, the Soviet Union's space program, and (post-)Soviet cinematic depictions of outer space. However, available inventories of the archives of Soviet and post-Soviet film production indicate that space-themed science fiction cinema, and films at the intersection of science fiction and popular science were especially well funded by the Soviet state in 1957–1992, resulting in approximately 2.4 Soviet-produced science fiction films per annum, around 50 percent of which were about space programs and space exploration.[16] Feature films about the history of the Soviet space program were significantly fewer; the website of the Russian Space Agency's television studio, TV Roscosmos, for instance, lists nine such films, produced in 1950–1991.[17] Moreover, space exploration and related tropes were not critically reflected upon in a humorous manner until the late1980s and the release of the cyberpunk space satire *Kin-Dza-Dza!*[18]

Post-Soviet cinema about outer space has demonstrated a sharp break with the outlined priorities, strategies, and tropes of the Soviet era.[19] In 1992–2010, for example, only eight space-themed films were produced, four of these in the genre of science fiction (around 50 percent of the overall production rate of science fiction films), and four historical dramas. All these films, even Fedor Bondarchuk's 2009 commercially produced science fiction blockbuster dilogy *The Inhabited Island I* and *II* were highly allegorical and employed irony as a powerful instrument of social critique, interrogating the mythology of the Soviet space program

as a sacred project and a symbol of the USSR's moral purity and technological superiority.[20]

In contrast, the following decade (2011–2021) was marked by both
the expansion of commercial film production in Russia and an increasingly rigid cultural policy advanced by Vladimir Putin's government,
which encouraged nationally affirmative films glorifying Russia's history
and present. In this context, the popularity of science fiction grew, testifying to the revival of the local film industry and considerable advances
in the VFX industry, but the fascination of science-fiction films with
space exploration dwindled, reflecting global trends.[21]

In contrast, Russian film-funding organizations' interest in the history of the Soviet space program and in Soviet science fiction increased;
incidentally, the five space films produced in the 2010s were all historical
blockbusters, supported by stakeholders like national television channels, national film-funding mechanisms, and the Russian Space Agency—Roskosmos. Four of these were based on figures or episodes that
marked the Soviet space program, and one was a science fiction production. Interestingly, all these big-budget productions employed humorous
elements, but to a different effect from their predecessors; instead of a
social critique, they contributed to a neocolonial revival of the grand
narrative of Soviet supremacy in the space race and the moral superiority
of the Soviet space program.[22]

Accordingly, two types of humorous approaches to the Soviet space
program coexist in post-Soviet Russian film production, populating the
landscape of nostalgic popular-cultural production about the Soviet
past. Svetlana Boym has argued that nostalgia for bygone epochs can
be categorized as either reflective (a positively charged critical kind of
remembrance of the past) or restorative (an uncritical glorification of
the past).[23] However, this chapter demonstrates that post-Soviet Russian
astroculture complicates this useful distinction.

IRONIC TAKES ON THE SPACE AGE IN RUSSIAN
POST-SOVIET CINEMA

While Soviet traditions of science fiction and popular science films
about the Space Age as well as related biopics and historical dramas have
been oriented toward the future, most thematically similar post-Soviet
film productions have thus far focused on historicizing this image of the
future. Although a new trend has been indicated by several productions
since 2017 (e.g., *The Attraction*), which are based on new, original screen-
plays, most science fiction films, such as the aforementioned *The Inhabited Island* dilogy, have either focused on the Soviet past (for example, *First

on the Moon, *Sputnik*) or have been new adaptations of Soviet science fiction literature and film (for example, *The Inhabited Island* dilogy and the animated film *Ku! Kin-Dza-Dza!*).[24]

The surge in non-science-fiction feature films about the Soviet space program since 2004 points in the same direction: outer space, once spontaneously associated with the future, appears to have become anchored in the Soviet past. The trend to historicize the of the twentieth century is not a Russian peculiarity (cf. Cuenca, chapter 11 in this book), but it is possible to argue that Russian cinematographers are especially invested in exploring whether the Soviet space program was as pure in its greatness, glory, and achievements as Soviet media reports and science fiction had suggested.

Transition and Internationalization in the 1990s

The first post-Soviet Russian production about the Space Age, Dmitry Astrakhan's *Fourth Planet*, produced in the context of economic hardship in the film industry, cofounded by both the national film studio Lenfilm and smaller, commercial stakeholders, was an intertextually rich production that combined existential themes from Andrei Tarkovsky's 1972 *Solaris* with realpolitik, depicting an Amero-Soviet crew encountering extraterrestrial intelligence, reminiscent of the sentient ocean from *Solaris*, on Mars.[25] Hallucinating severely due to the activities of the extraterrestrial presence, the crew reflects on the Soviet past through tropes from Soviet film, some directly reminiscent of optimistic, future-oriented science fiction films from the 1960s.

Although the film was formally set in the post-Soviet early 1990s, its message had little to do with the present. At best, it offered a strategy for coming to terms with the idealism and unfeasibility of Soviet dreams about a bright future of interplanetary communism under the auspices of the USSR. In this sense, the somewhat melancholic *Fourth Planet* was also satirical; it highlighted and scrutinized the incongruity between the post-Soviet present and haunting older visions of this present.

Fourth Planet alludes to the Soviet vision of the Space Age future using the popular Soviet song "Apple Trees Will Bloom on Mars." This song, composed in 1962 by Vano Muradeli to the lyrics of Evgeny Dolmatovsky, was produced for the Soviet science fiction film *Toward a Dream*, which explores a young engineer's dream about the nearing Space Age and contact with intelligent and benevolent aliens.[26] There, "Apple Trees Will Bloom on Mars" is played when the spectator is transposed into (and, in the end, out of) the fictional world of the engineer's dream. The refrain professes: "To live and to believe is wonderful. Unprecedented ways lie before us. Cosmonauts and dreamers claim that apple trees will bloom on Mars, too!" This refrain encourages the spectator to identify with the young engineer and his dream. In *Fourth Planet*, however, its effect was remarkably different; embedded into a post-Soviet cosmonaut's hallucinations, it sounded both satirical and melancholic.

The described melancholic sentiment picked up on and explored in *Fourth Planet* came to dominate over a decade of Russian film production about the Space Age spanning the 1990s and early 2000s, preceding the more general and more positive atmosphere of nostalgia for socialism that came to characterize a significant segment of Russian film production in the twenty-first century. In contrast to other aspects of the Soviet past and Soviet history, which were openly and heavily criticized in Russian cinema of the 1990s and early 2000s, the Space Age was reflected upon as a semifictional construct—a memory of a complex fairy tale, which should have had a beautiful ending.

Incidentally, this assessment of the Space Age was also captured in Andrei Ujica's 1996 German-produced film *Out of the Present*—one of the first (documentary or fiction) feature films in history that included footage shot in the Earth's orbit.[27] These shots, coordinated from Earth by Andrei Tarkovsky's cameraman Vadim Yusov, were made in 1991–1992, during the collapse of the USSR, when the last Soviet cosmonaut Sergei Krikalev was aboard the Mir space station. Ujica later edited them into a pensive documentary, aligning Krikalev and his colleagues' daily activities at the space station, such as exercising, performing scientific experiments, eating, and grooming, with shots of Earth from space, documentary footage from Moscow, and animated science fiction sequences. The soundtrack, consisting of dialogues from the space station, later comments given by the astronauts and cosmonauts, and an eclectic musical score, ranging from techno to classical music, was also created in postproduction.

Out of the Present captures a novel, international dimension of (post-)Soviet astroculture. In contrast to the standards of Soviet space-themed cinema intended to glorify the USSR, its space program, and its actors, *Out of the Present* exhibits documentary footage of the quotidian activities of the last Soviet cosmonaut, stranded in orbit during the collapse of the USSR—a human being in a precarious situation rather than a superhero. This film was probably the first to offer a novel, markedly ironic cinematic take on the proverbially serious topic of space exploration—and especially on the Soviet space program—irreverently combining documentary footage with science-fiction-style editing and critical sociopolitical reflection.

Out of the Present is difficult to classify in terms of genre; it delivers a cryptic message about the nature of time and space. Cooped up aboard the space station, the cosmonauts and astronauts seem almost unusually calm, as if attuned to the slow rhythms of galactic change, and too far away to witness the chaotic, quick transformations taking place on Earth. Accordingly, when Krikalev is asked to comment on the political events "below," and about his fate as the last Soviet cosmonaut, he famously replies: "For a cosmonaut, the speed with which day and night and the seasons fly by is more impressive."

Suspended in orbit due to political circumstances that delay his return to Earth for six months, Krikalev is literally jolted "out of the present": out of a linear and anthropocentric temporality.[28] The film thus questions the very sensibility of humanity's preoccupation with the future evident in science fiction history and space programs, both of which are hinted at by the animated sequences and the soundtrack. Ujica's juxtaposition of life as seen from space and life as seen from Earth foreground two different, misaligned temporalities, producing a satirical effect and exposing as fiction the heroic narrative of Soviet cosmonautics as a "road to the stars" and to a bright communist future.

Smiling about Past Potentialities in the Early 2000s

In the early 2000s, satire became a popular device used by filmmakers such as Aleksey Uchitel', Aleksey German Jr., and Aleksey Fedorchenko to interrogate the future-oriented temporality of the (Soviet) space program in a series of innovative films about the history of the Soviet space program: *Dreaming of Space, First on the Moon*, and *Paper Soldier*.[29] Fedorchenko's mockumentary *First on the Moon*, cofunded by a national film studio and the Ministry of Culture, is the most significant of the three for the purposes of this chapter, because it relies most heavily on its humorous aspects.[30]

Like *Fourth Planet* and *Out of the Present, First on the Moon* marvels at the history of the Soviet space program as a sociopolitical construct. Specifically, this film sets out to showcase the various mechanisms that have allowed the production and reproduction of certain space-oriented myths. To avoid ambiguity, the mockumentary, which relies on the skillful simulation of black-and-white pseudo-archival footage, examines the memory of the "early days of the Soviet space program," which included fictional top-secret experiments from the 1930s and the fictional cosmopilot Ivan Kharlamov's flight to the Moon, which accidentally ends somewhere in Chile.

The film, which received the award for Best Documentary at the 2005 Venice Film Festival, confused various groups of spectators, especially in Russia, because it was the first film to directly mock the Soviet nation-building mythology of Soviet supremacy in space. Moreover, it created an unprecedented liminal position, refusing to slip either into parody or into a serious conspiracy theory.

First on the Moon exploits the conventions of the documentary and insists on the truthfulness of all testimony—for instance, of the first Soviet cosmopilot who allegedly flies to the Moon, has an accident that lands him in Chile, is admitted to a psychiatric ward in Chita as he makes his way back to Moscow, and ends up playing the part of Prince Alexander Nevsky in a circus. In fact, *First on the Moon* deconstructs the very mechanism of mythmaking, possibly inspiring other productions, such as *Houston, We Have a Problem!*[31] While the deconstruction offered by such productions tends to be humorous, the humor employed in *First on the Moon* is not very sympathetic to the characters involved. Presented as elements of a heroic narrative and, later, as "surviving specimen," the protagonists of *First on the Moon*—mostly former cosmopilot candidates—appear completely mythological, defined by their contribution to that narrative, rather than by simply being human.

Moreover, the narrative has a carnivalesque quality: it presents a dance of bizarre, eccentric figures who had once participated in a common task and are now once again brought together in the name of this task. While the characters are rendered powerful, since, as witnesses, they speak the truth, they remain powerless at the same time. Namely, the truth that they speak in the mockumentary is nonsense, because they are caught up in empty signifiers that drive the narrative. This carnival is the core dynamic of the film. The visuals are accompanied by Soviet patriotic songs and tones reminiscent of Soviet electronic music that are common in Soviet science fiction films. The shots, on the other hand, present a different kind of bricolage: historical data are shown alongside pseudo-documentary montage and footage from Soviet space films. The only footage in *First on the Moon* that is not ambivalent are images from Vassily Zhuravlev's 1936 fiction feature *Cosmic Voyage*, also examines the possibility of flying to the Moon.

The only firm anchor of *First on the Moon* is therefore, ironically, fiction, which, in postmodernist terms, does not replace the reality that is lost and perhaps never was, but appears as an image, pointing to a void behind closed curtains. Accordingly, the affirmatively fictional shots from *Cosmic Voyage* are an especially important element of *First on the Moon*: they guarantee that there remains a little something to be trusted: at least some of the fiction in this mockumentary is truly fictional. Other shots mainly maneuver between fiction and reality—for example, footage of the cameras supervising the cosmopilots. Even excerpts from advertisements for these cameras are present in the movie, making the film even eerier: everything is shown to be contrived. Brian Willems has convincingly argued that *First on the Moon*'s aim is to demonstrate the necessity of exploring the past as an arena of potentialities. In the context of this interpretation, the film reexamines what it meant to claim that the mythology of the Soviet Space Age future materialized in an unanticipated way and generated unanticipated pasts.[32]

The three examples of post-Soviet cinematic takes on the Soviet space program examined in this section demonstrate how humor provided a powerful decolonial trajectory for filmmakers who sought to reflect on the history of the Soviet space program, fragmenting its triumphalist master narrative in at least three respects. *Fourth Planet* pioneeringly reflects on the end of an era dominated by the Soviet version of linear, modernist temporality, as it mixes science fiction tropes and elements of space history, aligning the two as two types of fictional narratives. While the history of Soviet science fiction cinema has abundant examples of films that are intended to prophesize, *Fourth Planet* outlines the limits of this "explanatory potential," hinting that, perhaps, the Soviet space program relied on science fiction more heavily than science fiction had relied on the space program. This was an emancipatory gesture for (post-)Soviet science fiction as a genre, freeing it of the prophetic function that it had carried for much of Soviet film history. *Out of the Present* and *First on the Moon* exhibit a similar preoccupation with the relationship between fiction and reality, the former adopting the highly original perspective of a cosmonaut as a worker on the Mir space station, and the latter stipulating the mythmaking mechanisms that the Soviet space program and its cultural memory had relied on.

HUMOROUS INTERPRETATIONS OF THE SOVIET SPACE PROGRAM IN RUSSIAN SPACE BLOCKBUSTERS

If Russian cinematography of the 1990s and 2000s tended to rethink the space-oriented notion of the Soviet future along the axes of fiction, genre, media, and various actors that had previously been considered as marginal, the 2010s signaled a shift in priorities. While the overall preoccupation with the history of the Soviet space program continued to dominate reflections about the future of space exploration, the perspective on this history changed dramatically.

Generously funded by state institutions, biopics about Yuri Gagarin (*Gagarin: First in Space*) and the chief designer Sergei Korolev (*Chief*), and history thrillers, such as *Spacewalk*, portrayed their protagonists as heroic, patriotic, courageous, strong-willed men, who persevered and followed their dreams for the better of humankind, against the backdrop of the horrors of World War II, and despite various hardships.[33] Their feats were framed within action-driven narratives situated in a highly stylized and idealized version of the Soviet past, populated by supportive characters, such as simple, but loving mothers and docile wives.

Formally, these box office blockbusters, often timed to premiere around important Space Age anniversaries, presented a mix of classical Soviet science fiction films, such as *Andromeda Nebula* and Western superhero narratives.[34] While their "historical" content seemed to indicate that Soviet history could well match up to Western fiction, it also reaffirmed the statement of their pensive predecessors: that the temporality of space exploration in the post-Soviet context is primarily oriented toward the past, and not the future.

Recent Russian space-history blockbusters share another trait with their satirical predecessors: their attitude toward the history of the Soviet space program was not as solemn as the one exhibited by Soviet films (science fiction, popular science, and biopics alike). In fact, Russian space-history blockbusters of the 2010s used humor as a mechanism for reinforcing, rather than destabilizing the master narrative of Soviet supremacy in the space race.[35]

Salyut 7, a unique film that addresses a period of Soviet space history that was no longer dominated by Sergei Korolev, taking place almost two decades after his death, is a representative case in point. *Salyut 7* focuses on a dramatic episode from 1985, when the Soviet cosmonauts Viktor Savinykh and Vladimir Dzhanibekov had to fly urgently into orbit to perform risky repairs on the unresponsive Salyut 7 space station.[36] The film was released around the fiftieth anniversary of Aleksey Leonov's 1965 spacewalk, outperforming *Spacewalk*, a film about this event, at the box office. Publicly available reviews reveal that *Salyut 7*'s lighthearted tone and humorous employment of stereotypes about the Soviet past were especially appreciated by the spectators.

In terms of structure, *Salyut 7* clearly draws inspiration from many other films on space exploration, highlighting the contrast between scenes "in space" and those "on Earth," such as *Out of the Present* and *Spacewalk*. However, while *Out of the Present* uses this contrast to demonstrate a misalignment between the cyclical temporality of space and the rushed, chaotic temporality of the human world on Earth, *Salyut 7* harnesses the juxtaposition between the two spheres to a different effect. If *Out of the Present* explores space exploration in relation to a certain present and to a certain ideal of the future, *Salyut 7* is completely focused on representing the Soviet past.

For example, at the beginning of the film cosmonaut Fedorov's romantic partner, the mother of his child, asks how he would describe his life in the USSR to the people of Madagascar if he accidentally landed on their island. "My daughter, football, and building communism," he responds, charting the coordinates of the Soviet world as timeless, and curiously set in a nostalgic, stylized version of the late 1950s and early 1960s, disregarding the fact that, "based on real events," *Salyut 7* is set in the 1980s, unlike films about Gagarin and Leonov.

In space, the protagonists, the cosmonauts Vladimir Fedorov (Vladimir Vdovichenkov) and Viktor Alyokhin (Pavel Derevyanko) are portrayed as longing for Earth—the Soviet past and its promises, including quiet family life, and the end of the Cold War (in the USSR's favor). They are part of a world built around an opposition between the USSR and the US, echoing *Spacewalk*, *Gagarin: First in Space*, and other Russian productions on the history of the Soviet space program. The "American" world remains inaccessible, even from the space station; its existence is only hinted at in conversation, through remarks about current events.

In contrast to standard science fiction and historical representations of the Soviet space program, however, *Salyut 7* portrays the Soviet past as a world that relied on autonomous, superhero-like cosmonauts. While this trope is tentatively explored in *Spacewalk* too, it remains clear that the cosmonaut, even a stubborn and opinionated one like Aleksey Leonov (as portrayed in *Spacewalk* by Evgeny Mironov) has to obey the mission control center headed by Sergei Korolev, the mythical mastermind of the Soviet space program. *Salyut 7*, however, is set after Korolev's death; the absence of an equally authoritative figure in mission control presented the director with an opportunity to reinvent the (male, heterosexual, patriotic, and tech-savvy) cosmonaut as an autonomous superhero.

Moreover, in contrast to most previous film representations of Soviet cosmonauts, the protagonists of *Salyut 7* appear entertaining: competent, but also amusing when looking for unconventional solutions. The film focuses on their ordeal at the unresponsive space station, diluting moments of despair with comedic insertions intended to paint the Soviet world and its inhabitants in positive—that is, warm, colorful, empathetic, and idealistic, if somewhat clumsy— brushstrokes.

Thus, the cosmonauts regularly overrule recommendations and even instructions given by the mission control center. They also act irrationally at times, for instance indulging in a cigarette, knowing that smoking wastes oxygen. For courage, they down a shot of vodka, remarking that there are no police in space to stop them. In the end, they still manage to single-handedly repair an unresponsive space station using a sledgehammer. Incidentally, the chief of mission control also finds a sledgehammer and uses it on the prototype of the station in his office, highlighting his own helplessness and uselessness—and the film's reference to *Armageddon*, which also includes a comic episode of a Russian cosmonaut desperately trying to use basic tools to repair space equipment.[37]

The frequently highlighted stereotypes of Soviet people, such as their affinity for alcohol and tobacco and their disregard of standard procedures, tie the heroism of the cosmonauts in *Salyut 7* to their Soviet provenance to the extent that they appear as ahistorical embodiments of a distinct national characteristic. Fedorov and Alyokhin, offbeat everyday heroes, are portrayed as embodied new communist men, committed to their mission, and capable of completing it regardless of the equipment at hand. In this sense, *Salyut 7* appears to respond to *First on the Moon*, which uses satire to question the veracity of such glorification by reaffirming it.

Salyut 7 ends on an ambivalent note. Alyokhin and Fedorov are shown sitting on top of the space station, waving at the American space vessel that is passing by. This comical image serves as a reminder that the cosmonauts have effectively risked their lives to avoid having to ask the US for help (and granting the Americans access to their top-secret technology). More observant spectators can also note a haunting reflection of the Moon in the cosmonauts' visors: the ultimate destination remaining out of reach for the Soviets. This scene produces a powerfully nostalgic effect, employing a comical situation that highlights the bravery and excellence of Soviet cosmonauts, combined with a hint at "what could have been" had the state not collapsed.

This way of using humor has been quite widespread in recent Russian films centered around a restorative nostalgia for the Soviet regime and the nation-building myths that supported it, corroborating Wilmes's observation about the complexity of modern cultural forms, which are highly intertextual and address audiences that are presupposed to be extremely sophisticated consumers. Modern Russian blockbusters "based on historical events" and supported by various national archives, the Russian Space Agency, and commercial film studios appear to address fans and skeptics of these historical events equally. Their preferred strategy is to convince through entertainment that is constructed around globally understandable, Hollywood-style action-driven narratives, while, at the same time, using props and tropes that reflect on both the material shortcomings and the broad spiritual horizons attributed to the Soviet past.

NOSTALGIA AND HORROR IN *SPUTNIK*

The post-Soviet films discussed throughout this chapter have mostly taken issue with whether the feats of the Soviet space program should be considered true or not, and to what extent they should be glorified, effectively using various registers (both auteur and genre film) to reflect on the complex relationship between space-oriented science fiction and real space programs.

Science fiction proper, on the other hand, appeared to have taken a back seat regarding discourses on outer space, at least until 2020 and the release of Evgeny Abramenko's debut feature, *Sputnik*—the first-ever space science fiction horror in the history of Soviet and Russian cinema. Like all the abovementioned post-Soviet films on outer space, *Sputnik*, a commercial project supported by the film industry magnate Fedor Bondarchuk and oriented toward international film festivals, is set in the Soviet past. It is a film about a fictional Soviet 1983 space mission gone wrong, with one cosmonaut dying upon landing, and the other bringing back a monstrous alien creature, loudly echoing more famous productions, such as Ridley Scott's *Alien*.[38] In the context of the present discussion, *Sputnik* matters especially in relation to the other films for its contribution to the nostalgic discourse on the Soviet Space Age, which dominates Russian popular culture today.

Specifically, *Sputnik* starts with several clearly nostalgic scenes from the unfortunate space vessel, with cosmonauts singing the iconic Soviet and Russian pop singer Alla Pugacheva's "Millions of Red Roses" (*Milliony alyh roz*), one of the biggest Soviet hits of 1983, the year in which the film is set. A cute matryoshka doll is shown flying around the cabin, while the cosmonauts casually discuss their desire to get home and take a hot shower. Outer space is thus linked to positive memories of socialism, such as camaraderie, friendship, and elements of popular culture.

As soon as the film moves on to the events on Earth, however, this atmosphere shifts completely, exposing the USSR in the 1980s as a cold, unfriendly, violent place ruled by military men and bureaucrats. The surviving alien-bearing cosmonaut Konstantin (Pyotr Fedorov) is portrayed as a Soviet "new man" who has abandoned his disabled child and sent him to an orphanage after his wife's death. The head of mission control, Colonel Semiradov (Fedor Bondarchuk), also hides dirty secrets behind his friendly smile.

In this company of antiheroes, the alien symbiote does not appear especially monstrous or evil. In fact, Abramenko's idea was to emphasize its alienness, that is, its radical alterity and specific goals, rather than its maliciousness. Effectively, the men running the Soviet state and its bureaucratic and military apparatus are presented as the "evil" side, more so than the cosmonaut-alien complex.

Sputnik's emphasis on the Soviet space program as a part of an extensive military apparatus veiled in secrecy and developed with the aim of increasing the state's international importance rather than protecting its own citizens, represents a significant departure from all the tropes related to the Soviet space program in films such as *Salyut 7* and *Spacewalk*. Its portrayal of the unfortunate cosmonaut who, realizing he could no longer live without his alien symbiote, takes his life, is also unprecedented. This plotline conjoins the two nostalgic impulses articulated by the reflexive auteur films of the 2000s and the action thrillers of the 2010s: the impulse to tell a space program story from the perspective of a non-celebrity, and the impulse to reinvent the cosmonaut as an autonomous agent.

Finally, *Sputnik* is a milestone in representations of women in (post-) Soviet space film.[39] The storyline includes a highly competent medical professional, Dr. Tatiana Klimova (Oksana Akinshina), called in by Colonel Semiradov to examine the case and get the alien creature under control. Dr. Klimova assumes a proactive role in the operation, and develops into a multilayered, complex character: an opinionated professional, the cosmonaut's potential romantic partner, his ally, and even a motherly figure. Indeed, at the end of the film, she adopts the dead cosmonaut's son. These characteristics are miles away from the Soviet representations of females in relation to space exploration: they have typically been featured rarely and have assumed silent, one-dimensional, passive human roles (mothers, wives, daughters) or dangerous alien ones.

THE COMPLEX IMPLICATIONS OF LAUGHING ABOUT THE SPACE AGE

The overview of narrative trends in space-themed (post-)Soviet cinema provided in this chapter indicates a great degree of intertextuality, which connects independent films to those financially endorsed by the state, and those popular with the public to those remaining on the margins of fandom and critical appraisal. The narrative and aesthetic choices adopted by film directors at a certain moment in time clearly resonate in later productions, where they are either supported and integrated or contested.

Accordingly, Russian space action thrillers and biopics from the 2010s acknowledged the specificity of the Soviet Space Age's significance as an object of nostalgic introspection proposed by a series of early post-Soviet films in the 1990s and 2000s. These earlier productions tended to regard the bygone Soviet Space Age as a fairy-tale-like construct that needed to be dismantled and reassembled in a way that would acknowledge its reliance on propaganda as well as the dependence of the entire endeavor of creating this "bright future" on numerous people, actions, and devices that were consistently left out of decision-making processes and memory politics. This critical way of exposing the limitations of Soviet-era narratives prevalent in science fiction and other modes of narrating the Space Age produced ironic and satirical effects along the so-called vertical axis of humor production: films like *First on the Moon* employed irony as a device of self-reflection, rather than as a means of humiliating an external adversary.

Blockbusters such as *Salyut 7* were a reaction to this perspective. While they also departed from the realization that the "Soviet space future" belonged to the past, they also uncritically embraced this past and its ideals, using anecdotal details about the USSR and its people mostly as a way of entertaining the spectator and highlighting the moral superiority of the heroes of the Space Age. In doing so, they used humor as a way of amplifying the figure of the cosmonaut—the new superhero of the Soviet Space Age, reinvented in post-Soviet productions as an independent agent, rather than an obedient tool of the Soviet mission control center.

Interestingly, both in critical reflections about the Soviet space program and in attempts to remythologize it, the humorous, laughter-inducing effect is commonly produced by the ambivalence of the nostalgic temporality. Critics have often found it difficult to pinpoint whether these films should be described as science fiction, historical drama (or action thriller), or even documentary, testifying to the interpretative openness of these cultural forms.

Space horror appears to be the newest addition to the palette of Russian space films, as indicated by *Sputnik*, analyzed as the final case study in this chapter. A commercial project, interested in the subversive potential of nostalgia, *Sputnik* offers a threefold innovation. In the context of (post-)Soviet space films, it has pioneeringly offered representations of femininity as a complex matrix of agency; of the USSR as a gloomy and dangerous place; and of alienness as a corporeal and willful aspect of one's subjectivity. These three points may strike the viewer as humorous, especially as they feed into several decades of intertextual dialogue that reflect advances in the post-Soviet popular-cultural imaginary of outer space and the Soviet space program. These include the replacement of the futuristic temporality with a nostalgic one; the gradual shift of focus from events and their architects, such as Korolev, to the laborers involved in producing them; and diverse perspectives on the relationship between science fiction, prognosis, and history.[40] *Sputnik* seems to suggest that there is a thin line between horror and humor; the horrifying events enacted in this film were, to a certain extent, just extrapolations of the humorous hints advanced in *First on the Moon* and *Salyut 7*.

In other words, while humor can certainly be used to distract the spectator from the inconsistencies of a triumphalist master narrative, this effect should not be seen as too profound. If the spectator is attentive enough to reflect on what exactly they found funny, they are bound to hit a roadblock, or, in other words, to find themselves wondering whether what they were smiling at is a laughing matter—or the prelude to an entirely different story.

PLANETARITY, PLANETIZATION, AND THE GLOBAL SPACE AGE

Genealogy and Prolegomena

ALEXANDER C. T. GEPPERT

The events we are witnessing and undergoing at this time are, without doubt, bound up with the general evolution of terrestrial life; they are of *planetary dimensions*. It is therefore on the planetary scale that they must be assessed.

–PIERRE TEILHARD DE CHARDIN, 1946

A "fragment," the *Oxford English Dictionary* informs us, is "a detached, isolated, or incomplete part; a (comparatively) small portion of anything; a part remaining or still preserved when the whole is lost or destroyed."[1] A "cosmic" fragment would then be nothing less than a small particle of the universe, understood as an ordered system of totality, found on Earth. As the contributions to this book demonstrate, albeit implicitly rather than systematically, over the course of the second half of the twentieth century such space particles spread and proliferated around the globe, like astrocultural modern-day equivalents to the age-old meteorites found on Earth.

That the power and allure of outer space as a modern fantasy, project, and product has never been confined to the two spacefaring superpowers alone has increasingly driven much historical research in roughly the past decade. One of the most important objectives of my ongoing inquiry into the history of astroculture, space thought, and sociotechnical imaginaries of extraterrestrial life in postwar Europe has thus far been the will to expand the historiography of outer space, spaceflight, and space exploration by pushing its geographical focus and temporal boundaries beyond the borders of the Soviet Union and the United States during the Cold War.[2] While this intricate task of double decentering is far from complete, it is becoming ever clearer to the rapidly growing number of space historians that the "promise of space" held enormous appeal in countries, regions, and places other than the USSR and the US. Equivalents of the so-called European space history paradox—understood as comprehensive space enthusiasm concomitant with decades-long abstinence from independent human spaceflight activities—were central to the national narratives of other nontraditional spacefaring nations, where outer space has constituted a similarly popular realm of futuristic thinking, other-worldly ambitions and military prowess.[3] Highly successful spaceflight programs, both human and robotic, currently pushed by the governments of China, India, and Japan show that the sheer capability to launch spacecraft continues to be considered the pinnacle of technological modernity worldwide. What is less clear, however, is how those "planetary dimensions" of outer space and global astroculture invoked in the epigraph should be understood. In the following, I offer an initial genealogy of various space "ages," a critique of prevailing periodization models, and a prolegomena to planetizing history.

SPACE AGES

The most fundamental and all-encompassing notion with which historians of science and technology have long operated to comprehend, analyze, and come to terms with human vertical expansion beyond the confines of Earth is the notion of the "Space Age." Usually neither theorized nor defined and hardly ever discussed, this ominous—or enthralling—period in human history is conventionally said to have begun with the launch of the first artificial satellite, Sputnik, on October 4, 1957. Whether—and if so, precisely when—the Space Age has ever ended is even less frequently a matter of debate. Concomitant with a growing realization that the likelihood of there being a future in the stars was increasingly diminishing, in the early 2000s historians and scholars in cultural studies did begin to treat the Space Age as past, yet the degree of conceptual and methodological self-reflexivity remained limited.[4]

Indirectly, the recent flurry of popular excitement about an impending "New" Space Age, a Space Age "2.0," or even an entire "New Space Race" seems to imply that the much-invoked "first" Space Age must definitively have ended, whether this was during the post-Apollo period or in more recent times. Possible end dates would include, but are not limited to, December 1972, with the last spacefarer leaving the Moon; February 1974, when the final Skylab mission returned to Earth; or the end of the US space shuttle program in July 2011. Other, non–US-related end dates would certainly be conceivable, but, again, a corresponding debate has not yet been initiated in public or among experts. For the time being, all "New Space Age" rhetoric remains historically unsubstantiated, their allusions to the "original" Space Age notwithstanding.[5]

One reason that such periodizations, despite the appeal of their clarity, are often grossly misleading is that they are built on historically shaky ground. Alluring, auspicious, and evocative as it may be, the term "Space Age" is less innocuous and more problematic than it appears. Contrary to expectations, it is not a product of what it stands for—that is, a clearly delineated period in human history when outer space was widely considered so central to a society's self-understanding and future development that it would serve to characterize the then-present. Rather, the term was a neologism created more than a decade before Sputnik and a formative element of a mission to realize precisely what it advocated. The "Space Age" is not only older than the proverbial Space Age itself, at least according to the conventional understanding, but—perhaps equally surprising—not of American origin either. It was the British aviation journalist Harry Harper (1880–1960) who seems to have used the term for the first time, as the headline for a celebratory feature article published in the British *Everybody's Weekly* in January 1946. Later that year, Harper authored a book-length study, *Dawn of the Space Age.*[6] For Harper, the promise of space and that of the future were tantamount, and the rocket was the ideal machine for the timely realization of both. "We have had an age of steam-power, an age of electricity and of the petrol engine, and an age of the air, and now with the coming of atomic power the world should, in due course, find itself in the space age," he proclaimed with a mixture of exuberant enthusiasm and pressing demand. While specific technological innovations determined the first two ages, the demarcation in the latter two cases was spatial rather than technical, with only the Space Age offering unprecedented fulfillment: "This should be the greatest age of all."[7] With the commercial success of Harper's book, both the term and the metaphor of its impending "dawn" were out in the world.

Why should a seemingly inconspicuous notion such as the "Space Age" then prove so convoluted? If one applies Reinhart Koselleck's classical distinction between historical concepts and historiographical categories, it instantly becomes clear that the "Space Age" is a residue, an integral part to be found in the source language rather than an analytical category construed and defined ex post, to describe and make sense of a paradigmatic rupture in our historical understanding.[8] The fact that the term has not been thoroughly conceptualized, let alone theorized, and thus has not been reviewed for its suitability as a historiographical master category and epoch designation would not necessarily have to be a problem—but in this case, it is. The signifier preceded the signified. The "Space Age" was introduced to proclaim and effectively help produce what it stood for and thus itself played a key role in the pro-space rhetoric and spaceflight propaganda of the 1950s and 1960s. The Sputnik moment was, therefore, not its prelude, but rather a much-awaited substantiation; it fulfilled expectations shaped much earlier. Still marked by the pre-scriptive-projective impetus ascribed to it in 1946, the term implies the far-reaching, yet hitherto unfulfilled promise of a limitless future in the stars, and it continues to do so up to the present day. Accordingly, there is no neutral use of "Space Age." Historians who do engage the term as a temporal marker take advantage of its vague, albeit elusive glitziness without deeper reflection on its precise outlook and pass off a historical concept as a historiographical category. Worse, they effectively engage in renewing the term's embedded futuristic promises.

To make things even more complicated, historians of that large-looming Age of Space—again, me included—have recently begun to ponder the notion of what a "global" incarnation of that Space Age could be.[9] Once again, such a conceptual innovation is not the result of a prior systematic debate. So far, the term does not stand for a clearly delineated and widely discussed research agenda. Nonetheless, the intention behind introducing a "Global Space Age" is obvious: to signal that there were other times and other places when and where outer space—or rather the prolonged promise of future expansion into space and the expected return benefit on Earth—was considered vital to a given society's self-understanding. What scholarship has highlighted for the "Global Cold War" applies here as well: by no means as bipolar in theory and practice as long assumed, from the outset the historical project of space exploration, what Harper called humankind's "greatest adventure," proved to be not only of worldwide appeal but also an endeavor of world-spanning dimensions.[10] Grounded in a deep-seated fascination with the extraterrestrial beyond, it was characterized by a complex interplay of international conflict and cooperation at many levels and across regions and regimes. Stating that "its interest is universal, world-wide. It fires the imaginations of people everywhere," as early as 1946 Harper recognized the global dimensions of what he was saluting as "the dawning of the great coming space age."[11]

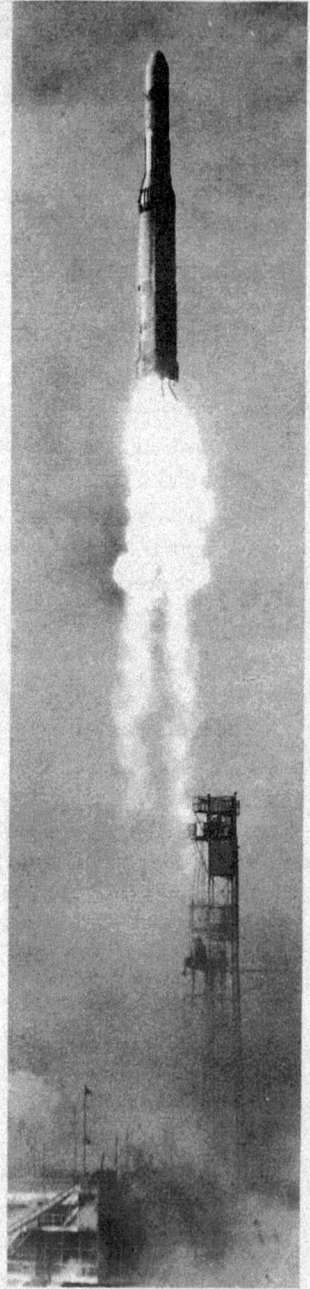

**Das Zeitalter
der Raumfahrt
hat begonnen**

Start der Europa-Rakete in Woomera (Australien) ▶

Fragment IV.1. "Das Zeitalter der Raumfahrt hat begonnen" (The Age of Spaceflight Has Begun). Source: *Mensch und Weltraum* 1 (1967), 9.

Yet the precise analytical yield of adding an adjective as ubiquitous and omnipresent as "global" to the in-and-of-itself already ill-defined "Age of Space" is far from self-evident. Would it be feasible to posit the coexistence of a multiplicity of staggered Space Ages, both dependent and yet interdependent? Could there have been several overlapping such ages in different parts of the world, defined by societal omnipresence and popularity of space thought and astroculture, together with an accompanying sense of future-tinged optimism? Or, alternatively, different local variants of one and the same world-encompassing Age of Space, coexisting and coeval, yet with different characteristics, at different times, and in different places?

Concrete periodization proposals depend, of course, on what one takes to be the deciding criterion of such an Age of Space. If an independently launched artifact in outer space is considered sufficient (as, for instance, with the first A4 launch in 1942 or that of Sputnik 1 in 1957), an Asian Space Age could be said to have set in as a consequence of the first Japanese and Chinese satellite launches in February and April 1970, respectively, the latter hailed in a contemporary pamphlet as "a good beginning in the development of China's space technology."[12] For the West European variant, the launch of the ESRO 2B research satellite in March 1968 could mark an equivalent starting point. "The Age of Spaceflight Has Begun," a German journal rejoiced, pointing to the first two-stage launch of a Europa rocket in Woomera, Australia, in August 1967—that is, an entire decade after Sputnik (see fig. Fragment IV.1). Alternatively, one could also argue for a takeoff point as late as Christmas Eve 1979, when the first Ariane 1 rocket was successfully fired from the European spaceport in Kourou in French Guiana.

If, however, an alternative key criterion is chosen, namely, that of human presence in space, such alternative periodization proposals shift again. The first person of Asian origin in space was the Vietnamese astronaut Phạm Tuân in July 1980; the first European was the Czech cosmonaut Vladimír Remek in March 1978; and the first West European was the French spationaut Jean-Loup Chrétien in June 1982. Their respective flights came two decades after those of Yuri Gagarin (1961) and John Glenn (1962), the first human and the first American, respectively, to orbit Earth. Whether one chooses the launch of a material artifact or that of a human body as the definiens of that foundational moment, both the Asian and the European Space Ages would not only have set in a decade, if not two, after Sputnik, but they would also coincide with a time when the "first," now sometimes also called "classical" Space Age, had effectively passed. According to numerous contemporary accounts, it was during the so-called post-Apollo period when the heyday of Western space enthusiasm came to a close, with that classical Space Age giving way to an era of widespread space fatigue—which, in turn, raises a third possibility: the Space Age as an actors' category, that is, when actors within history self-consciously referred to their ideas and actions as part of a "Space Age."[13]

In short, juxtaposing these fragmented ages and setting them in relation to each other demonstrates why the idea of a "Global" Space Age must be considered a useful and welcome addition to space history's otherwise neither very rich nor particularly well-developed conceptual toolbox. Such an immense geographical extension and consequent widening of the historiographical gaze requires conceptual clarification. As indicated, trying to identify multiple Space Ages worldwide inevitably raises complex periodization issues. At the same time, it also opens hitherto unthought-of comparative perspectives. Scholars in search of material artifacts or cultural constellations potentially indicating the centrality of space and spaceflight to a particular society's self-understanding need to clarify their concepts, categories, and terms of analysis before going global.

Yet taking such a conceptually controlled and more reflective view does not necessarily answer previous questions about the much-vaunted "classical" Space Age. Reraising these old issues in a new context makes their critical reassessment not less but effectively more central: Is it reasonable to cling to the Space Age as a meta-category? If so, when and where did that first Space Age begin, how long did it last, and what could be described as its main characteristics? How does it relate and compare to other technologically induced "ages," such as the Jet, the Atomic, the Information, and the Global Age or, more simply, the Age of Speed? Furthermore, most importantly, what kind of outer spaces did the Space Age produce?[14]

In all too brief a form, the remaining parts of this chapter introduce and discuss one potentially productive route to expand, decenter, and indeed globalize the histories of outer space, spaceflight, and astroculture. Driven by the conviction that the majority of global historians, with very few exceptions, err in not realizing the significance of overcoming Earth's spatial limits for the constitution of their very object of study, I argue that it is high time for historians of the Space Age—and be that one or several possible incarnations, including American, Russian, European, Asian, and global—to step in (and up) and turn their attention to the making of the world into a planet by way of outer space. From such a perspective, the "cosmic fragments" of the Global Space Age would by no means embody dislocation, decay, or decline. Rather, these particles of the universe, to be detected in the most unexpected places

hier machen
wir „Weltgeschichte"!

Fragment IV.2. "Hier machen wir 'Weltgeschichte!'" (Here We Are Making "World History!"), as proclaimed in this photomontage published in a popular German astronomy book in 1946. It is not clear whether the arrow is pointing to the Milky Way, our solar system, or planet Earth, according to the author our "dwelling-star" (*Wohnstern*). Source: Bruno H. Bürgel, *Der Mensch und die Stern*e (Berlin: Aufbau, 1946), 20.

around the world, would have to be understood as building blocks of an emergent planetary system, as scattered ingredients of an overarching historical process that I, borrowing from Pierre Teilhard de Chardin, suggest calling "planetization."

PLANETIZATION

Harper's was not the only voice to note a fundamental shift in the world's configuration during the 1940s. Long before Sputnik, observers across the political spectrum sensed that something was about to happen "out there" that would unfold a lasting impact on the human condition. In a brief book originally composed as a bedtime story for his daughter and published in the middle of World War II, philosopher Carl Schmitt diagnosed an ongoing "planetary spatial revolution," leading to the opening up of the third dimension as an entirely "new elementary area of human existence."[15] Similarly, the sociologist Helmuth Plessner considered the inclusion of "the vertical" into humankind's directions of movement a harbinger of a nascent planetary unity. "In this new world," Plessner anticipated in a radio lecture in 1949, an outside standpoint would be required to comprehend Earth as an entity, for "man will experience the unity of the Earth for the first time, because he will be able to detach himself from it."[16] Additionally, an illustration in a popular 1946 astronomy book pointed to a—presumably randomly chosen—celestial object within an unspecified part of the Milky Way, declaring "Hier machen wir 'Weltgeschichte!'" (Here we are making "world history!") (see fig. Fragment IV.2). Whether that object was supposed to represent the Milky Way, the Sun, or planet Earth itself, what is remarkable about this photomontage is not so much the fact that it seemingly anticipated the much-celebrated *Pale Blue Dot* photograph, the first ever "portrait" of Earth taken by Voyager 1 in February 1990 from the frontiers of the solar system, by half a century. Rather, it was the imagined gaze reversal, the act of self-distancing complete with a slightly amused, relativizing undertone that stands out.[17] Coming out of a global war of unprecedented existential dimensions, both the notion of "the planetary" and the idea of "planetarity" itself as a consequence of the already anticipated spaceflight revolution proved surprisingly popular, not only in philosophy, sociology, and astronomy but also in popular culture.

"Planetization," the term suggested here with a view to studying the production of outer space and the transformation of the world into planet Earth as mutually contingent historical processes, stems from the same temporal context, the transitional period of the early to mid-1940s. Intellectually, conceptually, and even topographically, however, it stands in a different tradition. Neither has "planetarity" formed an explicit part of the promise of space put forward by its agile and transnationally well-connected advocates in Europe, the United States and Russia since the mid-1920s nor has "planetization" constituted one of their magic "new words for a new world," as a Space Age glossary would later have it.[18] Unlike "astronautics" (coined in 1925), the "conquest of space" (1916/1931), the "final frontier" (early 1950s), or the already discussed "Space Age" (1946), planetization does not come from the lobby groups' conceptual treasure chest, otherwise so rich in alluring neologisms.

According to the present state of knowledge, it was the French Jesuit philosopher and paleontologist Pierre Teilhard de Chardin (1881–1955) who first wrote of "planetization" in the early 1940s, during his twenty-year-long exile in China, but only put it in print half a decade later.[19] Prohibited from preaching because of his evolutionist convictions and sent abroad in 1926, Teilhard traveled widely and engaged in archaeo-logical research, including the discovery of the Peking Man during that time. Even within Teilhard's comprehensive oeuvre, the term's geneal-ogy is intricate. Teilhard did introduce the underlying idea of an immi-nent unification of humankind, which he called a "méga-synthèse," in his main work, *Le Phénomène humain*. Begun in 1938 and completed in 1940, the book was stopped by censure in Rome and banned from pub-lication until 1955. Verbatim, however, "planétisation" was used a mere two times in the entire book, and without assigning it particular signifi-cance. "Une 'planétisation,' pourrait-on dire . . . de l'humanité," he noted for instance in passing, yet otherwise did not pay too much attention to his own, newly minted term.[20]

That would change in 1945/1946. In a private letter written on Oc-tober 10, 1945, Teilhard raved about the "'planetization' of Mankind all around our round Mother Earth" as one of his current "pet ideas."[21] For the first time and in much more detail than in *Phénomène humain*, albeit not entirely systematically either, he then outlined the contours of the argument in two shorter articles published in May and August 1946. The former was based on a lecture delivered to diplomats at the French Embassy in Beijing on March 10, 1945, whereas the latter, more explicit piece was dated December 25, 1945. Together these two brief articles mark the notion's moment of origin, both within and beyond Teilhard's body of work. "Planetization" and the "Space Age" are then of exactly the same age, with Harry Harper first writing of a "Space Age" in Janu-ary 1946, albeit 15,000 kilometers and a continent away.[22]

Complementary as they are, Teilhard's two essays need to be read in parallel. Both highlighted the planetary dimensions that human activi-ties on Earth had reached after six years of worldwide warfare. The first set the stage. Titled "Vie et planètes: que se passe-t-il en ce moment sur la Terre?" ("Life and the Planets: What Is Happening at this Moment on Earth?"), it offered a sweeping assessment of the place, significance, and importance of "our" planets within the universe, and that of humankind on Earth. The subtitle of the second piece, "Un grand événement qui se dessine: la planétisation humaine" ("A Great Event Foreshadowed: The Planetisation of Mankind"), pointedly provided an answer to the titular question: "The Planetisation of Mankind." For Teilhard, the postwar moment of an unprecedented global crisis did not signal an increasing advancement of the further disintegration of humankind, quite to the contrary. With "mankind, born on this planet and spread over its en-tire surface, coming gradually to form around its earthly matrix a sin-gle, major organic unity, enclosed upon itself," it rather gave rise to its "ever-increasing unification." It was this centripetal process of gradual, yet wide-reaching coalescence that he termed "planetization."[23] "When everyone else was talking about the end of the war, Teilhard was talking about the end of the world," one of his many biographers has remarked—and the world's subsequent transformation into a planet, one could add.[24]

What has sometimes been misunderstood as an early synonym of globalization or *mondialisation*, albeit avant la lettre, is in fact quite different, despite the terms' shared emphasis on expanding integration. Planetization à la Teilhard describes the gradual emergence of a global consciousness, whereas globalization usually denotes a capital-driven process of increasing worldwide socioeconomic entanglement.[25] While historians of science, technology, and religion have intensively studied and worked with Teilhard's notion of a "noosphere"—possibly better known due to its sister concept "biosphere"—the idea of an ongoing planetization of the world as a central component of his so-called cosmic vision has hitherto found little, if any, resonance.[26] Why should planetization be considered a key category and potentially a novel paradigm to historicize the Global Space Age? Admittedly, at least two problems must be addressed before elevating the term within the history of outer space, spaceflight, and astroculture at large and making it the conceptual cornerstone of this nascent body of literature: the teleological bent of Teilhard's theology, combined with his mistrust of space exploration.

First, when Teilhard describes planetization as the growth of a planetary consciousness, with humankind gradually coalescing into one global organic unit, there is unmistakably an underlying telos at work, reaching far beyond the postwar moment. In Teilhard's view, planetization was a central component of an evolutionary process, a necessary step humankind had to undertake before arriving at the noosphere and what he termed the Omega point. Even though Teilhard, a convinced and outspoken evolutionist, had not been permitted to teach in the Jesuit order for quite some years by 1946, he was, after all, not only a scientist and scholar, but first and foremost a Catholic theologian and ordained priest. To this day many of his critics continue to be put off by what they consider Teilhard's convoluted mysticism and an improper amalgamation of science and religion. The controversies his work triggered in the mid-twentieth century reverberate into the present. For some space historians, this powerful, yet deeply religious-spiritual surplus might already disqualify Teilhard's work, despite its original character. They might find such a conceptual apparatus too abstruse and mystically loaded to be useful and, accordingly, rather shy away from the theoretical labor required.

Second, it is far from clear what, if any, role the emerging spaceflight revolution, in 1945/1946 long underway on both sides of the Atlantic, played historically within the process of planetization as originally conceived and outlined by Teilhard de Chardin. Outer space did have a pronounced presence in his thought, and some scholars have gone so far as to attribute to him a "cosmic vision."[27] The "Vie et planètes" essay, for instance, deals with sweeping questions of scale and temporality, proximity and distance across the universe, and Teilhard showed an impressive command of astronomical knowledge. Carefully situating the Earth and the human life it houses vis-à-vis the Sun, the solar system, and even the Milky Way, the notion of "planetary dimensions" of current events introduced in the very first paragraph (and quoted in the epigraph to this chapter) proved by no means a metaphor. Yet, when compared to Schmitt, Plessner, later Hans Freyer, Hannah Arendt, Martin Heidegger, Maurice Merleau-Ponty, and many other contemporaneous mid-century intellectuals, philosophers and hommes de lettres, there is no explicit discussion of spaceflight, rocketry, or any technology-enabled departure from Earth, let alone a both imminent and desirable "conquest" of space. In a short paragraph toward the end, worth quoting in its entirety, Teilhard briefly pondered the possibility of leaving Earth behind: "Others seek to reassure us with the notion of an escape through space. We may perhaps move to Venus—perhaps even further afield. But apart from the fact that Venus is probably not habitable (is there water?) and that, if journeying between celestial bodies were practicable, it is hard to see why we ourselves have not already been invaded, this does no more than postpone the end," only to reject the idea outright, together with the potential existence of alien life.[28] In "Un grand événement qui se dessine," the second of the two essays, he juxtaposed such an "external" with an "internal" planetization of mankind. The Catholic left no doubt that he advocated for the latter: escape from Earth as spiritual metamorphosis rather than by means of technology. For Teilhard, the unfolding, internal planetization of humankind on Earth ruled out the conquest of the heavens.[29]

In short, the differences between Teilhard's original conceptualization of "planétisation" and the one developed and suggested here are stark. Quite some conceptual work is necessary before turning it into a useful category, potentially even a paradigm, when historicizing the Global Space Age. Yet is such an import and subsequent adjustment not only possible but also historiographically "permissible"? Following Koselleck again, the matter is a technical problem rather than one of principle, and the answer is a simple yes. Time and again, Koselleck reassures us, any ambiguities between a historical concept—planetization as developed, coined, and propagated by Teilhard himself—and a historiographical category—planetization as the making of the world into a planet by way of outer space, as suggested here—do not have to be off-limits, if only the differences are made sufficiently clear.[30]

PLANETARITY

Adoption, then, requires adaptation. However, is it also worth all the conceptual effort? Why should the import and subsequent adjustment of a notion originally coined by Teilhard de Chardin in 1945/1946 promise to open fresh perspectives to understand the history of outer space, astroculture, and extraterrestrial life on a global scale? Three final remarks, in the form of three preliminary theses, must suffice.

First, planetization evokes an extraterrestrial point of view and literally heralds a gaze reversal. It draws attention to the making of the planet as an imagined unit, based on reports, photographs, and data from space, and addresses the comparatively recent processes that led to its creation. Neither Earth nor outer space are set, unwavering entities. Rather, the transformation of our world into a planet and the production of outer space have been mutually contingent historical processes, and the category of planetization helps to bring this codependence into view. In so doing, it pushes us to rethink, recalibrate, and reformulate the idea behind the notion of a "Global Space Age." That the prospect of leaving Earth behind and expanding into outer space resonated in other parts of the world to the same degree as it did in Russia, Europe, and the United States—either at the same time or on different timescales—is all too obvious. What is less obvious, however, is that all individual space efforts, regardless of their technical specificities and precise geographical point of departure, share one common feature. They all come down to what Hannah Arendt described in 1963 as the never-ending, incompletable search for an Archimedean point, a point outside Earth "from which it would be possible to unhinge, as it were, the planet itself."[31] Every single launch of a spacecraft, whether human or robotic, civilian or military, governmental or private, anywhere around the world reinvokes this age-old quest to overcome the confines of the home planet; this is what explains much of the apparently inherent and never-ending fascination of spaceflight. To put it more crudely, the globe becoming global has been part and parcel of the Global Space Age from its very beginning.

Second, historicizing planetization and planetizing history is not tantamount to historicizing globalization and globalizing history, only with a refreshed vocabulary. As outlined above, the process I suggest terming "planetization" fundamentally differs from that of globalization, usually described as a multilayered and asymmetric process of worldwide entanglement driven by capitalism. With very few exceptions, historians of globality seem to be unaware of the constitutive role that the ongoing and infinite search for Arendt's Archimedean point has played in the constitution of their very subject matter. "*Global* . . . points in the direction of space; its sense permits the notion of standing outside our planet and seeing 'Spaceship Earth.' . . . This new perspective is one of the keys to global history," one of these rare exceptions, Bruce Mazlish, observed over a quarter century ago, when most historians were just beginning to learn how to think globally.[32] As recent cracks in the established globalization narrative—but not in that of the planetary—make clear, distinguishing between the globe and the planet is not simply adding to the otherwise long overabundant "environmental planet-speak" but rather an urgent necessity, analytically as well as politically.[33]

Third, planetization helps to demetaphorize and ground "planetarity." Together with concurrent debates on the Anthropocene and the climate catastrophe, a heterogeneous body of literature has rapidly sprung up and massively grown in recent years that invokes "planetarity" in all kinds of configurations and settings, usually reducing the term to an opaque, space-free, and ahistorical metaphor. Whether true or not, Gayatri Spivak claims ownership. Supposedly used for the very first time by her in Vienna in 1999, "planetarity", she tells us, is a word set apart from notions of the planetary, the planet, the Earth, the world, the globe, globalization, and the like in their common usage—yet without clearly indicating what precisely marks the difference.[34] Planetarity's ill-defined fuzziness, however, has not hampered the term's meteoric rise across the humanities, but quite to the contrary. Literary scholars have identified a "planetary turn," sociologists have begun to develop "planetary social thought," the historian Dipesh Chakrabarty has suggested provincializing Earth during the "planetary age," and so the list continues, amounting to a veritable fad in the making.[35] Ironically, histories of outer space, spaceflight technology, and global astroculture play hardly any role in this emerging literature, with pictorial clichés such as NASA's celebrated *Earthrise* (1968) and *Blue Marble* (1972) photographs the exception to the rule. It is time to amalgamate global and space history by developing a new field of historiography: planetary history.

Finally, what are the consequences of all this with a view to expanding and decentering the historiography of outer space, spaceflight, and space exploration? Where does this leave the newly proclaimed Global Space Age? It is clearly possible and certainly meritorious to identify and historicize hitherto unrecognized "space particles" and "cosmic fragments," unsystematically spread all around the globe, as the contributions to this book do. However, there may be another, alternative possibility: to understand them as building blocks of a much broader, world-encompassing process that helps us to understand the coproduction of outer space and the global present during the second half of the twentieth century. At a geopolitical moment when the extraterrestrial grip on Earth is more ubiquitous and omnipresent than ever, it is incumbent upon space history to provide the missing link. The notion of "planetization," developed by Pierre Teilhard de Chardin and adapted here for such a purpose, offers precisely this. As a Berlin 1930 radio broadcast proclaimed at the onset of the spaceflight revolution, "Planetarians of the world, unite!"[36]

AFTERWORLDS

Space in the Age of Ruins

EDWARD JONES-IMHOTEP

Consider two artworks from the past two decades, both created in the twilight temporal horizons of this book.

The first—called *Cosmic Thing*—is a disassembled 1983 Volkswagen Beetle, cast as an exploded view diagram (see fig. E.1). Created by the artist Damián Ortega and exhibited at the 2003 Venice Biennale, the sculpture pulls apart the component parts of the iconic car and holds them in suspension, like an insect trapped in amber, complicating our ideas about the object and revealing elements and relations usually hidden from view.[1] The second work is even more recent. *Anatomy of an AI System* is "an exploded view of a planetary system across three stages of birth, life and death" (see fig. E.2).[2] Created by Kate Crawford and Vladan Joler, it is a large-scale map illustrating the human labor, data, planetary resources, and extractive processes required to build and operate an Amazon Echo. As the artists note, "the scale of this system is almost beyond human imagining."[3] First shown at London's Victoria and Albert Museum in 2018, and now part of the digital collections of the Metropolitan Museum of Art in New York, the high-resolution diagram is designed to be printed at a minimum scale—roughly eight feet by ten and a half feet—otherwise its details are lost. Even beyond the vast spheres of artificial intelligence, it suggests the spiraling and poten-

E.1. Damián Ortega, *Cosmic Thing*, 2002, Malmö Konsthall, photograph © Helene Toresdotter/Malmö Konsthall.

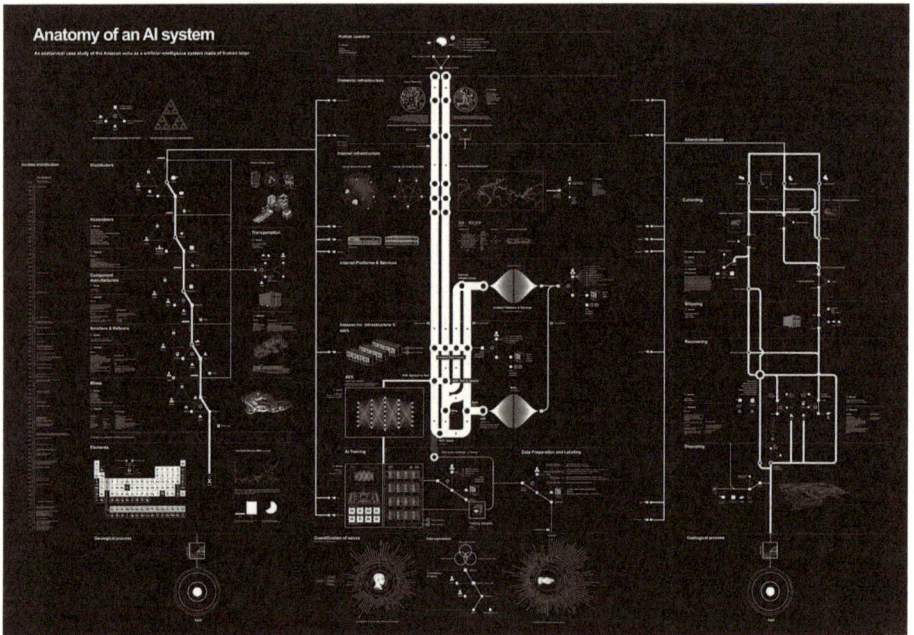

E.2. Kate Crawford and Vladan Joler, *Anatomy of an AI System*, 2018, MoMA.

tially infinite cosmographies that we could trace for each component of Ortega's sculpture.[4]

In their invocations of the cosmic and the planetary, both works suggest ways of thinking about perspective, fragmentation, totality, the deep past, and the ever present that are central concerns of this book. We might think of them as constellations. We might also think of them as complementary methods for writing a global history of space exploration through its fragments.

This book, after all, looks to dismantle. It works to take apart the neatly assembled narratives and the "normative fetishization of machines, men, and manifest destiny" that characterized earlier histories of space exploration.[5] The chapters function as fragments at the boundaries of those earlier stories—"forgotten, ignored, or invisible until now, but existing at edge-sites where the unitary narratives no longer apply."[6] Like an exploded sculpture, this book holds those elements in suspension, suggesting a larger totality and revealing what the assemblage hides. Like *Anatomy of an AI System*, the book aims to map the intricate genealogies of extraction, violence, ideology, hope, deception, and aspiration that lie behind each component. Together, the chapters both explode and excavate. They not only trace critical histories of space exploration through landscape, empire, waste, and rupture; they suggest what those histories mean for our own time. Even more than a constellation, they constitute a cosmogram of the space age.[7]

In this brief epilogue, I would like to suggest a few additional fragments for our collection: some drawn from this book, others from beyond its pages. And I would like to add four additional themes—emptiness, estrangement, expendability, and remains—to suggest ways to think about how the fragments of space history are produced and how they tie past and present together.

EMPTINESS

A stretch of Florida's Space Coast, occupied by humans for 12,000 years, illustrates how space activities have made "emptiness" central to their terrestrial ways of seeing.

Cape Canaveral—the launch site for countless space missions, from Project Mercury to SpaceX—occupies a landscape of palmetto-covered lowlands, barrier islands, coastal lagoons, sheltered estuaries, and pine flatwoods of Eastern Florida. Tom Wolfe described this terrain as, "one of those bleached, sandy, bare-boned stretches where the land that any sane man wants runs out [like the Mojave Desert and Edwards Air Force Base] . . . and the government takes it over for the testing of hot and dangerous machines."[8] He could have been channeling the Australian writer Ivan Southall on the Woomera Test Range: "Here it was, one of the greatest stretches of uninhabited wasteland on earth, created by God specifically for rockets."[9]

Landscapes are not innocent.[10] The "myth of the empty land" has antecedents stretching to the biblical. But it has also helped furnish the tabulae rasa of broader utopian visions at least as old.[11] What space exploration contributes to that longer mythology, both on and beyond the Earth, is an identity between "empty" spaces and advanced, potentially catastrophic machines.[12]

The claimed emptiness occupied by missile ranges, launch facilities, and advanced test sites is as much about time as about terrain. It aims to create what Alice Gorman calls a "chronological sequence of archaeological time periods" rather than a temporal and cultural mosaic.[13] Linking the late "emptiness" of Cape Canaveral to ancient human artifacts that predate it—flint-head missiles of early human, colonial, and precolonial inhabitants of the Cape—allows NASA to portray its former human inhabitants "at once as primitive squatters on land destined for greater uses, and as the innovative arrow-firing progenitors of the steely-eyed missile man."[14] These juxtapositions of the primitive and the modern, across the global sites of space activity, are striking in their consistency: the French guided missiles set against the stone tools of the Colomb-Béchar test range; the satellites and the rock art of the Woomera Prohibited Area (itself named after the Aboriginal spear-throwing technology). This deep time chronology outlines a Space Age/Stone Age narrative that casts space facilities, as Eleanor Armstrong suggests in her discussion of the pastoral, as spaces of an idealized and primitive past, a fallen present, and a possibly utopian future.[15] Socially, politically, and culturally, they mark out the space-faring—the ones who control the earth in order to leave it—from the "backward," the earthbound, and the dispossessed.

That chronology also masks the contemporary acts of dispossession that make space activities possible. These are spaces that have been *emptied*, albeit on different timescales and through different processes. The local experiences of the Mescalero Apache at White Sands or the Doui-Menia of Colomb-Béchar, differ profoundly from the Pitjantjatjara of South Australia or the Yanadi of Sriharikota.[16] But space activities have developed a generalized and sophisticated technical apparatus that supports the creation of specific kinds of terrestrial emptiness in the present, precisely because its spaces of action are anything but empty. They have mobilized not only a "logic of location" that justifies displacement, but a set of technical supports to enact it: legal frameworks that underwrite the condemnation and appropriation of lands; formulas for blast radii that define exclusion zones and risk; and the material and logistical integration of places like Sriharikota or Merritt Island with other, less visible, activities on the Bay of Bengal or Cape Canaveral.[17]

That emptiness works to erase not only the recent human occupants of these areas but also their inconvenient technical genealogies. The ostensibly civilian Kennedy Space Center on Cape Canaveral developed literally across the river from, and in deep sympathy with, the launch head for one of the most important military aerospace facilities of the Cold War, the Atlantic Missile Test Range—a five-thousand-mile missile-testing corridor created by President Harry Truman in 1949 and stretching from Cape Canaveral over the Bahamas, across the Lesser Antilles and into the South Atlantic, with its densest concentration of radar stations and military bases surrounding Barbados.[18] Like Vandenberg did for the Pacific, that test range shifted the risk from rocket failures onto non-US citizens—Pacific Islanders and African diaspora populations in the Caribbean.[19] The image of Cape Canaveral as a center for peaceful exploration of outer space was, among other things, a trick of perspective, created by placing other histories, infrastructures, and people just outside the frame of Merritt Island.[20] The launch facilities there were both a product and an actor in what we might call the "machinic orders" of space activities: the selective and malleable genealogies that linked specific technologies in particular contexts, for particular purposes, and not others.[21] They connected spacecraft, launch infrastructures, and command and control centers to the long histories of humankind, even while they severed those other machines, right next door, that might destroy it. Combined with the logic of location, they remind us that making space activities relatable and familiar is not their only gambit for public support. Much of that support is won by making these activities and infrastructures appear deeply exceptional: by justifying what happens there precisely because it happens in only a few places on the planet.

ESTRANGEMENT

Histories of space exploration are, in many ways, about displaced empires—empires going underground, living on in private enterprises, turning themselves into something palatable for a postcolonial world.[22] As Priya Satia beautifully puts it in a different context, these are empires that barely exist on paper. For many, though, they produce a familiar strangeness.[23]

The San Marco project was a two-decade collaborative venture between NASA and Italian researchers to launch scientific satellites into equatorial orbits from mobile seaborne platforms off the coast of Kenya.[24] This kind of transnational project would seem tailor-made for ideas of "circulation" and global knowledge flows. But as Asif Siddiqi has taught us elsewhere, emphasizing dynamic flows, imperial logistics, and porous boundaries can make us dispose too quickly of nation-states, their identities, and the hierarchies at work within space activities.[25] As a result, we miss the frictions and displacements that emerge when national interests and global mobility collide, and when state actors labor to control and disrupt flows of potentially mutinous knowledge, materials, and people.

What we need, Siddiqi teaches us, is not so much a way of tracking travel, movement, and flow, but a way of pausing on incoherence, conflict, and location. Against its public portrayals, the San Marco site was not neatly bounded. It was *dispersed*—like a stand of quaking aspen or the troops of matsutake mushrooms Anna Tsing describes in *The Mushroom at the End of the World*.[26] In contrast to the "project," the site included research and support facilities in Rome, offices in the United States, equatorial earth orbits, and aspects of Kenyan territory and coastal waters—the airport, its infrastructure and personnel, the Royal College, the harbor at Mombasa, marine vessels, the base camp at Ngomeni, and the city of Malindi. The Italian workers' home away from home was also a leisure, drug- and sex-tourism destination for their compatriots. That messy assemblage meant the San Marco site was not only dispersed, but fragmented.[27]

What world, then, emerges from these fragments of empires, even invisible ones? Fragmentation in the case of San Marco meant that the various elements of the site "came into violent friction with the entrenched human geographies and political imperatives of a newly independent and powerful postcolonial state, Kenya.[28] Those different elements created different worlds, different ontologies, and conflicting meanings for the space project. Kenyans made meanings out of San Marco that ran against Italian or American understandings of the project.[29] We are tempted to reach for an older language in the history of technology and science and technology studies and call this a kind of "interpretive flexibility."[30] But Siddiqi's point is more subtle and more powerful: it refuses to posit a stable reality underpinning multiple interpretations, but suggests that San Marco generated different realities and experiences at different points. This leads to a more general claim that sites of transnational knowledge, as sites of struggle and conflict, are generally (and possibly always) unstable. They are only partially about the site itself, and much more about the projection of a distant power, first onto these local sites, and from there into space. This vision of conflict and friction gets us to a more contested, more violent, and more critical vision of Peter Galison's trading zones, with its irenic connotations of lively, consensual, and mutually beneficial relations.[31]

What San Marco and the other dispersed and fragmented sites reveal is how space activities extended not only imperial relations, infrastructures, and ideologies but also a central experience of empire that crossed the spaces of North Africa, Oceania, the Caribbean, and even the Kazakh steppe. Years ago, Frantz Fanon pointed to the estrangement that is the consequence and the logic of colonialism: the feeling that we are never quite at home, even in our own skin.[32] Resigning his post at the psychiatric hospital in Blida-Joinville, Algeria, where he worked as a psychiatrist from 1953 to 1956, Fanon spoke of "the Arab, permanently an alien in his own country."[33] The more general experience of alienation is echoed across this book. It runs through the poetry of Oodgeroo Noonuccal, a member of the Indigenous Kaurna people of the Adelaide Plains, who lamented: "We are strangers here now, but the white tribe are the strangers. / We belong here, we are the old ways."[34] It lurks behind the Nigerian prime minister Abubakar Rafawa Balewa, flanked by the US chargé d'affaires and watched over by a portrait of John F. Kennedy while aboard the USNS *Kingsport*, anchored in the harbor of his country's largest city.[35] And it inspires the Kazakh activist Marat Dauletbayev's description of how the hybrid governance of the "sacrifice zone" of Baikonur, the world's oldest spaceport, "makes Kazakh people feel like foreigners in their own country."[36] Again and again, as the dispersion and distribution of space activities expanded, they continually served to remind disparate peoples that they ultimately did not belong in these spaces of the future.

EXPENDABILITY

The boundary separating spacecraft that fall back to earth from the ones that continue out into space represents an event horizon, shaped by a singularity in the South Pacific.

Point Nemo is a geospatial artifact located about 2,700 kilometers from the nearest landmass, Antarctica. Also known as the "ocean point of inaccessibility," it is defined by being the point on the earth's oceans farthest from any point of land. Surrounding this point, and four kilometers underwater, is a zone called the South Pacific Ocean Uninhabited Area. It is outside the legal jurisdiction of any country and is the resting place of almost 300 splintered spacecraft, including 145 of Russia's autonomous Progress resupply ships and roughly 25 tonnes of the Mir space station.[37] Any newly launched orbiting technologies are designed either to burn up on reentry, or to end up here—after a brief luminescence in the atmosphere, back in the cold and dark.

Viewed from the perspective of waste, it is difficult to see space activities as anything but a history of the expendable—of things meant to be sacrificed, used up, then abandoned or destroyed. That expendability has been part of the *human* experience of space activities. The resistance movements mobilized against launch sites and ground stations around the globe have partly been a refusal by people to accept the implications of their disposability. Opposition to Japanese research activities in the north of Honshu was shaped by the recent histories of Hiroshima and Nagasaki, the bombing of Akita (the last bombing mission of World War II), the irradiation of the fishing ship *Fukuryū Maru* during nuclear testing in the Pacific, and the mercury poisoning at Minamata.[38] The launch facilities on Michikawa Beach pointed to the legacies of the Pacific War, the militarized geography of islands it helped create, and the imperial occupations it supported, inspiring Akita's residents to implicitly ask: "What more should we sacrifice in their name?"[39]

Beyond the earth's atmosphere, the notion of space waste in particular trades on a very specific double notion of decay: one about functional degradation and failure, and the other about deteriorating orbits. Space junk is created by a temporal disjuncture between the longevity of orbits (comparatively long) on the one hand, and the longevity of the technologies designed to operate there (comparatively short) on the other. What kinds of meanings and histories emerge in that gap? How has decay been not just a historical force shaping the process and meanings of globalization, but a particular kind of historical object, created in part by space activities? Like globalization itself, why and how did it come to matter, and to matter differently, across the globe?

The South Pacific Ocean Uninhabited Area makes clear how the line between human and natural expendability and the waste of the Space Age has helped constitute both the local and the global. Space activities expressed what Steven Jackson has flagged as the productivist innovate-and-discard cycle that has contributed so disastrously to the dangers of the Anthropocene—obscene budgets, multimillion-dollar single-use components, storehouses of liquid propellants, sacrificeable animals, and fragile populations.[40] According to the European Space Agency's 2022 annual environmental report, "Ever since the start of the space age on the 4th of October 1957, there has been more space debris in orbit than operational satellites."[41] Point Nemo, with its underwater graveyard of space debris from Russia, the United States, Japan, and China, among other countries, forms the other face of what David Kaiser and Hunter

Heyck have called the federated world of Cold War science and technology.[42] The chapters in this book illustrate how transnational communities of failure and decay emerge, how the issues of falling satellites, space tourism, developing economies, nuclear attack, postcolonial sovereignty, exo-atmospheric liability, solar-terrestrial physics, and limited orbital paths converged after the 1960s to make waste a source of global concern.[43] Recent space tourism takes this to the limiting case of the globe itself, where the planet can be enjoyed and then abandoned by the people wealthy enough to escape its ruin.[44]

If taken too far, though, the focus on waste and expendability hides the enormous amounts of repair and maintenance that space activities have also required. As Réka Patrícia Gál makes clear, those techniques and practices have been central to the human experience of space activities. She writes, "Unsurprisingly, in their autobiographies astronauts who have served on various space stations repeatedly highlight maintenance work as a fundamental facet of life in space."[45] Although Gál focuses on the life-sustaining built environment of Skylab, she also points to the feminist media theorists and scholars who have signaled the centrality of care work, maintenance, and repair performed mainly by women, poorly paid immigrants, and people of color.[46] Her intervention reminds us that the high-speed consumption of materials and resources has always worked to mask the durable social structures and unequal relations that have characterized labor before and through the Space Age.

REMAINS

Space activities communicate through many media—sound, images, and text. They also communicate through their answers to the question: Who decides what persists and what disappears?

Operation Morning Light was a cleanup mission launched after a Soviet nuclear-powered satellite, Cosmos 954, slipped from orbit in late January 1978 and disintegrated during earth reentry, spreading radioactive debris over hundreds of thousands of square kilometers above northern Canada.[47] The cleanup operation lasted nearly nine months as Canadian and US military personnel, guided by airborne sensors and dressed in protective suits, fanned out over the debris field and into isolated communities of people who were still unaware of the crash itself.[48] Official accounts hailed the operation as a success, reassuring residents about the negligible risks while underlining southern Canadian ingenuity and ignoring the traditional and suddenly hazardous land uses of the Dene, the Métis, and Inuit people who traveled and hunted on the Barrens.[49]

The material response drew on existing apparatuses of disaster—radiation suits, Geiger counters, and surveillance overflights—that illustrate how the infrastructure of nuclear warfare also ministered to technological catastrophes, blurring the lines between accidental and intentional violence, between the mundane and the apocalyptic. But the experiences of Indigenous people were also shaped by a series of disjunctures that called into question the very nature of Operation Morning Light— shielded military personnel versus exposed children; government avowals versus public uncertainty; the airtightness of official coverup stories against the permeability of Indigenous bodies; the urgent military response against the languid half-life of U-235; the well-delimited government event versus the ongoing local experience of illness, contamination, and concern.[50]

Adding to the uncertainty was the larger geography of risk in the region. Former uranium, zinc, and lead mines made it difficult to distinguish the effects of Cosmos 954 from the broader contamination of the postwar Canadian North and especially their effects on Dene and Metis life. Here, the terrestrial language of contamination and decay was more than metaphorical. There were deep affinities between space activities and terrestrial environments. Decaying orbits and decaying isotopes merged in the bodies of northern residents and uncertainties about their health.[51]

How do we think about the media of space activities? In their various ways, the final chapters of this book address the historical conditions surrounding what it means for something to be spacial: the delicate dependence on public opinion that shaped cooperative ventures like the Apollo–Soyuz Test Project; the popular culture that channeled NASA's public relations messages into *Star Trek* and *2001: A Space Odyssey*; the complex mix of irony, incongruence, and subversiveness of late Soviet space cinema; even the shifting meaning of "planetization" in the thought of Pierre Teilhard de Chardin.[52] Tracking their respective ruptures, reversals, and mutations, the chapters argue that the categories of the spacial, the planetary, and the global would not stay still.

But it is useful to remember that before Teilhard's "planetization" had a name, it already had an ethos. In 1937, the philosopher-paleontologist-priest asked, "What fundamental attitude . . . should the advancing wing of humanity take to fixed or definitely unprogressive ethnical groups? The earth is a closed and limited surface. To what extent should it tolerate, racially or nationally, areas of lesser activity?"[53] Rooted in a conviction that "not all ethnic groups had the same value," Teilhard had called the previous year for "official recognition of: (1) the primacy/priority of the earth over nations; (2) the inequality of peoples and races."[54] The ruptures, mutations, and disappearances that eventually sanitized his view of the planetary were also taking part on the same "closed and limited surface" he was conceptualizing.

These are complex geographies. It is useful to think through their issues of visibility and power by considering what William Rankin has called "intangible artifacts"—a class of phenomena that includes radio waves, odors, noises, and toxins. Rankin explains that these artifacts are characterized by two common properties: they are temporally fleeting and therefore not always mapped; and they are difficult to contain. They are not just irreverent when it comes to regional and national boundaries, spilling across borders that halt other objects in their tracks. They embody a profound misalignment between their transient geography and more traditional political delineations of territory, like borders. Together, Rankin argues, those two properties give these intangible artifacts an insidious power, especially when used by states: they provide permanent occupations that masquerade as temporary events.[55]

We might consider another side of that relationship—how the claimed impermanence of things like toxicity and contaminations are used to justify the *absence* of the state and the abdication of its responsibilities in something like Operation Morning Light.[56] For the Canadian officials coordinating the operation, making contamination "disappear" was easy. They simply had to ground surveillance flights, recall military personnel, and turn off Geiger counters. Cosmos 954 calls attention to the striated geographies and official practices of neglect that made the impermanence of state involvement and responsibility possible. It puts in a new light what it meant for Teilhard's planetary vision to place certain peoples among those being left behind.

SPACE IN THE AGE OF RUINS

The history of space activities is a history of things not meant to survive—coastal scrublands, political antagonists, spent rocket boosters, early satellites, animal test subjects, local resistances, and even the globe itself. These are the fragments we write about. The work of seeing them as fragments is perhaps made easier at this moment because so many things have fallen apart.

Ruins are no more innocent than landscapes.[57] But they are what we are left with. One of the challenges of the chapters collected in this book has been to write backward from the ruins, to make clear that this is not where things must end. They confront us with the irony that the cosmic fragments, the exploded objects, and hidden genealogies, most excluded from the history of space activities (and its writing) are the ones best placed to help us understand its legacies. They are also the best vantage point from which to see its alternatives. As Ambrose Akinmusire puts it, not only the "the savage histories, brutal legacies, illusory democracies, feudal tendencies, rendering twisted souls, twisted like vines in the jungle" but also something just as essential: *"The hunger. The hunger for light."*[58]

NOTES

INTRODUCTION. INTO THE COSMIC (AGAIN)

1. For earlier surveys of the literature on space exploration, see Siddiqi, "American Space History"; and Launius, "Historical Dimension."

2. In this vein, see McDougall, *The Heavens and the Earth*; Launius and McCurdy, *Spaceflight*; Logsdon, *John F. Kennedy*; Bilstein, *Stages to Saturn*; and Johnson, *Secret of Apollo*. For Hughes, the obvious reference is his *Networks of Power*.

3. Hughes himself was a cheerleader for a kind of "American technological sublime." On American technological exceptionalism, see Hughes, *Rescuing Prometheus* and *American Genesis*.

4. An early articulation of American exceptionalism as a driving force in the space program can be found in McCurdy, *Space*.

5. The literature on colonial science is vast and uneven. For some useful surveys, see the special issues: MacLeod, "Nature and Empire," especially its "Introduction" (1–13); Osborne, "Social History of Science, Technoscience and Imperialism" and its "Introduction" (161–70); and Schiebinger, "Focus" and its "Forum Introduction" (52–55). See also Seth, "Putting Knowledge in Its Place." For useful surveys on the literature on postcolonial technoscience, see Anderson, "Introduction"; and Arnold, "Europe, Technology, and Colonialism."

6. Newell, *Destined for the Stars*.

7. Redfield, *Space in the Tropics*; S. Mitchell, *Constellations of Inequality*; Messeri, *Placing Outer Space*; Vertesi, *Seeing Like a Rover*; Olson, *Into the Extreme*; Gorman, *Dr Space Junk vs the Universe*; and Klinger, *Rare Earth Frontiers*.

8. A brief listing of some of this new historical work includes Rand, "Falling Cosmos; Bimm, "Andean Man"; Muir-Harmony, *Operation Moonglow*; Siddiqi, "Science, Geography, and Nation"; Siddiqi, "Dispersed Sites"; Siddiqi, "Competing Technologies"; Maher, *Apollo*; Tribbe, *No Requiem*; and Launius, *Apollo's Legacy*.

9. Krige, *Fifty Years*; Siddiqi, *Challenge to Apollo*; and Geppert, *Imagining Outer Space*.

10. Siddiqi, *Red Rockets' Glare*; Jenks, *Cosmonaut*; Andrews and Siddiqi, *Into the Cosmos*; Maurer et al., *Soviet Space Culture*; and Kohonen, *Picturing the Cosmos*.

11. McCray, *Visioneers*; Neufeld, *Von Braun*; Buss, *Willy Ley*; and MacDonald, *Escape from Earth*. See also Geppert, "Rocket Stars."

12. Holt, *Rise of the Rocket Girls*; and Shetterly, *Hidden Figures*.

13. Nochlin, *Body in Pieces*, 23–24.

14. Tronzo, "Introduction," in Tronzo, *Fragment*, 3.

15. For some recent scholarship, see Rieppel, Lean, and Derringer, "Science and Capitalism"; and Lucier, "Capitalism and Science."

16. Works on science and socialism in the twentieth century include L. Graham, *Science in Russia*; Kojevnikov, *Stalin's Great Science*; Gerovitch, *From Newspeak to Cyberspeak*; J. Smith, *Works in Progress*; Siddiqi, *Red Rockets' Glare*; Schmalzer, *Red Revolution*; and Wang, "Science and the State." Although much earlier, David Joravsky's work still offers many key insights. See Joravsky, *Soviet Marxism*.

17. For the question of universalism in science, see Somsen, "History of Universalism."

18. Muir-Harmony, *Apollo to the Moon*; and Parker, "Capitalists in Space."

19. Lenin, *Imperialism*.

20. Dickens, "Cosmos," 67.

21. Harraway, "Anthropocene." See also Trischler, "Anthropocene."

22. Pritchard and Zimring, *Technology*, 8.

23. Pritchard and Zimring, *Technology*, 9.

24. Jones-Imhotep, *Unreliable Nation*, 13, 216.

25. Maher, *Apollo*; Henry and Taylor, "Re-thinking Apollo"; and Cosgrove, "Contested Global Visions."

26. Olson, *Into the Extreme*.

27. Henry and Taylor, "Re-thinking Apollo," 198–201. See also Messeri, *Placing Outer Space*.

28. Gál is drawing from the work of Crosby and Stein, "Repair."

CHAPTER 1. "LOOSE IN SOME REAL TROPICS"

1. Bipartite and Butler, *History of the Kennedy Space Center*, 28.

2. Ferdinando, "Atlantic Ais," 10–15.

3. Bipartite and Butler, *History of the Kennedy Space Center*, 29.

4. Namakas, *Liquid Landscape*.

5. Navakas, *Liquid Landscape*, 2.

6. Navakas, *Liquid Landscape*, 126.

7. Navakas, *Liquid Landscape*, 126.

8. Benson and Faherty, *Moonport*.

9. Lipartito and Butler, *History of the Kennedy Space Center*, 37.

10. Launius, *NASA*, 15.

11. Paul and Moss, *We Could Not Fail*, 49.

12. Lipartito and Butler, *History of the Kennedy Space Center*, 58.

13. Faherty, *Florida's Space Coast*, 27.

14. See Marx, *Machine in the Garden*.

15. Mazlish, *Railroad and the Space Program*.

16. Mazlish, *Railroad and the Space Program*, 209.

17. Mazlish, *Railroad and the Space Program*, 216.

18. G. Harris, *Kennedy Space Center Story*, 6.

19. The development of the recreational potential of the area was an important policy focus for Brevard County in the 1960s; see Weaver and Anderson, "Some Aspects of Metropolitan Development."

20. G. Harris, *Kennedy Space Center Story*, 7.

21. G. Harris, *Kennedy Space Center Story*, 1.

22. For a legal history of the refuge system in the United States, which is distinct from the National Parks system, see Fischman, "National Wildlife Refuge System."

23. G. Harris, *Kennedy Space Center Story*, 1.

24. Arnold, *Problem of Nature*, 142–43. See also Arnold, *Tropics and the Traveling Gaze*; and Stepan, *Picturing Tropical Nature*.

25. S. Mitchell, *Constellations of Inequality*.

26. Redfield, "Beneath a Modern Sky," 260. See also Redfield, *Space in the Tropics*.

27. Redfield, "Beneath a Modern Sky," 261.

28. Redfield, "Beneath a Modern Sky," 259.

29. Maher, *Apollo*, 97.

30. "Cape's Old Lighthouse Has Yet to Go into Orbit," *Spaceport News*, February 7, 1963, 3.

31. "Veteran of 100+ Launches Recalls Early Cape Days," *Spaceport News*, February 28, 1963, 6.

32. NASA did get around to taking down many of these structures, but

not until the late 1960s. A committee was formed in 1966 to survey temporary facilities and recommend them for preservation or disposal, but many were not slated for disposal until 1969. Joseph Hester, NASA Suggestions memo, December 8, 1966. Memo to the Director of Administrations, re Disposition of Temporary Buildings, March 11, 1969 (National Archives, Atlanta. KSC files, box 2, Ad Hoc Committee on Temporary Facilities 1969).

33. "NASA Girl's Family MILA Pioneers," *Spaceport News*, March 7, 1963, 1.

34. "NASA Girl's Family," 6.

35. "KSC Story: From Marshland to Spaceport," *Spaceport News*, October 10, 1968, 2–3.

36. "KSC Story," 2.

37. "KSC Story," 4.

38. "Headquarters Pond Has 'Gator—You Name It and Win a Prize," *Spaceport News*, August 28, 1969, 3.

39. "Native Plants, Imports Add Beauty," *Spaceport News*, March 27, 1969, 8.

40. "Native Plants."

41. Lozano, "Race, Mobility, and Fantasy," 807–8. See also Lozano, *Tropic of Hopes.*

42. Cox, "Selling Seduction," 205.

43. Ring, "Inventing the Tropical South," 622.

44. Spence, *Dispossessing the Wilderness*, 5. On the role of the memory and representation of conflict with Indigenous people in the larger culture of postwar America, see also Engelhardt, *End of Victory Culture.*

45. "Counting Down with the Editor," *Spaceport News*, September 26, 1968, 8.

46. "Counting Down with the Editor."

47. Bullen, Bullen, and Bryant, *Archaeological Investigations*, 1. I would like to thank Dr. Kathleen Sheppard for help with materials related to the archaeological history of this area.

48. Bullen, Bullen, and Bryant, *Archaeological Investigations*, vii.

49. Bullen, Bullen, and Bryant, *Archaeological Investigations*, 27.

50. Letter from Elbert Cox, Regional Director Department of Interior to Clarence Bidgood, Director of Facilities Engineering, Kennedy Space Center, February 27, 1964 (National Archives, Atlanta. Kennedy Space Center Files. Directorate of Design Engineering, Real estate Branch 1963–1970, box 2, Ad Hoc Temporary Facilities-Cape Data Collection Annex-Archaeological Sites. Archaeological Sites Ross Hammock).

51. "Proposed Conditions and Restrictions to be Attached to National Park Service Permit to University of Florida for Archeological Survey, Excavation and Collection at KSC, April 18, 1966 (National Archives, Atlanta).

52. "Proposed Conditions."

53. See Mozingo, *Pastoral Capitalism*.

54. Mailer, *Of a Fire on the Moon*, 50.

55. Tom Wolfe, *Right Stuff*, 128.

56. Tom Wolfe, *Right Stuff*, 128.

57. Robert G. Whalen, "Visit to Three Cape Kennedy's," *New York Times*, December 13, 1964. See also Maher, *Apollo*, 97–103.

58. Al Rossiter Jr., "Huge Spaceport Right on Schedule," *Chicago Tribune*, July 26, 1964.

59. Jonathan Amos, "James Webb Space Telescope Lifts Off on Historic Mission," BBC News, December 25, 2021, https://www.bbc.com/news/science-environment-59782057.

60. Edgar Sandoval and Richard Webner, "A Serene Shore Resort, Except for the SpaceX 'Ball of Fire,'" *New York Times*, May 24, 2021, https://www.nytimes.com/2021/05/24/us/space-x-boca-chica-texas.html.

CHAPTER 2. FRAGMENTED HISTORIES OF A SATELLITE COMMUNICATIONS EARTH STATION

1. Rushdie recorded his experiences in a travel narrative, *Jaguar Smile*.

2. Earth stations, or ground stations, are the large technical installations required for sending and receiving radio signals to and from an Earth-orbiting satellite, as well as decoding and improving signal quality and redistributing the signals via ground-based radio networks.

3. While Intersputnik was initially limited to Soviet allies in Eastern Europe plus Mongolia and Cuba, the socialist telecommunications officials who created it envisioned it from the beginning as a genuinely global network. For more on Intersputnik see Stovbun, "Osobennosti Mezhdunarodnoi organizatsii"; Downing, "Intersputnik System"; and Evans and Lundgren, *No Heavenly Bodies*.

4. On Rushdie's self-fashioning as a political tourist in *Jaguar Smile*, see Moynagh, *Political Tourism*, 177–212.

5. Rushdie, *Jaguar Smile*, 36.

6. Trouillot, *Silencing the Past*, 72.

7. Edgerton, "From Innovation to Use."

8. *NASA, CR-144672*, sec. 10, p. 20.

9. Nicaragua joined Intelsat in 1969 under an interim agreement and was among the original signatories of the permanent agreement on August 20, 1971. Executive Office of the President, *Seventh Annual Report*, 4.

10. "Official Inauguration of the Earth Station," American Embassy Managua to Secretary of State, Telegram 2614, June 1, 1973, Tel Nic, Subject-Numeric File (SNF) 1970–73 Economic, Record Group (RG) 59: General Records of the Department of State, National Archives at College Park (NACP).

11. "Official Inauguration of the Earth Station."

12. "Official Inauguration of the Earth Station."

13. One striking example is the telephone conversation between President Nixon and King Hassan II of Morocco, during the opening of the first African Intelsat station in Morocco on January 7, 1970. Evans and Lundgren, *No Heavenly Bodies*, 109.

14. Letter from Jack Brooks to Henry Kissinger, May 4, 1976, box 30, folder "Shelton, Turner (4)," John Marsh Files, Gerald R. Ford Presidential Library, https://www.fordlibrarymuseum.gov/LIBRARY/document/0067/7773962 .pdf; see also Steinmetz, *Democratic Transition*, 107.

15. International Teleommunication Union (ITU), *Eleventh Report*, 101.

16. Valenzuela, "Policy-Making," 64–68.

17. Valenzuela, "Policy-Making," 68–69.

18. "Space Telecommunications: GON and COMSAT to Build and Operate Earth Station as Joint Venture," American Embassy Managua to Department of State, Airgram A-139, October 18, 1970, SNF 1970–73 Economic, RG59, NACP.

19. "Space Telecommunications." Decisions about where to build an Earth station were often made regionally because existing radio relay networks and satellite footprints generally crossed national borders. For small countries, having an Earth station carry traffic for multiple countries was believed to be essential for economic viability. See Early, Kumins, and Baer, "Business Forecasting." For an overview of regional development in Latin America, see Rizzoni, "Overview."

20. Lee, "De-Centring Managua."

21. Star, "Ethnography of Infrastructure," 380, 382; and Larkin, "Politics and Poetics," 336.

22. Our thanks to Asif Siddiqi for this astute observation.

23. Reid, "Symbolism of Postage Stamps," 223.

24. This stamp also does not include the red-and-black Sandinista flag alongside the blue-and-white Nicaraguan flag, which was typical of Sandinista-era stamps according to Jack Child. See Child, *Miniature Messages*, 184.

25. The examples discussed in this chapter are typically so-called commemorative (as opposed to definitive) stamps, depicting historical landmarks "with strong political iconic messages." Child, *Miniature Messages*, 17. For an overview of Soviet, US, and European astronomy and space-related stamps, see Poznakhirko, "Spravochnik."

26. See Anna Reser, chapter 1 in this book. It is worth noting, nonetheless, that the imagery of a high-tech installation against an uninhabited natural landscape was also typical of stamps produced by countries like Iceland.

27. ITU, "Eleventh Report", 100–101.

28. Russian State Archive of Economics (RGAE), f. 3527, Ministerstvo

Sviazi, op. 72, d. 665 "Ekspertnoe zakliuchenie no. 3358 po rabochemu proek-
tu na stroitel'stvo kosmicheskoi stantsii sistemy 'Intersputnik' v Nikaragua,"
March 16, 1984, l. 1–3.

29. Costs for telephone and television transmissions via Intersputnik were
significantly lower than those for Intelsat. Downing, "International Communications," 115–16.

30. RGAE, "Ekspertnoe zakliuchenie," l. 1–3.

31. RGAE, "Ekspertnoe zakliuchenie," l. 4.

32. RGAE, "Ekspertnoe zakliuchenie," l. 4.

33. See, for example, Osokina, *Stalin's Quest for Gold*; and S. Harris, "Dawn
of the Soviet Jet Age."

34. *K stoletiiu so dnia rozhdeniia.*

35. *K stoletiiu so dnia rozhdeniia,* 69.

36. Stanek, *Architecture.*

37. *K stoletiiu so dnia rozhdeniia,* 53–54.

38. Sanchez-Sibony, *Red Globalization.*

39. "RGAE, f. 3527, op. 72, d. 579, "Dokumenty ob uchastii MinSviazi v
rabote MOKS Intersputnika," l. 1.

40. Mark, Kalinovsky, and Marung, *Alternative Globalizations,* 2.

41. "Inauguration of Intersputnik Satellite Ground Station in Nicaragua,"
BBC Monitoring Summary of World Broadcasts, July 19, 1986.

42. "Pri sodeistvii SSSR," *Pravda,* July 19, 1986, 1.

43. Rushdie, *Jaguar Smile,* 35–36.

44. Evans and Lundgren, *No Heavenly Bodies,* 106.

45. "Soviet Intelsat Accord Enters Force," *Sel'skaia zhizn',* October 4, 1985,
3, in FBIS [Foreign Broadcast Information Service] no. 203, NASA archives,
record no. 15738, Series: International Cooperation and Foreign Countries,
Subseries "International Cooperation," folder "Intelsat."

46. "Soviet Intelsat Accord Enters Force."

47. "Nicaragua Closes Airspace, Report on Military and 'Espionage' Activities," *BBC Monitoring Summary of World Broadcasts,* July 17, 1986.

48. William R. Long, "US Accused of Spying for the Contras," *Los Angeles
Times,* July 18, 1986, https://www.latimes.com/archives/la-xpm-1986–07–18
-mn-16555-story.html.

49. Radio Sandino Managua broadcast,19:38 GMT, July 17, 1986.

50. The detailed report led US observers to believe that Wheelock and the
Sandinista Popular Front had active support from the Soviet Union, employing
an "advanced East Block radar system," in their efforts to gather this intelligence. Julia Preston, "Sandinistas Say U.S. Intelligence Overflights Are Frequent," *Washington Post,* July 18, 1986, A24.

51. Duncan Campbell, "Inside Echelon: The History, Structure and Func-

tion of the Global Surveillance System Known as Echelon," July 25, 2000, https://www.angelfire.com/retro/antropologos/echelon.pdf; and Duncan Campell, "My Life Unmasking British Eavesdroppers," *Intercept*, August 3, 2015, https://theintercept.com/2015/08/03/life-unmasking-british-eavesdroppers.

52. For a discussion of the dual use of space infrastructure, see Peldszus, "Architectures of Command."

53. Leire Ventas, "La enigmática estación satelital que Rusia instaló en Nicaragua para 'combatir el narcotráfico' y que inquieta a Estados Unidos," *BBC Mundo*, June 23, 2017, https://www.bbc.com/mundo/noticias-america-latina -40352903.

54. "La enigmática estación satelital rusa en Nicaragua," *Cuba en Cuentro*, June 24, 2017, https://www.cubaencuentro.com/internacional/noticias/la-enigmatica-estacion-satelital-rusa-en-nicaragua-329817. On the promise of infrastructures, see Anand, Gupta, and Appel, *Promise of Infrastructure*.

55. Joshua Partlow, "The Soviet Union Fought the Cold War in Nicaragua: Now Putin's Russia Is Back," *Washington Post*, April 8, 2017 ; LeireVentas, "La enigmática estación satelital rusa en Nicaragua," *Cuba en Cuentro*, June 24, 2017, https://www.cubaencuentro.com/internacional/noticias/la-enig matica-estacion-satelital-rusa-en-nicaragua-329817.

56. Ventas, "La enigmática estación satelital rusa."

57. Marcelo Cantelmi, "La enigmática base satelital rusa en Nicaragua que toma una nueva dimensión," *Clarín Mundo*, January 14, 2022, https://www .clarin.com/mundo/enigmatica-base-satelital-rusa-nicaragua-toma-nueva-di mension_0_Spul4676t.html.

58. Siddiqi, "Another Space"; Parks, "Global Networking"; Alonso and Palmarola, "NASA in Chile"; and Redfield, *Space in the Tropics*.

59. Parks, "Global Networking"; and Siddiqi, "Another Space."

60. Parks and Starosielski, "Introduction," 2–3. See also Schwoch, *Wired into Nature*.

61. Siddiqi, "Another Space," 23.

62. Banks, d'Avignon, and Siddiqi, "Introduction," 3.

63. Banks, d'Avignon, and Siddiqi, "Introduction," 3.

64. Siddiqi, "Shaping the World."

65. Geppert, "Post-Apollo Paradox," 11–12.

CHAPTER 3. "PRACTICALLY NO HABITATION"

Many of my ideas in this essay were sharpened during my stay in 2021–2022 at the Davis Center for Historical Studies at Princeton University. I wish to thank especially David Bell, Michael Gordin, Durba Mitra, and Peter Redfield for their invaluable insights. I also want to thank Shaik Basheer with the Association for Rural Development in Andhra Pradesh.

1. "PSLV Launch: What Made Modi Speak in English," *Times of India*, July 1, 2014, https://timesofindia.indiatimes.com/india/PSLV-launch-What-made -Modi-speak-inEnglish/articleshow/37543003.cms.

2. Bruce Bonta, "Police Remove Yanadi before Modi Visit," *Peaceful Societies*, July 10, 2014, https://peacefulsocieties.uncg.edu/2014/07/10/police -remove-yanadi-before-modi-visit/.

3. "Blanket of Security around SHAR, Nellore Coastline ahead of Modi Visit," *New Indian Express*, June 24, 2014, https://www.newindianexpress.com/ states/andhra-pradesh/2014/jun/24/Blanket-of-Security-Around-SHARN ellore-Coastline-ahead-of-Modi-Visit-628057.html.

4. Bonta, "Police Remove Yanadi."

5. "Blanket of Security around SHAR."

6. "Poverty Haunts Challa Yanadi Tribal Families," *Hindu*, June 21, 2014, https://www.thehindu.com/news/national/andhra-pradesh/Poverty-haunts -Challa-Yanadi-tribalfamilies/article11623704.ece.

7. A slew of recent semiofficial and independent works and memoirs on the Indian space program make no mention of the Yanadi community. These include semiofficial works such as Rao and Radhakrishnan, *Brief History of Rocketry*; and Rao, *From Fishing Hamlet to Red Planet*; independent works such as Singh, *Indian Space Programme*; and Maharaj, "Space for Development"; and memoirs such as Kalam and Tiwari, *Wings of Fire*; Rao, *India's Rise*; Aravamudan and Aravamudan, *ISRO*; and Sandlas, *Leapfroggers*.

8. "India's Space-Based 'Revolution,'" *BBC News*, November 3, 2013, https://www.bbc.com/news/scienceenvironment-24772147.

9. In 2023 India became the first country to set down a robotic probe near the south pole of the Moon.

10. Sankar, *Economics of India's Space Programme*, 292.

11. U. Tejonmayam, "Isro Takes Copy of Gita and PM Modi's Photo to Space," *Times of India*, February 28, 2021, https://timesofindia.indiatimes .com/india/isro-takes-copy-of-gita-and-pm-modis-photo-tospace/articleshow/ 81254647.cms.

12. The figure of ten thousand total displaced comes from revenue records of the Office of the Sub-Collector in the town of Gudur, cited in Reddy, *Displaced Populations*, 31.

13. Rodgers and O'Neill, "Infrastructural Violence." The literature on infrastructure is vast, but Brian Larkin's offers an excellent overview. See Larkin, "Politics and Poetics."

14. Redfield, *Space in the Tropics*.

15. Similar (although not identical) considerations are folded into the siting of astronomical observatories as underscored in the excellent work of Tiffany Nichols. See Nichols, "Finding Stillness."

16. See Siddiqi, "Another Space."

17. For biographies, see Chowdhury and Dasgupta, *Masterful Spirit*; and Shah, *Vikram Sarabhai*.

18. Sarabhai, *Science Policy*, 3.

19. Not coincidentally, these were all technoscientific fields principally identified with either the United States or the Soviet Union, and not Great Britain whose technical acumen was identified with such systems as the railroad, telegraph, naval ships, and so on, and thus still marked as instruments of colonial domination. For the reconfiguration of scientific imaginaries across the divide of 1947 in South Asia, see Prakash, *Another Reason*.

20. Siddiqi, "Science, Geography, and Nation."

21. Rostow, *Stages of Economic Growth*.

22. Besides Rostow, we might also include the work of Daniel Lerner, Rowan Gaither, Douglas Ensminger, Chadbourne Gilpatric, and Roger Evans. See Latham, *Modernization as Ideology*; Gilman, *Mandarins of the Future*; and Sackley, "Passage to Modernity."

23. Bhabha, "Science and the Problems of Development."

24. INCOSPAR was set up in February 1962. See Siddiqi, "Science, Geography, and Nation," 432.

25. Siddiqi, "Science, Geography, and Nation."

26. Department of Atomic Energy. See also Siddiqi, "Whose India?"

27. Sarabhai, *Science Policy*, 21–27.

28. Department of Atomic Energy, *Atomic Energy and Space Research*, 28.

29. Narayanan and Ram, *Ready to Fire*.

30. Narayanan and Ram, *Ready to Fire*.

31. Ramaswamy, *Passions of the Tongue*.

32. Reddy, *Displaced Populations*, 114.

33. Raj, *Reach for the Stars*, 47–48.

34. Government of India, *The Annual Report of the Department of Atomic Energy 1968–69*, Bombay, 1969, 70.

35. "AP Launching Station," *Times of India*, December 6, 1969, 1.

36. "Rocket Station in Nellore District," *Times of India*, April 30, 1969, 1.

37. "Crore" is the English term, common in South Asia, denoting ten million.

38. Singh, *Indian Space Programme*, 194.

39. Government of India, *The Annual Report of the Department of Atomic Energy 1969–70*, Bombay, 1970, 70. Emphasis added.

40. Atomic Energy Commission, *Atomic Energy and Space Research*, 34. Emphasis added. The same phrase appears in a press account of the selection of Sriharikota from 1970. See "Satellite to Be Launched in 1974 from Sriharikota," *Times of India*, July 21, 1970, 11.

41. "Satellite to Be Launched," 11.

42. Reddy, *Displaced Populations*, 27.

43. Reddy, *Displaced Populations*, 28.

44. The total population of Yanadi in Andhra Pradesh in 1971 numbered 239,409. They were the second largest "scheduled tribe" in Andhra (after the Koyas). Nearly 50 percent of them lived in the Nellore district, which contained Sriharikota Island. See Agrawal, Rao, and Reddy, *Yanadi Response to Change*, 13–14. By 1981, this number had decreased to 205,925. See Jaya Kumar, *Tribals*, 4.

45. Cushing, "Sriharikota."

46. Cushing, "Sriharikota," 19–20.

47. Cushing, "Sriharikota," 20.

48. Reddy, *Displaced Populations*, 101.

49. Aravamudan, "Space Saga," 18.

50. Reddy, *Displaced Populations*, 117. For conversion rates, see https://www.bookmyforex.com/blog/1-usd-inr-1947-till-now/.

51. Reddy, *Displaced Populations*, 120.

52. Reddy, *Displaced Populations*, 119.

53. Pratik Rakshit, "Yanadi Tribals of Nellore Are Caught in Vicious Cycle of Bondage," February 16, 2015, https://ballsoffury2811.wordpress.com/tag/sriharikota/.

54. Agrawal, Rao, and Reddy, *Yanadi Response to Change*, 20.

55. Agrawal, Rao, and Reddy, *Yanadi Response to Change*, 14.

56. Reddy, *Displaced Populations*, 127.

57. Agrawal, Rao, and Reddy, *Yanadi Response to Change*, 40.

58. Raj, *Reach for the Stars*, 62.

59. The original offer had come from P. Sudhakara Reddy. Unable to study the Yanadis on Sriharikota, he completed an extensive study of the Yanadi who had left Sriharikota, in 1977–1978, the results of which were published nearly two decades later. See Reddy, *Displaced Populations*.

60. Agrawal, Rao, and Reddy, *Yanadi Response to Change*, 128–29.

61. Siddiqi, "Science, Geography, and Nation."

62. The social science literature on Adivasis is vast, the historical canon less so. For some primarily historical accounts, see Das Gupta and Basu, *Narratives from the Margins*; Sen, *Indigeneity*; Nilsen, *Adivasis*; and Radhakrishna, *First Citizens*.

63. Bijoy, "Adivasis of India."

64. Bijoy, "Adivasis of India."

65. Ashish Kothari, "India Colonises Its 'Frontiers,'" *Wall Street International*, September 13, 2021, https://wsimag.com/economy-and-politics/66908-india-colonises-its-frontiers; and Shrivastava and Kothari, *Churning the Earth*.

66. Kothari, "India Colonises Its 'Frontiers.'"

67. Balagopal, "Land Unrest."

68. Padel and Das, *Out of This Earth*.

69. Bijoy, "Adivasis of India."

70. Ethnographical studies of the Yanadi include Agrawal, Rao, and Reddy, *Yanadi Response to Change*; Reddy, *Displaced Populations*; Rao, *Ethnography*; Jaya Kumar, *Tribals*; and Ravikumar, "Ethnographic Profile."

71. Rakshit, "Yanadi Tribals of Nellore"; Bonta, "Police Remove Yanadi."

72. Sk Basheer, email message to author, July 28, 2021.

73. Arnold, "Nehruvian Science." See also Prakash, *Another Reason*.

74. Siddiqi, *Red Rockets' Glare*; and McCurdy, *Space*.

75. For space and extractive political economies, see Klinger, "Environmental Geopolitics."

76. Rory James, "Anger after Indonesia Offers Elon Musk Papua Island for SpaceX Launchpad," *Guardian*, March 9, 2021, https://www.the guardian.com/world/2021/mar/10/anger-after-indonesia-offers-elon-musk -papuanisland-for-spacex-launchpad.

77. Although not explicitly historical, Sean T. Mitchell's ethnography of the politics surrounding the displacement of 1,500 Afro-Brazilians by the construction of a space launch site at Alcântara offers a trenchant overview how space infrastructure can produce unequal and racialized political arrangements in postcolonial settings. See S. Mitchell, *Constellations of Inequality*.

78. Agrawal, Rao, and Reddy, *Yanadi Response to Change*, 129.

FRAGMENT I. GROWING NEW NARRATIVES IN SCIENCE COMMUNICATION IN OUTER SPACE

1. Some of the foods popular on the space station currently—including the tacos that contained space-grown chilies—are themselves foodstuffs of Indigenous, often Global South communities. The complicated tensions of imperialist impulses are visible within these the foodstuffs—not only are US agencies expropriating the foods of Mexico and Central America, but there are cultural imbalances in, for example, the Han-majority foods that dominate the Chinese meals; and the locations that the ISRO's foods are drawn from within India; and the *gurpi* (cured reindeer meat) Fuglesang hoped to take is a foodstuff of the colonized Sami communities. Even in displaying these foods in the UK Museum, the exclusive showcasing of US foods rather than Russian equivalents such as jellied beef tongue or Riga bread, as is done in the Smithsonian, in the US, also re-creates international hierarchies in the galleries. Further integration of the work on culinary colonialism, food sovereignty, and food supply chains into the work of social studies of outer space would reveal other ways that capital-colonialism is integrated into the fabric of space projects but is beyond the scope of this fragment. For further information on the foods mentioned here, see: CNN, "China's Space Menu Revealed, October 10, 2003, http://edition.cnn .com/2003/TECH/space/10/10/china.space.food/; Dagens Nyheter, "Chris-

ter Fuglesang redo för rymden efter 14 års träning," October 1, 2007, https://web.archive.org/web/20071001090818/http://www.dn.se/DNet/jsp/polopoly.jsp?d=2597&a=589321&previousRenderType=6; Japan Aerospace Exploration Agency, "Candidates for Japanese Space Food Selected," September 5, 2014, https://global.jaxa.jp/press/2014/09/20140905_sfood.html; "Food in Space: From ISRO's Gaganyaan Mission to NASA, What Astronauts Eat," *Indian Express*, https://indianexpress.com/article/technology/science/food-in-space-from-isros- gaganyaan-mission-to-nasa-what-astronauts-get-6208429/; "What Do Astronauts Eat in Space?" *Chinadaily*, October 20, 2016, https://www.chinadaily.com.cn/china/2016-10/20/content_27118311.htm.

2. See, for example, Armstrong, "Exploring Space(s)"; Gorman, "Cultural Landscape of Interplanetary Space; and Small, "Space Museums."

3. Empson, *Some Versions of Pastoral*.

4. Garrard, *Ecocriticism*.

5. Kilgore, *Astrofuturism*, 128.

6. Pak, *Terraforming*, 58.

7. Kessler, "Pretty Sublime," 58.

8. Kat Deerfield, "When Is Space?" YouTube video, February 1, 2013, https://www.temporalbelongings.org/presentations3/kat-deerfield-cardiff-university (1:19).

9. Parker and Bell, *Space Travel & Culture*, 4.

10. Data on Exploring Space collected by Armstrong in 2019. For more details, see Armstrong, "Exploring Space(s)."

11. Fürst, "Cooking and Femininity."

12. Cairns and Johnston, *Food and Femininity*, vii.

13. Charnell Chasten Long, "The Black Women Food Scientists Who Created Meals For Astronauts," *Lady Science*, 2019, https://www.ladyscience.com/essays/black-women-food-scientists-who-created-meals-for-astronauts.

14. Shetterly, *Hidden Figures*.

15. NASA Fact Sheet, "Human Needs: Sustaining Life during Exploration," April 16, 2007, https://web.archive.org/web/20201205090814/https://www.nasa.gov/vision/earth/everydaylife/jamestown-needs-fs.html.

16. "Visions of the Future," https://www.jpl.nasa.gov/galleries/visions-of-the-future.

17. Lisa Messeri, "The Image of Exoplanets and the Imagination of Worlds," lecture at Pontificia Universidad Católica de Chile, September 29, 2020.

18. Koch, "Whose Apocalypse?"

19. Pearson, Hollinger, and Gordon, *Queer Universes*.

20. At the time, this blog also appeared on the Smithsonian *Air and Space* Magazine but is no longer archived there.

21. Don Pettit, "Letters to Earth: Diary of a Space Zucchini," April 3, 2012,

Letters to Earth (blog), NASA, https://blogs.nasa.gov/letters/2012/04/03/post_1333471169633/.

22. Valentine, "For the Machine."

23. Westling, *Green Breast.*

24. See, for example, Schama, *Landscape and Memory.*

25. Buell, *Environmental Imagination.*

26. Sarah Baggs, *I'm Mad They Left Poop on the Moon (And You Should Be Too)*, Melbourne, Australia: zine printed by the author, 2021.

27. Baggs, *I'm Mad*, 4.

28. Andrew Stanton, dir., *WALL-E* (Walt Disney Pictures/Pixar Animation Studio, 2008).

CHAPTER 4. "WE ARE AS STRANGERS HERE NOW"

Acknowledgments: My views may not represent those of the Indigenous groups named here, and I acknowledge the importance of discussions with Aboriginal people, which have informed my perspective, particularly Andrew Starkey (Kokatha). I would like to thank Asif Siddiqi for including my chapter in this book, and for his encouragement and extraordinary patience; and Brenton Griffin for his invaluable assistance with references.

1. A bora ground is a ceremonial space consisting of complex circular earthworks and paths.

2. Noonuccal, *We Are Going.*

3. Wright, "Weapon of Poetry," 21.

4. Brewster, "Oodgeroo," 99.

5. McGregor, *Imagined Destinies.*

6. Bawaka Country et al., "Dukarr Lakarama"; Alice C. Gorman, "To Boldly Go toward New Frontiers, We First Need to Learn from Our Colonial Past," *The Conversation*, October 7, 2016, https://theconversation.com/to-boldly-go-toward-new-frontiers-we-first-need-to-learn-from-our-colonial-past-65568; Schwartz, "Myth-Free Space"; and Treviño, "Cosmos Is Not Finished."

7. Gao, "Early Chinese Migrants to Australia."

8. Tom McKay, "Elon Musk: A New Life Awaits You in the Off-World Colonies—for a Price," *Gizmodo*, January 17, 2020, https://gizmodo.com/elon-musk-a-new-life-awaits-you-on-the-off-world-colon-1841071257.

9. Roberts, Howkins, and van der Watt, "Antarctica," 14.

10. Urey, *Planets*; and Watson, Murray, and Brown, "On the Possible Presence of Ice."

11. Byrne, "Nervous Landscapes."

12. Cait Storr, "Could Corporations Control Territory in Space? Under New US Rules, It Might Be Possible," *The Conversation*, June 2, 2022, https://theconversation.com/could-corporations-control-territory-in-space-under-new-us-rules-it-might-be-possible-138939.

13. Byrne, "Nervous Landscapes"; and Byrne, "Counter-Mapping."

14. Byrne, "Nervous Landscapes."

15. Byrne, "Nervous Landscapes."

16. Gorman, "Cultural Landscape of Interplanetary Space"; Gorman, *Dr Space Junk vs the Universe*; and Schwartz, "Myth-Free Space."

17. See, for example, Carnes and Luis, "Deep Time"; Cornum, "Outer Space Future"; and Rifkin, *Fictions of Land and Flesh*.

18. Bowden, *Pitt Rivers*.

19. Morgan, *Ancient Society*.

20. Gorman, *Dr Space Junk vs the Universe*.

21. Morton, *"Fire across the Desert."*

22. Morton, *"Fire across the Desert."*

23. Gorman, "Cultural Landscape of Interplanetary Space"; and Gorman, "Beyond the Space Race."

24. Bowen, *Original Sin*.

25. "Half-Castes May Work at Woomera," *Advertiser* (Adelaide, South Australia), July 7, 1954, 16; Alice C. Gorman, "A Colonial Space: Women and Rockets in Australia," poster presented at "Space Science in Context," online conference, University College London, May 14, 2020.

26. Gorman, "Archaeology of Space Exploration."

27. Gorman, "Archaeology of Space Exploration."

28. Kristeva, *Powers of Horror*.

29. Lyle, "Highlights," 122.

30. Michael Fletcher, "Another Hidden Figure: Clyde Foster Brought Color to NASA," *American Experience*, https://www.pbs.org/wgbh/americanexperience/features/chasing-moon-another-hidden-figure-clyde-foster-brought-color-nasa/.

31. Hamilton Bims, "Rocket Age Comes to Tiny Triana," *Ebony*, March 1965, 106–19.

32. Gorman, "Space Cowboys"; and Gorman, "La Terre et l'espace."

33. Gorman, "Space Cowboys."

34. Gammage, *Biggest Estate on Earth*.

35. Southall, *Woomera*, 3.

36. Wikipedia, s.v. "Mining," https://en.wikipedia.org/w/index.php?title=ining&oldid=1183346487.

37. Edgington, *Range Wars*, iv.

38. Morton, *"Fire across the Desert"*; and Mann, Clarke, and Gostin, "Surveying for Mars."

39. Gorman, "Archaeology of Space Exploration."

40. Blenkinsopp, "Bible, Archaeology and Politics."

41. See, for example, Furniss, "Indians"; Lu and Schönweger, "Great Expectations"; and van Eeden, "Surveying the 'Empty Land.'"

42. Lucke et al., "Questioning Transjordan's Historic Desertification."

43. Blenkinsopp, "Bible, Archaeology and Politics"; and Cezula and Modise, "Empty Land Myth."

44. Carroll, "Myth of the Empty Land"; and Cezula and Modise, "Empty Land Myth."

45. Alice C. Gorman, "The Sustainable Management of Lunar Natural and Cultural Heritage: Suggested Principles and Guidelines," *Report to the Global Expert Group on Sustainable Lunar Activities*, 2023, https://moonvillageassoci ation.org/gegsla/documents/gegsla-reference-documents/; and Gorman, "How We Let the Moon Die."

46. P. O'Brien and S. Byrne, "Cooler than Cool: Doubly Shadowed Regions at the Lunar Poles," paper presented at the fifty-third Lunar and Planetary Science Conference, March 7–11, 2022, Woodlands, Texas, https://www.hou.usra .edu/meetings/lpsc2022/pdf/2372.pdf.

47. I. Crawford et al., "Managing Activities."

48. Fredengren, "Nature"; and Harrison, "Beyond 'Natural' and 'Cultural' Heritage."

49. Enric Volante, "Navajos Upset after Ashes Sent to Moon: Nasa Apologizes," *Spokesman-Review*, January 15, 1998, https://www.spokesman.com/ stories/1998/jan/15/navajos-upset-after-ashes-sent-to-moon-nasa/.

50. Bawaka Country et al., "Dukarr Lakarama," 6.

51. Ceridwen Dovey, "Making Kin with the Cosmos," *Humans and Nature Website* (2021) https://humansandnature.org/making-kin-with-the-cosmos/.

52. Freud, "Uncanny."

53. Jorgenson, "Middle America."

54. Jorgenson, "Middle America," 187.

55. Jorgenson, "Middle America," 187.

56. Freud, "Uncanny."

57. Castaneda, *Journey to Ixtlan*, 264.

58. Tsing, *In the Realm*, 154, quoted in Byrne, "Counter-Mapping."

59. Mudrooroo, "Poetemics of Oodgeroo."

60. Fuller, Hamacher, and Norris, "Astronomical Orientations."

61. Alexis Wright, "In Times Like These, What Would Oodgeroo Do?" *The Monthly*, December 2020–January 2021, http://www.kooriweb.org/foley/news /2000s/2020/monthly_dec_2020b.pdf.

62. Furaih, "'Let No One Say the Past Is Dead,'" 170.

63. Byrne, "Counter-Mapping," 257–58.

64. Noonuccal, *My People*, 22.

CHAPTER 5. INVENTING SYNCOM

Acknowledgments: I am grateful for the guidance and support of my coadvisers at the Department of History at Princeton University, Angela N. H. Creager and Hendrik Hartog, and to numerous colleagues for their feedback, especially Michael Gordin, Erika Milam, Jennifer Rampling, Katharina Schmidt, Keith Wailoo in the History of Science Program Seminar at Princeton; Lisa Larrimore Ouellette, Aileen Robinson, Kara Swanson, and Steven Wilf at the Stanford Center for Law and History "Working with IP" conference, alongside Brent Salter and Amalia Kessler (for recognizing my work with the Graduate Student Paper Prize); Skyler Gordon and Saumyashree Ghosh at the Colonialism and Imperialism Workshop at Princeton; Kamari Clarke, Mary X. Mitchell, and Simon Stern at the Innovation Law and Policy Workshop at the University of Toronto Faculty of Law; Stephen Garber at the NASA History Division; and Edward Jones-Imhotep, Asif Siddiqi, Christine E. Evans, Lars Ludgren, and Julie Michelle Klinger at the "Cosmic Fragments" workshop. I am also indebted to many archivists: Robert Beebe at the National Archives at Kansas City; Gene Morris at the National Archives at College Park; Elizabeth Suckow, Colin A. Fries, and Sarah H. Jenkins at the NASA Historical Reference Collection; and Elizabeth C. Borja at the Archives Division of the National Air and Space Museum's Steven F. Udvar-Hazy Center.

1. Hughes Aircraft Company, Space Systems Division, "Syncom 2 Performance during First 60 Days in Orbit," November 1963, folder 16509, box 3/14/5, HRC (hereafter "Syncom 2"); *Syncom Engineering Report, Vol. I*, NASA Technical Report no. R-233, Syncom Projects Office, Goddard Space Flight Center, March 1966 (hereafter "Report").

2. Edgar Morse, "Preliminary History of the Origins of Project Syncom," NASA Historical Note no. 44, Goddard Space Flight Center, September 1, 1964, folder 16507, box 3/13/4, HRC (hereafter Morse, "Preliminary"); Hughes Aircraft Company, "Syncom: The Synchronous-Orbit Communications Satellite by Hughes," 1962, folder 006421, box 3/13/4, HRC (hereafter "Brochure"); Clarke, "Extra-Terrestrial"; and Pierce and Kompfner, "Transoceanic Communication."

3. Morse, "Preliminary"; and Whalen, *Origins of Satellite Communications*.

4. *Astronautics and Aeronautics, 1964*, 332–33.

5. Morse, "Preliminary"; Whalen, "Billion Dollar Technology"; and Communications Satellite Act, 76 Stat. 419 (1962).

6. Gavaghan, *Something New*, 199–222.

7. Whalen, "Billion Dollar Technology."

8. Novak, "'Myth'"; and Scheiber, "Review."

9. McDougall, *The Heavens and the Earth*.

10. Hurst, *Law*.

11. Trubek and Galanter, "Scholars"; and Kennedy, "Three Globalizations of Law."

12. Tani, *States of Dependency*, 242–43; and Seo, "New Public."

13. Reich, "New Property."

14. See Immerwahr, *How to Hide an Empire*; cf. Paul Kramer, "How Not to Write"; A. Macpherson, "A Caribbean Historian Extends Scholarly Critiques of *How to Hide an Empire: A History of the Greater United States*," March 15, 2020, https://s-usih.org/2020/03/a-caribbean-historian-extends-scholarly-critiques -of-how-to-hide-an-empire-a-history-of-the-greater-united-states. While historians of law and technology write extensively about international law, diplomacy, and transnational phenomena, they seldom link US domestic legislation and regulation to extraterritorial conduct during the Cold War. For exceptions, see Oldenziel, "Islands"; Black, *Global Interior*; Dudziak, *Cold War Civil Rights*; Swanson, *Banking on the Body*, 84–119; M. Mitchell, "Land, Culture, and Marshall Islanders"; Creager, *Life Atomic*, 107–42; and Grisinger, "'South Africa.'"

15. I approach the new idea of US empire as a generative concept, as Talal Asad describes of "the West," or a sociolegal imaginary, as employed by Samuel Moyn. Asad, "Introduction"; and Moyn, "Imaginary Intellectual History."

16. Harold A. Rosen and Donald D. Williams, Santa Monica, CA, "Changing the Orientation and Velocity of a Spinning Body Traversing a Path," Patent no. 3,394,344, December 27, 1966, Serial no. 862,921 (filed December 30, 1959, later issued to NASA as Patent no. 3,398,920) (hereafter Rosen-Williams Patent); and Donald D. Williams, Inglewood, CA, assigned to Hughes Aircraft Company, Culver City, CA, "Velocity Control and Orientation of a Spin-Stabilized Body," Patent no. 3,758,051 (filed August 21, 1964, issued September 11, 1973) (hereafter Williams Patent).

17. *Rosen and Williams and Hughes Aircraft Company v. The National Aeronautics and Space Administration*, US Patent Office Board of Patent Appeals, No. 8/64 (September 30, 1966) (hereafter *Rosen*)

18. *Williams v. Admin., Nat. Aero. Space*, 463 F.2d 1391 (C.C.P.A. 1972) (hereafter *Williams*).

19. *Hughes Aircraft Co. v. U.S.*, 226 Ct.Cl. 1, 640 F.2d 1193 (1980) (hereafter *Hughes I*).

20. *Hughes Aircraft Co. v. U.S.*, 29 Fed.Cl. 197 (1993) (hereafter *Hughes II*).

21. Slotten, "Satellite Communications."

22. Reich, "New Property," 736–37.

23. I treat the public/private boundary, alongside the idea of the global, as a social construct, following methodologies in technopolitics and legal history. T. Mitchell, "Limits of the State"; G. Hecht, *Radiance of France*; and Gordon, "Critical Legal Histories."

24. Gavaghan, *Something New*, 201; and Whalen, "Billion Dollar Technology," 95.

25. *Hughes II*.

26. *Rosen*; *Hughes II*; and NASA Act, §305.

27. Another difference is that the Williams Patent described control of the satellite's orientation or attitude (rotational position in orbit, or angle of the

satellite's spin axis), whereas the Rosen-Williams patent described control of the satellite's translational position (movement closer to or farther from Earth, or forward/backward along its orbital path).

28. *Hughes I.*

29. *Rosen.*

30. *Hughes I.*

31. *Hughes I,* 1208.

32. Whalen, *Origins of Satellite Communications*; Whalen, "Billion Dollar Technology"; and Gavaghan, *Something New.* This is remarkably similar to reliance by contemporary companies, such as SpaceX, on government ground infrastructure. (Credit to Asif Siddiqi for drawing the comparison.) On government support for contemporary aerospace corporations, see Space Angels, *U.S. Government Support of the Entrepreneurial Space Age,* June 20, 2019, https://www.spacecapital.com/publications/us-government-support-of-entrepreneurial-space-age-nasa-jpl; cf. Eric Berger, "A New Contract Highlights the Difference between 'New' and 'Old' Space," *Ars Technica,* June 1, 2017, https://arstechnica.com/science/2017/06/the-us-military-is-still-paying-a-spacex-competitor-for-rocket-upgrades/.

33. Morse, "Preliminary."

34. *Hughes I* (Skelton, Byron, dissenting).

35. *Rosen.*

36. *Rosen*; and NASA Act.

37. *Rosen,* 10–12.

38. *Rosen,* 1; Patent Application File, Rosen-Williams Patent, 75–91.

39. *Williams.*

40. Gavaghan, *Something New,* 211–20.

41. See Shapin and Schaffer, *Leviathan,* 225–82.

42. *Rosen,* 13–16.

43. T. Mitchell, "Limits of the State."

44. Hughes, Brochure, 1, 9, 3.

45. Hughes managers believed that Syncom would be cheaper than AT&T's Telstar, thus available to "have not" nations, a lever in "this vital race for men's minds." Slotten, "Satellite Communications," 342–43.

46. Hughes, Brochure.

47. Brian McNeil, "Nigeria and the New Frontier," *H-Net,* July 10, 2013, https://lists.h-net.org/cgi-bin/logbrowse.pl?trx=vx&list=H-Diplo&month=1307&week=b&msg=HVVuSrt4fIZfwzUuWW3cvg&user&pw. Stations in South Africa had been used since the late 1950s but did not meet public controversy in the United States until the late 1960s. See L. Ezell, *NASA Historical Data Book,* 406.

48. Conversation with Prime Minister Balewa of Nigeria by Means of the Syncom Communications Satellite, August 23, 1963, White House Audio Re-

cordings, 1961–1963, White House Audio Collection, John F. Kennedy Presidential Library and Museum, Boston, https://www.jfklibrary.org/asset-viewer/archives/JFKWHA/1963/JFKWHA-211–003/JFKWHA-211–003.

49. Edmond Buckley (director, Tracking and Data Acquisition), "Memorandum for the Record: Recommended Position on SYNCOM Ship in Nigeria," March 21, 1962, folder 006422, box 3/13/4, NASA Historical Reference Collection, Washington, DC (hereafter HRC), warning *Kingsport* officials about "tightening . . . Nigerian attitude toward our Mercury Station" in Kano; "Nigerians Call Off General Strike, but in No Mood to Celebrate," *Guardian*, October 1, 1963; Clyde Sanger, "Chances for Nigeria's Left Wing: New Party Gathers Support," *Guardian*, November 10, 1963.

50. *Astronautics and Aeronautics, 1966*, 134–35; "Demonstration of Syncom II Satellite with Trans Oceanic News Conference," NASA News Release, November 8, 1963, folder 006424, 28–29, box 3/13/5, HRC.

51. *Extraordinary Administrative Radio Conference to Allocate Frequency Bands for Space Radiocommunication Purposes* (Geneva, October 7–November 8, 1963) (hereafter EARC-63 Documents).

52. Slotten, "International Telecommunications Union," 323.

53. Bronfman, *Isles of Noise*; and Altshuler, "From Shortwave and Scatter to Satellite."

54. Minutes of the Fourth Plenary Meeting, November 6, 1963, Document no. 236-E, November 12, 1963), 9 (Cuban delegate's statement), EARC-63 Documents (Geneva, 1963).

55. This was a choice to conceal the reservation of frequencies for military applications. Slotten, "International Telecommunications Union."

56. Minutes of the Ninth Plenary Meeting, November 8, 1963, Document no. 241-E, November 14, 1963), 7 (Cuban delegate's statement), EARC-63 Documents (hereafter "Cuban dissent").

57. Slotten, "International Telecommunications Union."

58. In the 1970s, lawyers and diplomats from Colombia led a group of equatorial nations to sign the Bogotá Declaration of 1976, dividing sovereignty over GEO segments and their associated frequencies according to equatorial territory below: GEO "frequency-orbits" above the high seas were the "common heritage of mankind," while equatorial countries held sovereignty over frequency-orbits above them. *Declaration of the First Meeting of Equatorial Countries*, Bogotá, Colombia, adopted December 3, 1976. When US lawyers complained that Colombia had never made such claims, the Colombian delegates "point[ed] out that it drew attention at the 1963 Space Conference in Geneva to the future implications of these matters which are now becoming topical and urgent." Documents of the *World Administrative Radio Conference for the Planning of the Broadcasting-Satellite Service in Frequency Bands 11.7–12.2 GHz (Regions 2 and 3) and 11.7–12.5 GHz (Region 1)*, Geneva, January 10 to February 13, 1977. When

the *Journal of Air and Space Law* published the declaration in a 1978 issue, a Polish space lawyer mocked the declaration, tracing the origin of these conflicts to the launch of Syncom II in 1963. Gorbiel, "Legal Status of Geostationary Orbit," 171.

59. J. J. Dougherty and Leonard Jaffe, NASA Memorandum: "SYNCOM Turnover to the DOD," November 25, 1964, folder 006422, box 3/13/4, HRC.

60. *Astronautics and Aeronautics, 1964*, 332–33.

61. Parillo, "Leviathan."

62. See, for example, Black, *Global Interior.*

63. Jessup and Taubenfeld, *Controls for Outer Space*, 4–5, 121, 281–82.

64. Schmitt, Nomos *of the Earth*, 283–85; and Jessup, "Monroe Doctrine."

65. Cuban dissent.

CHAPTER 6. BAIKONUR AS A SACRIFICE ZONE

1. This research was funded by the Science Committee of the Ministry of Science and Higher Education of the Republic of Kazakhstan (Grant no. AP14870269), "Memory Infrastructure: Revitalization of Cultural Landscapes of Kazakhstan."

2. G. Hecht, *Entangled Geographies*, 3–4.

3. The name "Baikonyr" is a more authentic spelling in Kazakh as opposed to the Russified version, "Baikonur," used historically since the Soviet period.

4. Morrison, "Introduction."

5. Peterson, *Pipe Dreams*, 3.

6. Kopack, "Baikonur 2.0," 98.

7. Dubuisson, "Whose World?" 410.

8. Dawson, *Eco-Nationalism*, 162.

9. Weinthal and Watters, "Transnational Environmental Activism," 782.

10. Rozsa, "Nevada Movement," 99.

11. Cederlöf and Sivaramakrishnan, *Ecological Nationalisms.*

12. Kirchhof and Meyer, "Global Protest," 177.

13. Redfield, *Space in the Tropics.*

14. Kopack, "Baikonur 2.0," 97.

15. Lerner, *Sacrifice Zones.*

16. De Souza, "'Sacrifice Zone.'"

17. Heptyl is the local name used for the compound known in the West as unsymmetrical dimethylhydrazine (UDMH).

18. Andrews, *Red Cosmos.*

19. Josephson, "Rockets, Reactors, and Soviet Culture."

20. T. Martin, *Affirmative Action Empire.*

21. Batykova and Medeuova, "'Cosmos,'" 152.

22. Gumbert, "Cold War Theaters," 240.

23. Siddiqi, "Competing Technologies," 431.

24. Bekus and Medeuov, "Aspirations and Challenges," 327.

25. Nazarbayev, *Kazakhstan's Way*, 310–12.

26. Högselius, "Hidden Integration," 223.

27. Cooley, "Imperial Wreckage," 100.

28. Based on data available for 2020.

29. Kopack, "Rocket Wastelands," 556.

30. Nazarbayev, *Kazakhstan's Way*, 317.

31. "Deputat predlozhil otdat gorod Baikonur v vedenie Kazakhstana," *Kursiv*, April 11, 2019, https://kursiv.kz/news/politika/2019–04/deputat-pred lozhil-otdat-gorod-baykonur-v-vedenie-kazakhstana.

32. For the text of the original Agreement (in Russian), see https://docs .cntd.ru/document/901941545.

33. Ivan Cherebko, "Startovyi kompleks Zenitov na Baikonure perekhodit k Kazakhstanu," *Izvestiia*, July 19, 2013.

34. Sergei Golubev and Elisaveta Pestova, "U vas chto-to s iurisdiktsiei: Kazakhstanskogo obshchestvennika osudili za klevetu iz-za zhaloby rossiiskomu prezidentu," *Mediazona*, March 6, 2017.

35. Petr Trotsenko, "Kritika ili Kleveta? Kak mery Baikonura sudilis s kazakhstanskim pravozashchitnikom," *Radio Azattyk*, April 11, 2019.

36. Baikonur city administration, http://www.baikonuradm.ru/index.php ?mod=1957.

37. Marat Dauletbayev, "For Termination of the Lease Agreement for the Baikonur Complex," Facebook, June 17, 2020, https://www.facebook.com/ photo?fbid=1163547377331251&set=a.253428031676528.

38. "Ocherednoe narushenie kontitutsionnykh prav grazhdan Kazakhstana na Baikonure," Facebook, November 19, 2019, https://www.facebook.com/pro file.php?id=100010279989030&fref=ts.

39. "Laborers of the Earth—Preparing Ships for Space," Facebook, August 22, 2017, https://www.facebook.com/permalink.php?story_fbid=47938147574 7848&id=100010279989030.

40. Analytical Information Portal Almakz.info 2018, https://almakz.info/ antigeptil.

41. Scott, *Domination*, 207.

42. Sabitov, *Kazakhstanskii Kosmos*, 14.

43. The Eurasian Economic Union was established in 2014 with the participation of Belarus, Kazakhstan, and Russia, and was joined in 2015 by Armenia and Kyrgyzstan.

44. "Kazakhstantsy vyrazili protest vlastiam Rossii v sviazi s padeniem rakety na Baikonure," *Tengrinews*, July 2, 2013.

45. Svetlana Glushakova, "Aktivisty Antigeptila prizvali ne oskverniat' kazakhskuiu zemliu," *Radio Azattyk*, October 19, 2013.

46. "Antigeptil otmetil god s momenta avarii 'Protona M' aktsiei u posol'st-

va Rossii," *Matriza.kz*, July 3, 2014, http://www.matritca.kz/news/11462-anti geptil-otmetil-god-s-padeniya-protona-m-akciey-u-posolstva-rossii.html.

47. Orken Zhoyamergen, "Piket uchastnika 'Antigeptila,'" *Radio Azattyk*, February 5, 2014.

48. Sergei Tsekhmisternko, "Mezhdu Moskvoi i Alma-Atoi prolegla baikonurskaia step," *Kommersant*, February 25, 1994.

49. A. Berkimbayev, "Kosmodrom Baikonur—novyi vzgliad na razvitie Kazakhstana," Internet News Portal Zonakz.net, July 7, 2009.

50. Bekus, "Outer Space Technopolitics," 354.

51. Merkhat Sharipzhanov and Andrei Sharogradkii, "Saigaki gibnut riadom s Baikonurom," *Radio Svoboda*, June 3, 2015.

52. Kock et al., "Saigas on the Brink."

53. Caroline Charlier and Pavel Moiseev, "The Baikonur Cosmodrome: Are There Legal Ways to Address the Danger?" *Legal Dialogue*, December 12, 2018, https://legal-dialogue.org/the-baikonur-cosmodrome-are-there-legal-ways-to-address-the-danger.

54. Chernykh and Fominykh, "Diskursy iadernogo nerasprostraneniia."

55. See https://ecomuseum.kz.

56. Turgai Alimbayev et al., "Ecological Problems of Modern Central Kazakstan: Challenges and Possible Solutions," E3S Web of Conferences, 2020, 157, https://doi.org/10.1051/e3sconf/202015703018.

57. Kopack, "Rocket Wastelands," 565.

58. Yessenova, "Tengiz Oil Enclave."

59. Klinger, "Environmental Geopolitics," 683.

60. Ormrod "Beyond World Risk Society?" 740–41.

61. Larkin, "Politics and Poetics."

62. Cresswell, *In Place/Out of Place*.

63. Dawson, *Eco-Nationalism*.

64. Abraham, *Making of the Indian Atomic Bomb*.

65. Svetlana Glushakova, "Ekspert OON vstretiltsia s aktivistami Astany," *Radio Azattyk*, March 28, 2015, https://rus.azattyq.org/a/26925195.html.

66. Latour, *Down to Earth*, 19.

FRAGMENT II: TERRESTRIAL CHAUVINISM

1. N. S. Kardashev, "On the Inevitability and the Possible Structures of Supercivilizations," *The Search for Extraterrestrial Life: Recent Developments; Proceedings of the Symposium*, Boston, June 18–21, 1984.

2. "John F. Kennedy Moon Speech—Rice Stadium," NASA, https://er.jsc.nasa.gov/seh/ricetalk.htm.

3. "The Man Trap," *Star Trek*, season 1, episode 1, premiere, September 8, 1966.

4. Bush, *Science*.

5. Launius, "Underlying Assumptions."

6. Siddiqi, *Red Rockets' Glare.*

7. Wukelic, *Handbook*, 456.

8. Kardashev, "Peredacha informatsii."

9. Kardashev, "On the Inevitability."

10. Kardashev, "On the Inevitability."

11. For histories of SETI in the Soviet Union and Russia, see Gindilis, "40 let SET v SSSR"; and Gindilis and Gurvits, "SETI in Russia."

12. It should be noted that while the intention was to send the message to Venus, it was later calculated in 2002 that the message was probably unintentionally sent in the direction of the star HD131336 in the Libra constellation.

13. Quast, "Profile of Humanity."

14. For general histories of SETI in the US context, see Dick, *Life on Other Worlds*; Squeri, *Waiting for Contact*; and S. Smith, "'Cosmic Rorschach Test.'"

15. M. Hart, "Explanation," 132.

16. National Security Agency, "Stonehouse: First U.S. Collector of [REDACTED] Signals," (undated but declassified September 2007), https://www.nsa.gov/Portals/70/documents/news-features/declassified-documents/cryptologic-spectrum/stonehouse.pdf; and "The Longest Search: The Story of the Twenty-One Year Portrait of the Soviet Deep Space Data Link and How It Was Helped by the Search for Extraterrestrial Intelligence," National Security Archives (undated, but produced after 1983, declassified September 2011), 2.

17. "Longest Search."

18. See especially Gorman (chapter 4), Siddiqi (chapter 3), and Bekus (chapter 6) in this book.

19. Carl Sagan, "The Common Enemy," *Parade*, February 7, 1988.

20. Sagan, "Common Enemy."

21. Sagan, "Common Enemy."

CHAPTER 7. WASTE MAKES THE FRONTIER

1. For a global overview of space expenditures by national governments, see Space in Africa, *Global Space Budgets.*

2. Seaton, "Guided by the Dark," 240.

3. See White, *Overview Effect.*

4. See Spector and Higham, "Space Tourism."

5. On climate doomerism, see, for example, Mary Helgar, "Home Is Always Worth It," *Medium*, September 19, 2019, https://medium.com/@maryheglar/home-is-always-worth-it-d2821634dcd9; on neo-Malthusian perspectives in contemporary space expansionism, see Deudney, *Dark Skies.* Marie Morales, "Blue Origin's Jeff Bezos Predicts Earth Will Be a 'Natural Resort' and Only a Few Will Be Allowed to Stay," *Science Times*, November 15, 2021, https://www

.sciencetimes.com/articles/34513/20211115/earth-natural-resort-future-blue -origin-s-jeff-bezos-predicts.htm.

6. Colonizing space by any means necessary to escape the failure to address crises on Earth is an idea that has been explored with increasing intensity in the past decade. See, for example, Joe Mascaro, "To Save Earth, Go to Mars," *Aeon*, May 11, 2016; and Tom Risen, "Selling Mars as Planet B," *Aerospace America*, May 31, 2017.

7. See, for example, Catherine Thorbecke and Jackie Wattles, "Here's Who's Flying on Thursday's Blue Origin Space Launch," *CNN*, March 28, 2022, https://www.cnn.com/2022/03/28/tech/blue-origin-launch-passengers-scn/ index.html.

8. See, for example, Christopher Ingraham, "World's Richest Men Added Billions to Their Fortunes Last Year as Others Struggled," *Seattle Times*, January 1, 2021, https://www.seattletimes.com/business/worlds-richest-men-added -billions-to-their-fortunes-last-year-as-others-struggled/; Sara Sirota and Ryan Grim, "Senate Preparing a $10 Billion Bailout Fund for Jeff Bezos Space Firm," *Intercept*, May 25, 2021, https://theintercept.com/2021/05/25/jeff-bezos-blue -origin-senate-bailout/; Christopher Jasper and Fabian Graber, "Virgin Atlantic Wins Creditor Backing for $1.6bn Bailout," *Aljazeera*, August 25, 2020, https:// www.aljazeera.com/economy/2020/8/25/virgin-atlantic-wins-creditor-back ing-for-1-6bn-bailout; and Shawn Tully, "As Bitcoin Soars to Near $50,000, Elon Musk's Profit Jumps by 250%," *Fortune*, August 24, 2021, https://fortune .com/2021/08/24/bitcoin-price-btc-elon-musk-profits/. See also Beery, "State, Capital and Spaceships."

9. On "social wage," see Estes, Gilmore, and Loperena, "United in Struggle."

10. A 2005 article in the *Economist* read: "Running out of exotic places to visit or more extreme thrills for an even bigger adrenaline rush? For those who can afford it, a wholly new travel experience is coming over the horizon. When it arrives, space tourism will offer the ultimate in bragging rights." *Economist*, May 12, 2005, https://www.economist.com/news/2005/05/12/the-ride-of-your -life. See also Liu and Li, "How Travel Earns Us Bragging Rights."

11. "We react to apocalyptic threats by either partying (assuaging our apocalyptic anxiety through increased consumerism, reasoning that if it all may be gone tomorrow, we might as well enjoy it today), praying (in hopes that divine intervention or mere time will allow us to avoid confronting the challenges before us), or preparing (packing 'bugout' packs for a quick escape or stocking up on gold, guns, and canned food, as though the transformative moment we anticipate will be but a brief interlude, a bad winter storm that might trap us indoors for a few days or weeks but that will eventually melt away)." Matthew Barrett Gross and Mel Gilles, "How Apocalyptic Thinking Prevents Us from Taking Political Action," *Atlantic*, April 23, 2012, https://www.theatlantic.com/politics/archive/2012/04/ how-apocalyptic-thinking-prevents-us-from-taking-political-action/255758/.

12. See Jonathan Watts, "The Sound of Icebergs Melting: My Journey into the Antarctic," *Guardian*, April 9, 2020, https://www.theguardian.com/world/ng-interactive/2020/apr/09/sound-of-icebergs-melting-journey-into-antarctic-jonathan-watts-greenpeace.

13. Estimates on CO_2 emissions vary. For one such estimate, see Champion Traveler Editors, "One SpaceX Rocket Launch Produces the Equivalent of 395 Transatlantic Flights Worth of CO2 Emissions," https://championtraveler.com/news/one-spacex-rocket-launch-produces-the-equivalent-of-395-transatlantic-flights-worth-of-co2-emissions/. On space colonies, see J. Phelps, "Gary Johnson Calls for Space Colonization, Vows to Stay in Race," *ABC News*, September 25, 2016, https://abcnews.go.com/Politics/gary-johnson-calls-space-colonization-vows-stay-race/story?id=42341834; and Matt Novak, "Libra: The 21st Century (Libertarian) Space Colony," *Smithsonian Magazine*, February 18, 2013, https://www.smithsonianmag.com/history/libra-the-21st-century-libertarian-space-colony-19119736/.

14. Liboiron and Lepawsky, *Discard Studies*.

15. Gorman, "La Terre et l'espace."

16. Klinger, *Rare Earth Frontiers*.

17. d'Avignon, *Ritual Geology*.

18. For a brilliant examination of the exhaustion trope in literature and fiction, see C. Miller, *Extraction Ecologies*. Quote is from Morales, "Blue Origin's Jeff Bezos Predicts."

19. See Gabrielle Hecht, "Residue," *Somatosphere*, January 8, 2018, http://somatosphere.net/2018/residue.html/. On "Eden under glass," see Hecht and Cockburn, *Fate of the Forest*.

20. On thanotourism and mining disasters, see Świtała-Trybek, "Memorial Tourism." On mine tourism, see Pretes, "Touring Mines."

21. The literature on reservations and parks as spaces of enclosure, explusion, confinement, and exclusion is deep and global in scope. See, for example, Biolsi *Power and Progress*; and Brockington, *Fortress Conservation*.

22. The conservationist advocacy for outer space is also several decades old. To be clear, despite my critique of how the conservationist ethos serves broader practices of enclosure, extraction, and abandonment, the conservation movement has been and continues to be an important force for staying the wholesale destruction of landscapes. The early concerns that runaway industrialism (and now, unfortunately, militarization it seems) could eventually lay ruin to the shared right to peaceful use and access of the near-Earth space environment were prescient and have only grown more important. For some influential examples, see Cockell and Horneck, "Planetary Park System"; and Elvis and Milligan, "How Much of the Solar System Should We Leave?"

23. On the co-constitution of Native American dispossession and the creation of national parks in the United States, Carolyn Merchant writes, "The parks were vast managed gardens in which the wild was contained for viewing.

People could have a wilderness experience in a protected environment." Merchant, "Shades of Darkness," 382. On the logics of enclosure that undergird high-end nature-based tourism and gated "wildlife estates" in South Africa, see Koot, Büscher, and Thakholi, "New Green Apartheid?"

24. For a global analysis of the majority incidence of potentially economic critical raw materials deposits on Indigenous and peasant lands, see Owen et al., "Energy Transition Minerals." For a discussion of critical or strategic mineral mining as a technology of territoriality, see Klinger, *Rare Earth Frontiers.* For an examination of the environmental impacts and politics around industrialized mine waste and e-waste processing, see Wolk, "Invisible Damage."

25. Siddiqi, "Dispersed Sites."

26. See the watershed work of Redfield, *Space in the Tropics.* On regimes of exclusion in northeastern Brazil, see S. Mitchell, *Constellations of Inequality.* On ranges and detention centers, see Gorman, "La Terre et l'espace."

27. See, for example, Woodburn, "Pushback in the Jet Age." See also S. Mitchell, *Constellations of Inequality.*

28. On Baikonur cosmodrome (Kazakhstan), see Christine Bichsel, "When Things Fall from the Sky," https://www.societyandspace.org/articles/when-things-fall-from-the-sky-understanding-rocket-stages-on-the-kazakh-steppe-as-imperial-debris; and Kopack, "Baikonur 2.0." On Algeria, see, for example, Hennaoui and Nurzhan, "Dealing with a Nuclear Past." On Kenya, see Siddiqi, "Dispersed Sites."

29. See, for example, Ryan et al., "Impact of Rocket Launch."

30. Small aircraft still use leaded gasoline. See, for example, Federal Aviation Administration, "Leaded Aviation Fuel and the Environment," United States Department of Transportation, 2019, https://www.faa.gov/newsroom/leaded-aviation-fuel-and-environment.

31. See Bumpus and Liverman, "Carbon Colonialism?"

32. See Liboiron, *Pollution Is Colonialism*, 6, 135.

33. See, for example, Brenner and Katsikis, "Operational Landscapes."

34. On graveyard orbit, see Saperstein, "Opening the Black Box."

35. See Rush, "Astronauts' Sensemaking."

36. Frank White, the philosopher who coined the term "the overview effect," quoted in an NPR story on William Shatner's experience of profound grief during his spaceflight on October 13, 2021. See Enrique Rivera, "William Shatner Experienced Profound Grief in Space: It Was the 'Overview Effect,'" *National Public Radio*, October 23, 2022, https://www.npr.org/2022/10/23/1130482740/william-shatner-jeff-bezos-space-travel-overview-effect.

37. See, for example, Carolin O'Donovan and Lauren Kaori Gurley, "Labor Board Rejects Amazon's Objections to Union Victory," *Washington Post*, September 1, 2022, https://www.washingtonpost.com/technology/2022/09/01/amazon-union-victory/.

38. Refer to Lisa Ruth Rand's forthcoming *Space Junk*.

39. See, for example, Garber and Rand, "Montreal Protocol?"

40. For a proof of concept that platinum group elements can be had from the dust in tailpipes and parking garages, see the performance piece by the artist Louize Harries: *ALL THAT'S SOLID MELTS INTO AIR (and the Domestic Mobile Mining Unit)*, 2018, https://vimeo.com/319012478 (at 4m, 14s). See Ingrid Burrington, "In Space No One Can Hear You Sell," *RIP Corp*, March 23, 2021, https://ripcorp.biz/episodes/in-space-no-one-can-hear-you-sell, 45m21s; and Andrew Glester, "The Asteroid Trillionaires," *Physics World*, June 11, 2018, https://physicsworld.com/a/the-asteroid-trillionaires/.

41. On the health and livelihood status of informal recyclers in Mongolia, see Uddin and Gutberlet, "Livelihoods and Health Status." On prison labor and trash handling in the United States, see, for example, Tim Darragh, "Dirty Details of How Much Trash N.J. Prisoners Are Clearing from Our Highways," *NJ Advance Media*, May 16, 2015, https://www.nj.com/news/2015/05/you_might_be_surprised_by_how_much_trash_was_picked_up_by_nj_prisoners_last_year.html.

42. Olson, "Political Ecology."

43. Perdue, *China Marches West*.

44. There is an immense literature on nomads and sedentarization. For recent theorizations with respect to state territoriality, see, for example, Levin et al., "Before and after borders."

45. NASA, "Planetary Protection," NASA Office of Safety and Mission Assurance, https://sma.nasa.gov/sma-disciplines/planetary-protection.

46. For the theory, see A. Cartwright, "The Venus Hypothesis," *arXiv: Earth and Planetary Astrophysics*, https://doi.org/10.48550/arXiv.1608.03074.

47. See, for example, O'Donovan and Gurley, "Labor Board Rejects Amazon's Objections"; Robert Reich, "In Space, No One Will Hear Bezos and Musk Workers Call for Basic Rights," *Guardian*, April 25, 2021, https://www.theguardian.com/commentisfree/2021/apr/25/elon-musk-jeff-bezos-space-moon-mars-workers-rights-unions; and Alexia Fernández Campbell, "Elon Musk Broke US Labor Laws on Twitter," *Vox*, September 30, 2019, https://www.vox.com/identities/2019/9/30/20891314/elon-musk-tesla-labor-violation-nlrb.

CHAPTER 8: FIXING SKYLAB

1. For the American context, see Kilgore, *Astrofuturism*; O'Neill, *High Frontier*; and Scharmen, *Space Settlements*. For the Russian context, see Scharmen, *Space Forces*; and Siddiqi, "Imagining the Cosmos."

2. On capitalism, see Dickens and Ormrod, *Cosmic Society*; and Valentine, "Exit Strategy." On climate change, see Tutton, "Sociotechnical Imaginaries."

3. Tutton, "Sociotechnical Imaginaries," 426.

4. A. Russell and Vinsel, "Make Maintainers."

5. S. Graham and Thrift, "Out of Order"; Henke, "Mechanics of Workplace Order"; Orr, *Talking about Machines*; and Strebel, Bovet, and Sormani, *Repair Work Ethnographies*.

6. Jackson, "Rethinking Repair."

7. Burrell, *Invisible Users*; Kesha Michelle Fevrier, "Race and Waste: The Politics of Electronic Waste Recycling & Scrap Metal Recovery in Agbogbloshie, Accra, Ghana," September 20, https://yorkspace.library.yorku.ca/xmlui/handle/10315/37961; Jackson, "Rethinking Repair"; and Jesse Peterson and Alex Zahara, "Anthropocene Adjustments: Discarding the Technosphere," *Discard Studies*, May 26, 2016, https://discardstudies.com/2016/05/26/anthropocene-adjustments-discarding-the-technosphere/.

8. Crosby and Stein, "Repair"; and Mattern, "Maintenance and Care."

9. Robinson, *Black Marxism*.

10. Debbie Chachra, "Why I Am Not a Maker," *Atlantic*, January 23, 2015, https://www.theatlantic.com/technology/archive/2015/01/why-i-am-not-a-maker/384767/.

11. Aronowsky, "Of Astronauts and Algae."

12. Garriott, "Skylab Report"; Kelly and Dean, *Endurance*; Linenger, *Off the Planet*; Parazynski and Flory, *Sky Below*; and Robert Thirsk, "Preparing for the Next Global Crisis," *Robert Thirsk* (blog), August 21, 2020, https://robertthirsk.ca/2020/08/21/preparing-for-the-next-global-crisis/.

13. Crosby and Stein, "Repair," 181.

14. Mindell, *Digital Apollo*.

15. Russell L. Schweickart, Oral History 2 Transcript, interview by Rebecca Wright, March 8, 2000, Johnson Space Center History Office, https://historycollection.jsc.nasa.gov/JSCHistoryPortal/history/oral_histories/SchweickartRL/SchweickartRL_3-8-00.htm, 13–14.

16. *MCC Performance Evaluation Reports for the Skylab Missions*, vol. 1, Skylab 1 and Skylab 2, Manned Phase, 1973. Johnson Space Center Archives.

17. *MCC Performance Evaluation Reports*, vol. 2.

18. Russell and Klaus, "Maintenance."

19. Garriott, "Skylab Report," 225.

20. These main solar arrays were to provide the station with an adequate amount of power, but thankfully, the Apollo Telescope Mount (ATM) also had four, albeit much smaller, solar array panels that provided the power to operate the scientific instruments of the ATM. These solar arrays sustained the station until the first Skylab crew unjammed the remaining solar array of the station.

21. Compton and Benson, *Living and Working in Space*, 257.

22. Garriott, "Skylab Report."

23. Jack Kinzler, Oral History 2 Transcript, interview by Paul Rollins, January 16, 1998, 14–45, Johnson Space Center History Office, https://history

collection.jsc.nasa.gov/JSCHistoryPortal/history/oral_histories/KinzlerJA/JAK_1-16-98.pdf.

24. "The First Space Stations," 2023, https://airandspace.si.edu/stories/editorial/first-space-stations.

25. Jackson, "Rethinking Repair," 227.

26. Chachra, "Why I Am Not a Maker."

27. Ahmed, Jackson, and Rifat, "Learning to Fix"; Jackson, "Rethinking Repair"; Mattern, "Maintenance and Care"; and Rifat, Prottoy, and Ahmed, "Breaking Hand."

28. Tichi, *Shifting Gears.*

29. Kinzler, Oral History 2 Transcript, 13–12.

30. Mika McKinnon, "This Is the Woman Who Replaced Skylab's Destroyed Sunshield," Gizmodo, May 3, 2015, https://gizmodo.com/this-is-the-woman-who-replaced-skylabs-destroyed-sunshi-1689346849. NASA Archives, "Space History Photo: Skylab Solar Shield," Space.com, January 22, 2014, https://www.space.com/24375-skylab-solar-shield.html.

31. Compton and Benson, *Living and Working in Space*; Hitt, Garriott, and Kerwin, *Homesteading Space*; Melanie Whiting, "Skylab 2: 'We Can Fix Anything!,'" Text, NASA, May 22, 2018, http://www.nasa.gov/feature/skylab-2-we-can-fix-anything; and NASA, "Seamstresses Stitch a Sun-Shade for Skylab," Text, NASA, 2008, http://www.nasa.gov/multimedia/imagegallery/image_feature_543.html; and Zimmerman, *Leaving Earth.*

32. Margalit Fox, "Jack Kinzler, Whose Ingenuity Saved Skylab, Dies at 94," *New York Times*, March 15, 2014, https://www.nytimes.com/2014/03/15/us/jack-kinzler-skylabs-savior-dies-at-94.html.

33. Kerwin, Oral History 2 Transcript, 12–40.

34. Wilde et al., "One Hundred US EVAs."

35. Kerwin, Oral History 2 Transcript, 12–42.

36. Kerwin, Oral History 2 Transcript, 12–43.

37. *MCC Performance Evaluation Reports for the Skylab Missions*, vol. 3, Skylab 3, Unmanned Phase, and Skylab 4, Manned Phase, 1974.

38. Gorman, "Sky Is Falling."

39. Lousma, "Role of Man in Space," 8.

40. Lousma, "Role of Man in Space," 8.

41. Jackson, "Rethinking Repair," 221.

42. Garriott, "Skylab Report," 225.

43. Lousma, "Role of Man in Space," 9.

44. Réka Patrícia Gál, "Climate Change, COVID-19, and the Space Cabin: A Politics of Care in the Shadow of Space Colonization," *Mezosfera*, October 2020, http://mezosfera.org/climate-change-covid-19-and-the-space-cabin-a-politics-of-care-in-the-shadow-of-space-colonization/.

45. Kaitlin McTigue et al., "Extreme Problem Solving: The New Challenges of Deep Space Exploration," in *SpaceCHI: Human-Computer Interaction for Space Exploration Workshop at CHI* 5 (2021), 1.

CHAPTER 9. EARTHLY DREAMS AND COSMIC AFTERLIVES

1. Latour, "From Realpolitik to Dingpolitik," 521–22.
2. See Dusinberre, *Hard Times*.
3. Masao, "Production of Weapons," 2.
4. Matogawa Yasunori, "Pencil Rocket Story," https://global.jaxa.jp/article/interview/sp1/episode-4_p1_e.html.
5. See Logsdon, "U.S.-Japanese Space Relations," 294–300; Sato, "Contested Gift of Power," 182–85; and Watanabe, "Japan-US Space Relations," 1.
6. Oda, *Uchū kūkan kansoku 30-nen shi*, 24.
7. Oda, *Uchū kūkan kansoku 30-nen shi*, 8.
8. "Kotoshi no hanashi," *AS Akita Supplement*, December 17, 1955.
9. "Iwaki chō Michikawa kaigun de—ken mo kyōryoku jissoku wa sanju ni nen kara," *AS Akita Supplement*, August 2, 1955.
10. Oda, *Uchū kūkan kansoku 30-nen shi*, 7; and "Iwaki chō Michikawa kaigun de."
11. Matogawa Yasunori, "Pencil Rocket Story," https://global.jaxa.jp/article/interview/sp1/episode-4_p1_e.html.
12. "Kotoshi no hanashi"; and "Suichoku jikendai mo tsukuru," *AS Akita Supplement*, November 14, 1955.
13. See Low, "Accelerators," 275–80, 286.
14. See George, *Minamata*.
15. Aldrich, *Site Fights*, 1–7.
16. Dusinberre, *Hard Times*, 157–64.
17. "Yozora ni hikari no o hiite—ju ku nichi kara yokan roketto,' *AS*, June 5, 1957; "Jikken han sanjūnin, junbi o hajimeru," *AS*, June 19, 1957; and "Mujōken de kyōryukō—gyogyōsha ga mōshiawasu," *AS*, September 12, 1957.
18. "Suichoku jikendai mo tsukuru."
19. See, for example, "Shin roketto o tsukuru," *AS Akita Supplement*, February 9, 1956; "Ichigatsu ni 'Kappa' jikken," *AS Akita Supplement*, December 28, 1955; and "Tōbu kaishū testo—17 nichi Kappa go-kei ni sonaete," *AS Akita Supplement*, September 13, 1957.
20. "Kōsō no kiryū, kaze o kansoku—kyō Michikawa de Kappa roku sei uchiage,' *AS Akita Supplement*, December 25, 1958.
21. "Kotoshi no hanashi."
22. "Roketto jikken—Michkawa Kaigan ga saiteki," *AS Akita Supplement*, September 19, 1955; and "Hikitsuzuki Michikawa kaigan de—hokano yoi basho wa nai," *AS Akita Supplement*, December 5, 1957.

23. "Roketto kansoku dōryōkukai tansei," *AS Akita Supplement*, July 25, 1957; and "Shigoto shiyasui yo—roketto kansoku dōryōku kai hassoku," *AS Akita Supplement*, July 30, 1957.

24. "Dai-2 ki was fuseikō—'Bebii-T kei no tesuto,'" *AS Akita Supplement*, September 20, 1955.

25. "Kotoshi no hanashi."

26. "Natsuzora ni hakuen hitosuji—roketto jiken testo wa jōjō," *AS Akita Supplement*, August 7, 1955; and "3-ban ki wa muji ni hassha," *AS Akita Supplement*, September 22, 1955.

27. "Kenbutsu jin kara mo kansei," *AS Akita Supplement*, September 25, 1956; Gōkana jikkenhan—Michikawa no roketto tesuto," *AS Akita Supplement*, December 4, 1956; "Jiken kaku han hyaku % no katsudo—aozora yogiru ookei roketto," *AS Akita Supplement*, April 25, 1957; and "Kappa-3 ki tetsu kan hikō jiken," *AS Akita Supplement*, June 24, 1957.

28. "24 nichi no kappa jikken ni sonaete," *AS Akita Supplement*, September 21, 1956.

29. "Eikoku terebi han mo satsuei," *AS Akita Supplement*, May 3, 1957.

30. "Kasei o kansoku shiyo—ganbaru tenbun kurabu in," *AS Akita Supplement*, September 4, 1956.

31. "Sensei to seito no gassoku—Yokote ni-chū ni yon tsubō no seiza," *AS Akita Supplement*, November 2, 1956.

32. "Jinkō eisei no X-masu kēki—Akita, onedan wa 20 man en," *AS Akita Supplement*, December 10, 1957.

33. "Kappa-3 ki tetsu kan hikō jikken 'Chōonsoku no hashiri ya'—Michikawa kaigan kenbutsu jin mo odoroku," *AS Akita Supplement*, June 24, 1957; and "Isshun ni kumoma ni—'Kappa' roketto jikken," *AS Akita Supplement*, December 9, 1956.

34. "Kappa-3 ki tetsu kan hikō jikken"; and "Iwaki chō Michikawa kaigun de."

35. "24 nichi no kappa jikken ni sonaete."

36. *AS Akita supplement*, "Tōbu kaishū testo—17 nichi Kappa go-kei ni sonaete," *AS Akita Supplement*, September 13, 1957.

37. Matogawa, "Survey of Rocketry," 211.

38. Oda, *Uchū kūkan kansoku 30-nen shi*, 7.

39. "Roketto jikken—Michkawa Kaigan ga saiteki."

40. Matogawa, "'Pencil' Rocket," 129.

41. "Roketto uchū kansoku—kyō dai-1 ken no jikken," *AS Akita Supplement*, August 6, 1955; and "Bebii-s' kei mo jōjō—kokusan roketto tesuto," *AS Akita Supplement*, August 24, 1955.

42. "Kōkei roketto bakuhatsu jikken—chūgakusei hitori shinu," *AS Akita Supplement*, September 25, 1957.

43. "Roketto asobi de jūshō," *AS Akita Supplement*, March 4, 1958.

44. "Chūgakusei roketto asobi de jūshō," *AS Akita Supplement*, January 30, 1959; and "Sanin ga jūkeishō, Hanawa," *AS Akita Supplement*, June 26, 1959.

45. "'Kinshi o seyō' to senmonka," *AS Kagoshima Supplement*, October 10, 1968; and "Isshun kieta shimpai—kurōhō irareta to yorokobi," *AS Kagoshima Supplement*, September 18, 1968.

46. Oda, *Uchū kūkan kansoku 30-nen shi*, 7.

47. Matogawa, "Survey of Rocketry," 211; and Oda, *Uchū kūkan kansoku 30-nen shi*, 53.

48. "Roketto jikken shippai de jimoto no fuan takamaru—michikawa kaigan," *AS Akita Supplement*, May 26, 1962.

49. "Roketto sentā o Noshiro e," *AS Akita Supplement*, June 27, 1962; "Noshiro e no iten gutaika—Michikawa no roketto sentā," *AS Akita Supplement*, July 8, 1962; "Noshiro e iten ga ketteiteki—Roketto sentā Tōdai seiken ga setsumei kai," *AS Akita Supplement*, July 12, 1962.

50. "San shū sanshō no kenbutsu kansei—Noshiro no roketto jikkenjō, kumitate tendō tawā nado," *AS*, December 12, 1964.

51. Nobuyoshi , "Eisei yowa," 2–4.

52. Middleton, "Space Rush," 261.

53. Siddiqi, *Challenge to Apollo*, 135.

54. Redfield, "Beneath a Modern Sky," 261.

55. Latour, "On Actor-Network Theory," 373.

56. Redfield, "Half-Life of Empire," 802, 804; and A.-S. Martin, "Kourou."

57. Takahashi Tetsuo, "Jikkenjō totomoni 50 nen," in ISAS, *Uchinoura uchū kūkan kansokujo no 50 nen 1962–2012* (Uchinoura uchū kūkan kansokujo no 50-nen' kinen-shi henshū iinkai, November 2011), 16.

58. "Tōdai uchū kūkan kansokujo o kikō—Kagoshima, Uchinoura," *AS*, February 3, 1962; "Honnendo nimo kōji susumu—roketto dōro," *AS Kagoshima Supplement*, May 7, 1964; "Jimoto ga gakkuri—Uchinoura, Tanegashima, "Yota yota roketto," *AS*, March 28, 1967.

59. JAXA, "Ibuki uchiage tokusetsu saito," https://www.jaxa.jp/countdown/f15/tane_guide/index_j.html#01; Kagoshima Prefecture Kumage Branch Office, *Reiwa gannendo Kumage chiiki no gaikyō*, Kagoshima Prefecture, March 2019, 217.

60. "Are hōdai no uchū sentā Tanegashima," *AS Kagoshima Supplement*, July 13, 1967; "Umetatechi no zōsei hajimaru," *AS Kagoshima Supplement*, August 13, 1969; and Minamitanechō City Hall, "Uchūgaoka kouen," http://www.town.minamitane.kagoshima.jp/institution/ucyugaokapark.html.

61. Kagoshima Prefectural Office, *Kakukai kokuseichōsaji no shichōsonbetsu jinkō no suii*, https://www.pref.kagoshima.jp/ac09/tokei/bunya/kokutyo/h27kokutyo/documents/55416_20161128120608-1.pdf.

62. Kagoshima Prefectural Office, *Dai-4 no jinkō dōkō*, http://www.pref.kagoshima.jp/ap01/chiiki/kumage/chiiki/documents/71700_20190409114440-1.pdf.

63. Seki Shūichi, *Chūsei Nihon ni okeru gairai gijutsu denrai no sho jōken: Kaijō kōtsū to no kanren kara (chūsei no gijutsu to shokunin ni kansuru sōgō-teki kenkyū)*, Bulletin of the National Museum of Japanese History 210, 251–52.

64. Wilson, "'New Paradise,'" 242–44.

65. "Furuki yoki Tanegashima o'—Uchugahama ni hakubutsukan o," *AS Kagoshima Supplement*, September 20, 1969.

66. Takenaka Yukihiko, "N-1 roketto kaihatsu no ayumi," *Nihon kōkū uchū gakkaishi* 23, no. 362 (1984), 127.

67. "Roketto kichi no meian—Tanegashima," *AS Kagoshima Supplement*, September 8, 1969.

68. Kimotsukicho Kentō iinkai, "Kimotsuki machi toshi keikaku masutāpuran' sakutei ni muketa chōmin ishiki chōsa hōkoku," Kimostukicho yakushō, Kimotsukicho, July 18, 2012, 21–23.

69. Oda, *Uchū kūkan kansoku 30-nen shi*, 7.

70. Oda, *Uchū kūkan kansoku 30-nen shi*, 18.

71. "Wasei shatoru jikken," *AS*, June 17, 1986.

72. University Space Engineering Consortium (UNISEC), "Report on the Current State of 'Japanese University Rocket Projects,'" Tokyo, October 2011, 4–7.

73. "Yume mo takadaka," *AS*, October 27, 2013.

74. "'Noshiro uchū ibento' asu kara," *AS*, August 18, 2010.

75. "San nin dake no chikara uchū e," *AS*, June 30, 2014.

76. Noshiro City Office, "Noshiro-shi kodomokan 2-kai uchū kan shōkai," https://www.city.noshiro.akita.jp/c.html?seq=8125.

77. Akita Prefectural Office, *Tsukurō! Tobasō! Famirii sui roketto gijutsukai*, Akita City, 2012.

78. Noshiro Uchū Ebento, *Setsuritsu shushi*, http://www.noshiro-space-event.org/about_sub01.html.

79. Noshiro Uchū Ebento, *Setsuritsu shushi*.

80. Noshiro Uchū Ebento, *Setsuritsu shushi*.

81. MEXT stands for the Ministry of Education, Culture, Sports, Science and Technology. Tatemori Seikō, "Ginga Renpō," *ISAS News*, no. 81, October 1987, 10.

82. "Ginga Renpō no kenkokushiki," *ISAS News*, no. 81, December 1987, 6. Uchū kagaku kenkyūjō, *Ginga Renpō kyōwakoku yūkō toshi "Iwate-ken Ōfunato-shi" e no Higashi Nihon daishinsai fukkō shien undō no kiroku*, Kimotsuki, Kagoshima-ken, March 2012, 10–12.

83. Noshiro City Office, "Ginga Renpō kaku kyōwakoku no shōkai," https://www.city.noshiro.akita.jp/c.html?seq=1993.

84. Kashiopea Renpō, *Kashiopea Renpō*, https://cassiopeia-iwate.jp/.

85. Taiki City, *Ginga Renpō Yūko kōryū 25 shū nen*, https://www.town.taiki.hokkaido.jp/soshiki/kikaku/kikaku/ginga-Renpō.data/ginga25.pdf; and

Kawaguchi Isao, "Uchukagakukenkyūjō ga musubu Ginga Renpō," *ISAS News*, no. 212, November 1998, 12.

86. "Noshiro no nebuta karite ka-nibaru—Sagamihara no shōtengai to jidai-kai," *AS*, October 3, 1993; "Ginga Renpō no ringo no ki, kotoshi no ōnā bōshū," *AS*, June 18, 1994; Kawaguchi Isao, "Uchukagakukenkyūjō ga musubu Ginga Renpō," *ISAS News*, no. 212, November 1998, 12; and "Ringo no ōnā bōshū," *AS*, June 10, 1988.

87. "Ginga Renpō kokki, uchū o kakemeguru," *AS*, January 18, 1996.

88. "Shika no akachan tanjō," *AS*, June 14, 1990.

89. "Uchū kagaku hito me de, Sagamihara de tenrankai," *AS*, November 4, 1990; Miura Hideo, "Ginga Renpō Noshiro samitto kaisai," *ISAS News*, no. 150, September 1993, 4; Ogawa Isao, "Ginga Renpō ni yume o yosete," *ISAS News*, no. 273, December 2003, 9; "Ringo no ōnā bōshū," *AS*, June 10, 1988; and "Chijō no 'tennyō' ga atsumaru—Ginga Renpō kōryūkai, asukara Sagamihara de," *AS*, July 28, 1994.

90. "Rikisaki zurari go man ten: 'Zōkei Sagami kazekko ten' Sagamihara," *AS*, November 3, 1991.

91. Tatemori Seikō, "Ginga Renpō," *ISAS News*, no. 81, October 1987, 10; and Kogawa Hisao, "Ginga Renpō ni yume o yosete," *ISAS News*, no. 273, December 2003, 9.

92. "Otoshiyori, sakuhinkai de 1-nen no seika happyō," *AS*, February 13, 1991.

93. "Petto botoru sairiyō no roketto asobi ninki kyūjōshō," *AS*, August 25, 1995.

94. "'Ginga Renpō' saigaiji mo renkei," *AS*, February 1, 1996.

95. Kimotsuki Town Office, *Ginga Renpō kyōwakoku yūkō toshi 'Iwate-ken Ōfunato-shi' e no Higashi Nihon daishinsai fukkō shien undō no kiroku*, Kimotsuki, March 2012, 5.

96. Kimotsuki Town Office, *Ginga Renpō*, 20–31, 33, 37.

97. Kimotsuki Town Office, *Ginga Renpō*, 60.

98. Taiki City, *Chokkin 1-nenkan no jinkō suii*, https://www.town.taiki.hok kaido.jp/population.html.

99. Taiki City, *Chokkin 1-nenkan no jinkō suii*.

100. Saku City, *Saku-shi no jinkō dēta*, https://www.city.saku.nagano.jp/shi sei/profile/tokei/jinkodata01.html.

101. Sagamihara City, *Jinkō to setai-sū no suii*, https://www.city.sagamihara .kanagawa.jp/toukei/1010325/1010767.html.

FRAGMENT III. RUPTURE AND RUINATION IN THE EMPYREAN EMPIRE

1. See Nelly Bekus, chapter 6 in this book.

2. The launch vehicle officially bore the name 8K71PS or "Sputnik-PS." Although "Sputnik" signifies the sphere in Western contexts, in Russia that name often refers to the rocket. To the Soviet specialists who worked on the Sputnik 1

launch, the enormous amount of time and resources devoted to the launcher's development—and potentially its overall significance—far eclipsed that of the simple sphere. See Siddiqi, "Iskusstvennyy sputnik zemli," 432.

3. Plenty of historical ink has been spilled on the aftermath of the Sputnik launch. See, for example, McDougall, *The Heavens and the Earth*; Bulkeley, *Sputniks Crisis*; Siddiqi, *Challenge to Apollo*; and Launius, Logsdon, and Smith, *Reconsidering Sputnik*.

4. Project Moonwatch bestowed the designation of "1957α3" to the nose cone, but the small, faint object was not registered in the satellite catalog maintained by the US Department of Defense. Leon Campbell Jr., "MOON-WATCH Observations," Smithsonian Astrophysical Observatory Bulletin for the Visual Observers of Satellites, March 1958, Smithsonian Astrophysical Observatory Records ca. 1954–1966, box 47, Moonwatch Bulletins, Smithsonian Institution Archives.

5. Maher, *Apollo*.

6. Roger Launius has argued that Sputnik's significance as a turning point in history may be overblown when considering unconcerned responses by American leadership, the state of in-progress satellite development in both the United States and the Soviet Union, and how quickly stories about the first Soviet moon fell out of American newspapers. Launius, "What Are the Turning Points?" 28–33.

7. Douglas, *Purity and Danger*; and Viney, *Waste*.

8. Armiero, "Case for the Wasteocene."

9. Tarr, *Search for the Ultimate Sink*; and Rand, "Falling Cosmos."

10. For more on the respatializing of inner and outer space, see Olson and Messeri, "Beyond the Anthropocene." On vertical, circumplanetary wasting practices, see Rand, "Falling Cosmos."

11. See Liboiron and Lepawsky, *Discard Studies*, 20–25.

12. Armiero, *Wasteocene*, 1.

13. For a concise overview of wasting relationships and the creation of wasted people and places, see Armiero, *Wasteocene*.

14. Deondre Smiles, "The Settler Logics of (Outer) Space," *Society and Space*, October 26, 2020, https://www.societyandspace.org/articles/the-settler-logics-of-outer-space.

15. Liboiron, *Pollution Is Colonialism*, 68.

16. Liboiron notes that colonialism is predicated on perpetual settler access to Land for settler objectives, regardless of whether or not settlers own, use, or "set foot" on said Land. Liboiron, *Pollution Is Colonialism*, 66–68.

17. Liboiron, *Pollution Is Colonialism*, 5.

18. Liboiron differentiates between settler notions of commodifiable land (lower case) and the active, specific relations that make up Land (upper case). See *Pollution Is Colonialism*, especially 39–79.

19. For examples of historical scholarship on resistance to space industry inequities, see Maher, *Apollo*; Wormbs and Rand, "Techno-Diplomacy"; Gerovitch, "'Why Are We Telling Lies?'"; and Gorman, "Cultural Landscape of Interplanetary Space."

20. As of 2015, most space debris recovered from land had been found in regions of the Global South. See Aerospace Corporation, "Summary of Recovered Reentry Debris," http://www.aerospace.org/cords/reentry-data-2/summary-of-recovered-reentry-debris/.

21. Nixon, "Anthropocene"; and Nixon, "Unequal Anthropocene."

22. For a full explanation of the Wasteocene as an era shaped by wasting relationships, see Armiero, "Case for the Wasteocene"; and Armiero, *Wasteocene*.

23. Bruno Latour uses the wreckage of sublunar space portrayed in the film *Gravity* to illustrate the transformation of the "human" into the "Earthbound"—confined to a planet reeling through the Anthropocene with no escape route. Latour, "Telling Friends from Foes"; and Gorman, "Anthropocene in the Solar System."

24. Green, *Vanguard*, 254–84.

25. Darrin and O'Leary, *Handbook of Space Engineering*; Walsh, "Protection of Humanity's Cultural and Historic Heritage."

26. Aparna Venkatesan and John C. Barentine, "Noctalgia (Sky Grief): Our Brightening Night Skies and Loss of Environment for Astronomy and Sky Traditions," arXiv, 2023, https://doi.org/10.48550/ARXIV.2308.14685; Barentine et al., "Aggregate Effects"; "Let There Be (Natural) Light"; and K. Smith et al., "Losing the Darkness."

27. Gastón Gordillo notes that since the twentieth century the use of aviation technology has extended the horizons of imperial control beyond prior horizonal limits to vision and mobility and now encompass a vertical frame of surveillance and command. The vertical breadth extends even further if one includes the satellite surveillance infrastructure within this frame. Gordillo, *Rubble*, 255.

28. Gordillo, *Rubble*, 253–69.

29. Stoler, "'Rot Remains,'" 5.

30. Gordillo, "Void," 246.

31. Viney, *Waste*, 129.

32. Gordillo notes the contradictions of Nazi practices of glorifying monumental ruins while simultaneously prioritizing the reduction of nations to rubble. Gordillo, *Rubble*, 253–69.

33. Schönle, "Ruins and History," 666–67.

34. Grigorii Revzin, "Modern dolzhen byt' razrushen," *Kommersant-vlast'*, June 14, 2004. Translation provided in Schönle, "Ruins and History," 667.

35. Lahusen, "Decay or Endurance?"

36. In line with the uneven contours of Space Age citizenship, Gabrielle

Hecht demonstrates that being nuclear means different things depending on where the nuclear individual or community is positioned within these global networks. Hecht, *Being Nuclear.*

37. Masco, "'Survival Is Your Business.'" See also Rose, *One Nation Underground.* For more on civil defense programs and American citizens' participation in nuclear governance during this period of nuclear ruination, see Robey, *Atomic Americans.* Acts of nuclear ruination of the post-1945 period were not contained to just the tenuous boundaries of the atmosphere, either. Both the United States and the Soviet Union began exo-atmospheric nuclear testing shortly after Sputnik's ascent to orbit and continued to do so until the 1963 Partial Nuclear Test Ban Treaty outlawed atmospheric nuclear tests of any kind. For more on one of the first exo-atmospheric nuclear tests, Project Argus, see Wolverton, *Burning the Sky.*

38. Gordillo, *Rubble*, 247. For more on ruins as positive structures that conceal voids of colonial conquest and destruction, see Gordillo, *Rubble*, 253–69.

39. Siddiqi, "Another Space."

40. I am gesturing here to William Cronon's *Changes in the Land*, which traces the shaping of what is now known as New England by Indigenous inhabitants and reassesses the colonial gaze that assumed providential provision of untouched, empty land for exploitation by Christian settlers.

41. Macaulay, *Pleasure of Ruins*, 40.

42. Macaulay, *Pleasure of Ruins*, 40.

43. Walsh and Gorman, "Method for Space Archaeology."

44. Gorman, "Cultural Landscape of Space."

45. Archaeology as a discipline has long valued discarded things as remarkably honest primary sources that provide evidence that can be missing from formal archives, including revealing information about individuals and communities excluded from such archives. For a primer on this approach, see James Deetz, *In Small Things Forgotten.* Notably, Deetz briefly designates "an interplanetary space vehicle" as a complex form of material culture (24).

46. On the temporal displacement of Indigenous peoples into a "pristine" past, see, for example, Ramírez, "Indigenous Latino Heritage," 155–56.

47. Gordillo, "Void," 247.

48. On twenty-first-century nostalgia for empire, see Stoler and Bond, "Refractions Off Empire."

49. Both the United States and the Soviet Union designed, built, and flew partially reusable space shuttle systems, but most satellites and space probes launched before, during, and after the Space Shuttle and Buran programs flew aboard rockets that were designed to be used only once. For more on the construction of the US Space Shuttle and Buran as partially reusable, see Heppenheimer, *Space Shuttle Decision*; and Hendrickx and Vis, *Energiya-Buran.*

50. Baiocchi and Welser, "Democratization of Space"; and Paat-Dahlstrom and Dahlstrom, "Democratizing Access to Space."

51. Rand, "Orbital Decay," 194–266. See also Réka Patrícia Gál, chapter 8 in this book.

52. On these themes, see Siddiqi (chapter 3), Bekus (chapter 6), Klinger (chapter 7), and Jones-Imhotep (epilogue) in this book.

53. See Tsing et al., *Arts of Living*; and Tsing, *Mushroom*.

54. Buck-Morss, *Dialectics of Seeing*.

CHAPTER 10. FROM SPACE RACE DRAMA TO HAPPY HANDSHAKE

1. Fabian Schmidt, "Internationale Kooperation statt Wettlauf ins All" [International Cooperation Instead of a Space Race], *Deutsche Welle*, July 4, 2013, https://www.dw.com/de/internationale-kooperation-statt-wettlauf-ins-all/a-16927559.

2. L. John Martin, *Analysis of Newspaper Coverage of the US in the Near East, North Africa and South Asia*. Prepared for the Office of Research, US Information Agency (R-2-76, January 22, 1976) (Washington, DC: US Government Printing Office).

3. "Beyond Soyuz–Apollo," *New York Times*, July 20, 1975, E14.

4. "Mezhdunarodnaia kosmicheskaia expeditsiia na montazhnom orbite" [International Space Expedition on the Mounting Orbit]. *Pravda*, July 17, 1975, 4.

5. "Neue Perspektive für Kooperation" [New Perspective for Cooperation], *Neues Deutschland*, July 25, 1975, 6.

6. John Noble Wilford, "Far in the Future, Cooperation on the Grand Scale Will Be Needed: Apollo–Soyuz May Be Only a Beginning," *New York Times*, July 13, 1975, 149.

7. Kenneth Gatland, "Space . . . Proving Ground for Detente," *Daily Telegraph*, July 11, 1975, 12–13.

8. Grampp, "Watching Television."

9. Siddiqi, "Sputnik 50 Years Later."

10. Krasnyak, "Apollo–Soyuz Test Project"; and Karash, *Superpower Odyssey*.

11. Jenks, *Collaboration in Space*.

12. Jenks, "Securitization and Secrecy"; Asif A. Siddiqi and Dwayne A. Day, "Handshakes and Histories: The Apollo–Soyuz Test Project, 45 Years Later," *Space Review*, July 20, 2020, https://www.thespacereview.com/article/3991/1.

13. "Von Braun Foresees More Soviet 'Firsts': Von Braun Assesses Missile Program," *Washington Post*, November 10, 1957, A1.

14. "Webb Concedes Soviet May Reach Moon First," *Washington Post*, May 7, 1963, A9.

15. See, for example, translated press articles collected in the Archive of the Russian Academy of Sciences, Moscow (ARAN), f. 1678 (Interkosmos), op. 1, d. 93, or the summary of an article from the *New York Times* in ARAN, f. 1678 (Interkosmos), op. 1, d. 142, l. 59–60.

16. "Zapiska rukovoditelei sovetskoi promyshlennosti i nauki v TsK KPSS" [Memo from the Leaders of Soviet Industry and Science to the Central Committee of the Communist Party of the Soviet Union], Ministerstvo obshchego mashinostroeniia, https://www.roscosmos.ru/media/files/history/voshod/04.pdf.

17. "Man in Orbit 'Soon,'" *New York Times*, August 25, 1960, 5. See also Siddiqi, "Sputnik 50 Years Later," 535.

18. See Logsdon, *John F. Kennedy*, 10–15; and for a more recent study, see McMahon, "To the Moon and Back."

19. Launius, "Public Opinion Polls," 167–68.

20. On the specific role, the mass media played in the Soviet political system, see Wolfe, *Governing Soviet Journalism*, esp. 16–19.

21. Davies, *Popular Opinion*, 183. See also Postoutenko, Tikhomirov, and Zakharine, *Media and Communication*.

22. Gerovitch, *Soviet Space Mythologies*, 140.

23. See, for example, Richers, "Himmelssturm, Raumfahrt und 'kosmische' Symbolik," 207; and Siddiqi, "From Cosmic Enthusiasm to Nostalgia," 285.

24. On the notion of the imagined global audience, see Werron, "On Public Forms of Competition."

25. Examples include the Soviet movie *Nebo Zovet* (*The Sky Beckons*, 1959, dir. Mikhail Karyukov and Aleksandr Kozyr, USSR), the episode "Mission to Mars" of the US TV series *Men into Space* (1960, prod. Ziv Television Programs), or the movie *Marooned* (1969, dir. John Sturges, US) based on a novel by Martin Caidin. See Volf, "Between Cooperation and Competition."

26. John F. Kennedy, Address before the Eighteenth General Assembly of the United Nations, September 20, 1963, https://www.jfklibrary.org/archives/other-resources/john-f-kennedy-speeches/united-nations-19630920.

27. William L. Laurence, "Joint Moon Trip? Kennedy's Proposal for Venture Presents Formidable Problems," *New York Times*, September 29, 1963, 177; Howard Simons, "Opinion Divided Here on Joint Moon Shot Plan," *Washington Post*, September 21, 1963, A9; and Robert C. Toth, "House Opposes Joint Moon Trip, Votes NASA Fund," *New York Times*, October 11, 1963, 1.

28. Rockwell, "They May Remake Our Image," 133.

29. Krige, "NASA," 208–9.

30. Geppert, "Post-Apollo Paradox," 9; and Maher, *Apollo*, 92–136.

31. Roy, Gresham, and Christensen, "Complex Fabric," 670.

32. Howard Rusk, "Moon Shot Benefits: Cost and Scientific Gains from Apollo Being Equated with Earth's Problems," *New York Times*, January 5, 1969, 58.

33. Space Task Group, *The Post-Apollo Space Program: Directions for the Future*, NASA Historical Reference Collection, History Office, NASA Headquarters, Washington, DC, 1969, https://www.nasa.gov/history/the-post-apollo-space-program-directions-for-the-future/.

34. Logsdon, *After Apollo?*

35. Georgy Petrov, "Zachem my shturmuem kosmos" [Why We Are Storming Space], *Izvestiia*, October 3, 1967, 6.

36. Boris Konovalov, "Iz pokoleniia tridtsatiletnikh" [From the Generation of Thirty-Year-Olds], *Izvestiia*, June 2, 1970, 3.

37. Bernard Gwertzman, "High Space Costs Backed in Soviet: Expert Says Big Outlays for Research Are Justified," *New York Times*, February 28, 1971, 20.

38. Siddiqi, *Challenge to Apollo*, 552.

39. Richard M. Nixon, Address by President Richard Nixon to the UN General Assembly, US Department of State, September 18, 1969, https://2009-2017.state.gov/p/io/potusunga/207305.htm.

40. Richard D. Lyons, "NASA Seeks Links to Soviet Science but Moscow Response to 7 Letters Has Been Tepid," *New York Times*, March 12, 1970, 9.

41. This is supported by the increased frequency of terms such as "cooperation in space" or "space cooperation" in 1969 in the print media—for example, in the *New York Times*, the *Washington Post*, and the German *Frankfurter Allgemeine Zeitung*.

42. "Vereinte Nationen fordern Raketenstopp" [The United Nations Calls for a Missile Ban], *Frankfurter Allgemeine Zeitung*, December 18, 1969, 5.

43. "Sovetsko-amerikanskoe sotrudnichestvo v razlichnykh oblastiakh issledovaniia kosmicheskogo prostranstva—spravka" [Soviet–American Cooperation in Various Fields of Space Research: A Reference], 1971, ARAN, f. 1678 (Interkosmos), op. 1, d. 110.

44. The second article was published in *Air & Cosmos* on February 7, 1970. Both translated articles are in ARAN, f. 1678 (Interkosmos), op. 1, d. 93.

45. "Spravka" [Reference], March 30, 1972, ARAN, f. 1678 (Interkosmos), op. 1, d. 224v.

46. Gatland, "Space."

47. Ezell and Ezell, *Partnership*, 9–10. Pacner, *Sojuz volá Apollo*.

48. The US personnel of the ASTP pointed out that they had proceeded so fast with the talks only thanks to the good working knowledge of English among Soviet participants. See NASA, Johnson Space Center, "Letter from Manager of Apollo Spacecraft Program to ASTP Director, Summary Report on Results of the ASTP Working Groups 2 and 4 Meetings," December 6, 1972, National Archives at Fort Worth, TX. RG 255-E267 (ASTP Transmittals), box 1.

49. Sovet po mezhdunarodnomu sotrudnichestvu v oblasti issledovaniia i ispol'zovaniia kosmicheskogo prostranstva Akademii nauk SSSR (Interkosmos), "Otchet o rezultatakh peregovorov delegatsii Akademii nauk SSSR po organizatsii sovmestnoi ekspozitsii 'Soiuz–Apollon' na 30-om Parizhskom salone 23 maia–3 iiunia 1973 goda" [Report on the Results of Talks Conducted by a Delegation from the Academy of Sciences of the USSR concerning the Organization

of a Joint Exhibition "Soyuz–Apollo" during the Thirtieth Paris Air Show from May 23 to June 3, 1973], March 23, 1973, ARAN, f. 1678, op. 1, d. 294.

50. NASA, Johnson Space Center, "Apollo/Soyuz Film: Memo from Assistant Administrator for Public Affairs (John P. Donnelly) to Program Director," September 20, 1973, National Archives at Fort Worth, RG 255–264 (Apollo Soyuz Test Program), box 7.

51. "Est' stykovka: 'Soiuz–Apollon'" [There Is a Docking: "Soyuz–Apollo"], *Pravda*, July 1, 1975, 1; and Bushuev, *Soiuz i Apollon*. In the GDR, for example, see Eyermann, *Sojus-Apollo*.

52. Minister obshchego mashinostroeniia, "Prikaz ministra obshchego mashinostroeniia no. 259" [Order from the Minister of General Machine Building no. 259], August 18, 1972, https://www.roscosmos.ru/28774/.

53. Nauchno-tekhnicheskii sovet MOM, "Zakliuchenie na eskiznyi proekt raketno-kosmicheskogo kompleksa Soiuz-M" [Conclusions on the Preliminary Design of the Soyuz-M Rocket and Space Complex], February 27, 1973, https://www.roscosmos.ru/28774/.

54. US Senate Aeronautical and Space Sciences Committee, *Soviet Space Programs, 1971–75 Overview, Facilities and Hardware, Manned and Unmanned Flight Programs, Bioastronautics, Civil and Military Applications, Projections of Future Plans (Staff Report)*, August 30, 1976, US Government Printing Office, 60.

55. "Abschied vom Apollo" [Farewell to Apollo], *Süddeutsche Zeitung*, July 26/27, 1975, 4.

56. Institut kosmicheskikh issledovanii (IKI, Space Research Institute), "Opredelenie osnovnykh napravlenii razvitiia kosmicheskikh issledovanii po nauchnym programmam mezhdunarodnogo sotrudnichestva—otchet po nauchno-issledovatel'skoi teme" [Definition of the Main Directions for the Development of Space Research in the Field of International Cooperation Programs: Report on Research Topic], 1976, ARAN, f. 1862 (IKI), op. 2, d. 41.

57. Peter Osnos, "Soviets Tally Their Gains," *Washington Post*, September 1, 1975, A1.

58. Beloglazova, *Sovershenno sekretnyi general*, 251, 300.

59. Sagdeev, *Making of a Soviet Scientist*, 174.

60. Interkosmos, "Stenogramma peregovorov o vkliuchenii predstavitelei sotsialisticheskikh stran v sostav ekipazhei sovetskikh kosmicheskikh korablei i stantsii" [Stenographic Record of Talks about Assigning Representatives of Socialist Countries as Crew Members aboard Soviet Spaceships and Space Stations], July 14, 1976, ARAN, f. 1678, op. 1, d. 733, l. 47–48.

CHAPTER 11. INTERNATIONAL COOPERATION, COMPETITION, AND CHANGING TRENDS IN AMERICAN SPACE CINEMA, 1997–2019

1. Harvey, Smid, and Pirard, *Emerging Space Powers*, esp. 1–140.

2. Catchpole, *International Space Station*, 1–4.

3. *Attitudes toward Space Exploration*, July 10, 2019, distributed by Ipsos on behalf of C-SPAN, https://www.ipsos.com/sites/default/files/ct/news/docu ments/2019-07/c-span-space-exploration-07-10-2019_for_release.pdf; Cary Funk and Mark Strauss, "Majority of Americans Believe It Is Essential That the U.S. Remain a Global Leader in Space," Pew Research Center, June 6, 2018, https://www.pewresearch.org/science/2018/06/06/majority-of-americans-be lieve-it-is-essential-that-the-u-s-remain-a-global-leader-in-space/. This enthu-siasm, however, does not always translate into support for more funding; see Weitekamp, *Space Craze*, 189.

4. Funk and Strauss, "Majority of Americans Believe It Is Essential."

5. McCurdy, *Space*, 6–7, 155, 309–10. See also Landis, "Spaceflight."

6. For scholarship on this topic, see M. Smith, "Selling the Moon"; Prelinger, *Another Science Fiction*; Kauffman, *Selling Outer Space*; and Byrnes, *Politics and Space*.

7. A similar study has been conducted on Russian space films by Natalija Majsova, "Neither Rupture nor Continuity."

8. Launius, "Public Opinion Polls," 170.

9. On the use of nostalgia in films more generally, see Sperb, "Specters of Film"; and Sprengler, *Screening Nostalgia*, 11–38.

10. Hickam, *Rocket Boys*.

11. Dufournaud, "'Our Common Community.'"

12. Boczkowska, *Impact of American and Russian Cosmism*, 140.

13. The film is based on Margot Lee Shetterly's *Hidden Figures*.

14. Nkrumah, "Problems of Portrayal."

15. Ikhsano and Jakarudi, "Representation of Black Feminism"; and Thomp-son, "'See What She Becomes.'"

16. Torras i Segura, "'First Man.'"

17. Willems, "Potential of the Past." See also Majsova, "Neither Rupture nor Continuity," 212–16; and Asif Siddiqi, "A Brief Guide to Russian Space Movies," *Air and Space Magazine*, April 9, 2021, https://www.airspacemag.com/ daily-planet/brief-guide-russian-space-movies-180977460/.

18. For a general history of the Mir space station, see Harland, *Story of Space Station Mir*.

19. For comparison and further discussion on space stations in popular cul-ture, see Weitekamp, *Space Craze*, 187–226.

20. Boczkowska, "Spaceflight."

21. For contemporary reviews and commentaries on the film, see Michael Benson, "Andrei Ujica, Director: *Out of the Present*," *Independent Film and Video Monthly*, April 1997; and David Rooney, "Out of the Present," *Variety*, March 10, 1996, https://variety.com/1996/film/reviews/out-of-the-present-1200445375/. The full film may be viewed on Vimeo: Miska Henner, "Out of the Present (1997)," https://vimeo.com/325601048.

22. Sagan, *Contact*. A more critical view of the feminist politics in *Contact*, as also compared to another film in our study, *Armageddon* (1998), can be found in J. Martin, "Anti-Feminism." See also Steinke, "Women Scientist Role Models."

23. Peter Kramer, "'Want to Take a Ride?'"; Peter Kramer, "Making Contact: Exploring Book Adaptations and the Young Woman Adventurer in Contemporary Hollywood Science-Fiction," *Pure Movies*, November 20, 2013, https://www.puremovies.co.uk/columns/making-contact/.

24. Webber, "Space Tourism."

25. Denise Chow, "Russia's Space Station Mir: The First Space Tourist Hotspot?" *Space.com*, April 26, 2011, https://www.space.com/11480-space -tourists-russia-space-station-mir.html.

26. On this phenomenon more generally, see Morris, *Japan-Bashing.*

27. United States Senate, *Hearings before the Committee on the Budget: United States Senate, Ninety-Eighth Congress, First Session* (Washington: U.S. Government Printing Office, 1983), 3.

28. Volnarovski, "*Deep Impact* vs. *Armageddon*"; and Kristen Lopez, "Why 'Deep Impact' Couldn't Top 'Armageddon,'" *Hollywood Reporter*, May 8, 2018, https://www.hollywoodreporter.com/movies/movie-news/why-deep-impact -couldnt-top-armageddon-1109247/.

29. Jorgensen, "States of Weightlessness," esp. 205–6.

30. Lawler, "Accident Clouds US Future."

31. Jorgensen, "States of Weightlessness," 206.

32. T. Miller, *Mars in the Movies*, 105–9; and Stanley et al., *Martian Pictures*, 21–28.

33. For a contrasting viewpoint, see Peter Sobczynski, "'Mission to Mars': Twenty Years Later," *Spool*, March 10, 2020, https://thespool.net/reviews/ movies/mission-to-mars-de-palma/.

34. Kosmos 419, for example, was intended to go to Mars in 1971, but failed to leave Earth's orbit. See Siddiqi, *Beyond Earth*, 99–100.

35. L. Graham and Dezhina, *Science in the New Russia*, 18–32.

36. "Red Planet: Trivia," *IMDb*, https://www.imdb.com/title/tt0199753/ trivia.

37. Pfarrer, Cobb, and Palmiotti, *Virus.*

38. Quarantelli, "Realities and Mythologies"; J. Mitchell et al., "Catastrophe in Reel Life"; Kakoudaki, "Spectacles of History"; Kakoudaki, "Representing Politics"; and Haney, Havice, and Mitchell, "Science or Fiction."

39. Laura M. Delgado, "The Commercialization of Space in Science Fiction Movies: The Key to Sustainability or the Road to a Capitalist Dystopia?" Paper presented at *AIAA SPACE 2010 Conference & Exposition*, August 30–September 2, Anaheim, CA (Reston, VA: American Institute of Aeronautics and Astronautics, 2010), 1–12, http://dx.doi.org/10.2514/6.2010-8654.

40. Matt Nisbet, "Evaluating the Impact," *Skeptical Inquirer*, June 16, 2004, https://skepticalinquirer.org/exclusive/evaluating-the-impact-of-the-day-after-tomorrow/.

41. Andrew Norton and John Leaman, "The Day After Tomorrow: Public Opinion on Climate Change," MORI Social Research Institute, London, May 2004, https://www.climateprediction.net/wp-content/schools/mori_poll.pdf.

42. Von Burg, "Decades Away?"; Stefan Rahmstorf, "*The Day After Tomorrow*: Some Comments on the Movie," Potsdam Institute for Climate Impact Research, (2004), https://www.pik-potsdam.de/~stefan/tdat_review.html.

43. Reusswig and Leiserowitz, "International Impact."

44. Livesey, "Climate Change."

45. *Attitudes toward Space Exploration*, C-SPAN/Ipsos, July 10, 2019; "Majority of Americans Believe It Is Essential That the U.S. Remain a Global Leader in Space," Pew Research Center, June 2018.

46. P. Hart and Leiserowitz, "Finding the Teachable Moment"; and von Mossner, "Facing *the Day after Tomorrow*."

47. Malone, "Catholic Movies," 140.

48. Josh Sandy, "*Sunshine* (2007) Review: Integrating Science and Religion," *High on Film*, July 22, 2021, https://www.highonfilms.com/sunshine-2007-review/.

49. Grant Watson, "'Every Time I Close My Eyes': *Sunshine* (2007)," *Fiction Machine*, July 10, 2014, https://fictionmachine.com/2014/07/10/every-time-i-close-my-eyes-sunshine-2007/.

50. For an early Marxist and feminist take on the film, see Kavanaugh, "'Son of a Bitch.'"

51. Jonathan Pile, "*Life* Review," *Empire Online*, March 21, 2017, https://www.empireonline.com/movies/reviews/life-9-review/.

52. Warren, "'Moon'"; and Sundvall, "Clonetrolling the Future."

53. For more on capitalism in *Moon*, see Pierson, "Speculative Finance."

54. "*Moon* Is Duncan Jones' Homage to Classic Sci-Fi," *Wired*, June 12, 2009, https://www.wired.com/2009/06/duncan-jones-moon/.

55. Llinares, "Idealized Heroes."

56. Anatoly Zak, "A Complete Chronology of ISS Missions," *Russian Space Web*, http://www.russianspaceweb.com/iss_chronology_flights.html.

57. Chelsea Gohd, "Why Jeff Bezos's Blue Origin Is So Reviled," *Scientific American*, August 26, 2021, https://www.scientificamerican.com/article/why-jeff-bezos-blue-origin-is-so-reviled/.

58. "Majority of Americans Believe It Is Essential."

59. S. Carroll, "Lost in Space."

60. For a general history of the Chinese space program, see Harvey, *China in Space*.

61. Jonathan O'Callaghan, "What If Space Junk and Climate Change Be-

come the Same Problem?" *New York Times*, May 12, 2021, https://www.nytimes .com/2021/05/12/science/space-junk-climate-change.html.

62. Rand, "Falling Cosmos"; Gorman, "Ghosts in the Machine"; and Damjanov, "Of Defunct Satellites."

63. For an excellent reading of this film through a feminist perspective (and the protagonist's use of "masculine" tools), see Palmer, "Untethered Technology." See also Purse, "Working Space."

64. Weir, *Martian*.

65. T. Miller, *Mars in the Movies*, 125–30.

66. For more on the labor issues in the film, see Moss-Wellington, "Individual and Collaborative Labour."

67. Weir, *Martian*, 147.

68. Weir, *Martian*, 368–69.

69. Sculos, *"Martian."*

70. Patrick Frater, "China's Bona Film Invests \$235 Million in Fox Movie Slate," *Variety*, November 4, 2015, https://variety.com/2015/biz/asia/ bona-film-fox-investment-1201633139/.

CHAPTER 12: IRONY, SLAPSTICK, GORE

1. Condee, *Imperial Trace*.

2. Wilmes, "Empire Reloaded," 49.

3. Wilmes, "Empire Reloaded," 56.

4. Geppert, *Imagining Outer Space*, 28.

5. Sandner, "Shooting for the Moon"; Schwartz, *Expeditionen*; and Beumers, "Special/Spatial Effects."

6. Lewis, *Cosmonaut*, 202–3.

7. Kohonen, *Picturing the Cosmos*.

8. Anonymous, "1958: Put' v kosmos," *Nauka i zhizn'*, no. 1 (1958), 40. *Doroga k zvezdam*, directed by Pavel Klushantsev (Lennauchfilm Studio, 1957).

9. Lewis, *Cosmonaut*, 202.

10. For a definition of the chronotope, see Bakhtin, "Forms of Time."

11. *Aelita*, directed by Yakov Protazanov (MezhrabpomRus, 1924); and *Kosmicheskii reis*, directed by Vassily Zhuravlev (Mosfilm, 1936).

12. Beumers, "Special/Spatial Effects."

13. Kohonen, "Space Race."

14. Pahl, "Pussy Riot's Humour," 68.

15. *Voyage dans la Lune*, directed by Georges Méliès (Star Film Company, 1902); and Solomon, *Fantastic Voyages*. Trick films, a popular film format until the mid-1910s, were silent shorts exploring special effects (the capacity of the camera and editing to create temporal and spatial illusions), for example, by experimenting with slow and fast motion or multiple exposure. See Solomon, "Up-to-Date Magic."

16. Web repositories such as kino-teatr.ru.

17. Tvroscosmos.ru web repository.

18. *Kin-dza-dza!* directed by Georgy Daniliya (Mosfilm, 1986).

19. For more on the evolution of Soviet and post-Soviet space-themed science fiction cinema, see Majsova, *Soviet Science Fiction Cinema.*

20. Gerovitch, *Soviet Space Mythologies*; and *Obitaemyi ostrov I and II*, directed by Fedor Bondarchuk (Art Pictures Studio, Non-Stop Production, and STS Channel, 2009).

21. Condee, *Imperial Trace*, 69–84.

22. Crilley and Chatterje-Doody, "From Russia with Lols."

23. Boym, "Nostalgia."

24. Sasha Senderovich,"Georgii Daneliia and Tat'iana Il'ina: *Ku! Kin-dza-dza* (2013)," *Kinokultura*, no. 43 (2014), http://www.kinokultura.com/2014/43r-ku-kindzadza.shtml. *Pervye na lune*, directed by Aleksey Fedorchenko (Sverdlovsk Film Studio, 2005); *Pritiazhenie*, directed by Fedor Bondarchuk (Art Pictures Studio, Russian State Film Fund, Vodorod Film Company, Columbia Pictures, 2017); *Sputnik*, directed by Evgeny Abramenko (Vodorod Pictures, Art Pictures Studio, Hype Film, National Media Group Studio, STS, 2020), and *Ku! Kin-Dza-Dza.*

25. For intertextuality, see Bakhtin, "Discourse in the Novel," 291. *Chetvertaia planeta*, directed by Dmitry Astrakhan (Lenfilm, Mikofilm, Astrakhan-Lumex Studios, 1995); *Solaris*, directed by Andrei Tarkovsky (Creative Unit of Writers and Cinema Workers, Mosfilm, Unit Four, 1972).

26. *Mechte navstrechu*, directed by Mikhail Kariukov and Otar Koberidze (Odessa Film Studio. 1963).

27. *Out of the Present*, directed by Andrej Ujica (Köln Filmverein, 1996). *State of Weightlessness* (*Stan niewazkosci*), directed by Maciej Drygas (Studio Filmowe Logos, ADR Production, Canal+, 1994) is another case in point, using footage from the same space mission.

28. His mission lasted 311 days (May 18, 1991–March 25, 1992). The USSR collapsed on December 26, 1991.

29. *Kosmos kak predchustvie*, directed by Aleksey Uchitel' (TPO Rok, 2005); *Bumazhnyi soldat*, directed by Aleksey German Jr. (Lenfilm, Fenomen Films, Rossiia TV Channel, 2008).

30. For more on *First on the Moon* in relation to the memory of the Space Age, see Majsova, *Soviet Science Fiction Cinema*, chap. 7; see also Kabanova, "Mourning the Mimesis."

31. *Houston, imamo problem!* directed by Žiga Virc (Studio Virc, Nukleus Film, Sutor Kolonko, 2016) is a Slovenian-based international coproduction about the myth of the Yugoslav space program.

32. Willems, "Potential of the Past."

33. *Gagarin: Pervyi v kosmose*, directed by Pavel Parkhomenko (Kremlin

Films, 2013); *Glavnyi*, directed by Yuri Kara (Master film, 2015); and *Vremia pervykh*, directed by Dmitry Kiselev (Bazelevs Production, 2017).

34. *Tumannost' Andromedy*, directed by Evgeny Sherstobitov (Dovzhenko Film Studio, 1967).

35. For more on *Gagarin: First in Space*, *Spacewalk*, and *Salyut 7*, see Majsova, *Soviet Science Fiction Cinema*, chap. 8.

36. *Salyut 7*, directed by Klim Shipenko (CTB Film Company, Globusfilm, Lemon Films Studio, 2017).

37. *Armageddon*, directed by Michael Bay (Touchstone Pictures, Jerry Bruckheimer Films, Valhalla Motion Pictures, 1998).

38. Nick Levine, "'Sputnik' review: Russia's Riff on Ridley Scott Is Light Years Away from 'Alien,'" *NME.com*, August 11, 2020, https://www.nme.com/reviews/film-reviews/sputnik-review-2725403. *Alien*, directed by Ridley Scott (20th Century Studios, 1979).

39. See Høgetveit, "Female Aliens."

40. See also Siddiqi, "From Cosmic Enthusiasm to Nostalgia."

FRAGMENT IV. PLANETARITY, PLANETIZATION, AND THE GLOBAL SPACE AGE

Teilhard de Chardin, "Vie et planètes," 145 (emphasis in the original); for an English translation, see Teilhard de Chardin, "Life and the Planets," 97. This chapter forms part of a book project tentatively titled *Planetizing Earth: An Extra-Terrestrial History of the Global Present*. Thanks are due to Martin Collins, Michael Hagemeister, Michèle Mateschk, Asif Siddiqi, Tilmann Siebeneichner, Brad Tabas, Bernd Weisbrod, and, as always, Anna Kathryn Kendrick.

1. See "fragment, n., 2a," *Oxford English Dictionary* (Oxford: Oxford University Press, 2021), www.oed.com/view/Entry/74114.

2. In this context, see the European Astroculture trilogy, consisting of Geppert, *Imagining Outer Space*; Geppert, *Limiting Outer Space*; and Geppert, Brandau, and Siebeneichner, *Militarizing Outer Space*.

3. Geppert, "European Astrofuturism," 10, 13, 19. For the "promise of space," see Johnson, "International Control," 721, and, of course, Clarke, *Promise of Space*.

4. See, for instance, Boym, "Kosmos"; Benjamin, *Rocket Dreams*; and Dick, *Remembering the Space Age*, especially the contributions by Emily S. Rosenberg (chap. 10), Martin Collins (chap. 11), and Sylvia Kraemer (chap. 20).

5. For a sharp critique of "New Space Race" rhetoric and the historically skewed analogy that the phrase implies, see Dwayne Day, "Racing to Where/What/When/Why?" *Space Review*, March 2, 2020, https://www.thespacereview.com/article/3893/1. A quick newspaper search reveals that both usages can be found earlier, for instance, in the *New York Times*, but that the term "New Space Age" has gained more critical prominence only within the past years. For

an earlier example highlighting the notion of a thirty-year-long "Space Shuttle Era," lasting from 1981 through 2011, see Dennis Overbye, "As Shuttle Era Ends, Dreams of Space Linger," *New York Times*, July 5, 2011, D1.

6. Harry Harper, "The Space Age," *Everybody's Weekly*, January 19, 1946, 8–9; and Harper, *Dawn of the Space Age*. For a detailed *Begriffsgeschichte*—or conceptual history—of "Space Age," see Geppert, "Die Zeit des Weltraumzeitalters," 226–28.

7. Harper, *Dawn of the Space Age*, 5.

8. Koselleck, *Vergangene Zukunft*, 349–50. Historians such as Robert Poole, Benjamin Lazier, and others have spoken of a "first" or "classical Space Age," lasting from Sputnik to Skylab, with the December 1968 *Earthrise* moment constituting an irreversible turning point where an outward-oriented expansionism turned inward.

9. To that end, I have recently founded a Global Astroculture Research Group, comprising junior and senior scholars from four different continents. A first result is "Rocket Stars: Astrocultural Genealogies in the Global Space Age," a special issue coming out in spring 2025 in the *British Journal for the History of Science*, on the making of technocelebrities and space personas in China, India, Sri Lanka, East and West Germany, and Cuba.

10. Westad, *Global Cold War*, 71.

11. Harper, *Dawn of the Space Age*, 4, 142.

12. "China Successfully Launched Its First Man-Made Earth Satellite," *Peking Review*, April 25, 1970.

13. On 1970s space fatigue, see Geppert, "Post-Apollo Paradox."

14. Combining all three criteria discussed above, I have argued elsewhere that the Space Age lasted for precisely three decades, from Saturday, October 3, 1942, 14:58 (GMT) to Thursday, December 14, 1972, 21:54 (GMT), from the first artifact that crossed the (arbitrarily defined) boundary between Earth's atmosphere and outer space, to the moment when the last human left the Moon and returned to Earth, basing the proposed periodization on a combination of physical artifacts reaching beyond the Earth's atmosphere and the prominence of Earth-based space thought and astroculture on the ground. See Geppert, "Die Zeit des Weltraumzeitalters," 250.

15. For "planetarische Raumrevolution," see Schmitt, *Land und Meer*, 44–49.

16. Helmuth Plessner, "Gedanken eines Philosophen zur Weltraum-Rakete," *Gedanken zur Zeit*, Bayerischer Rundfunk, October 13, 1949, 22:45–23:00, reprinted in *Soziologie der Weltraumfahrt*, ed. Joachim Fischer and Dierk Spreen, 197–201 (Bielefeld: transcript, 2014), 200–201 (my translation).

17. Bürgel, *Der Mensch*, 20. There is no information as to how this image was produced and whether the arrow points at Earth or the solar system as such. As is well known, Carl Sagan took the *Pale Blue Dot* photograph—about which surprisingly little of critical substance has been written—as an occasion

to formulate an astrofuturistic manifesto for the 1990s; see Sagan, *Pale Blue Dot.*

18. Simmons, Diamond, and Hubbard, *Language of a New World*, 1.

19. The literature on Teilhard is massive. A bibliography published in 1981 lists 4,317 items, in addition to 621 works by Teilhard himself; see McCarthy, *Pierre Teilhard de Chardin*. Biographies include Speaight, *Teilhard de Chardin*; Schiwy, *Teilhard de Chardin*; King, *Spirit of Fire*; and Boudignon, *Pierre Teilhard de Chardin*.

20. Teilhard de Chardin, *Phénomène humain*, 280, 234.

21. Teilhard de Chardin, *Letters to Two Friends*, 163.

22. Teilhard de Chardin, "Vie et planètes"; and Teilhard de Chardin, "Un grand événement." Both articles were republished in his *L'Avenir de l'homme*, 127–56, 157–75, and translated for the English version; see Teilhard de Chardin, "Life and the Planets"; and Teilhard de Chardin, "Great Event Foreshadowed."

23. Teilhard de Chardin, "Life and the Planets," 115. See also "planétisa- tion," in Baudry, *Dictionnaire Teilhard de Chardin*, 69.

24. Speaight, *Teilhard de Chardin*, 266. A comprehensive *Begriffsgeschichte* of *planétisation* would exceed the scope of this think piece. However, the term's sudden rise in popularity (and subsequent fall) in French, as evidenced both in *Le Monde* and in general parlance, from the late 1950s through the early 1970s and thus in complete compliance with the Space Age as conventionally defined, is remarkable; see https://www.lalanguefrancaise.com/dictionnaire/definition/ planetisation. Thanks to Brad Tabas for the reference.

25. Nor is this the place for a detailed conceptual history of "globalization." The OED dates its first use to 1930, with "global" taking on its contemporary theoretical connotations in the early 1970s. See "globalization, n.," *Oxford English Dictionary* (Oxford: Oxford University Press, 2021), www.oed.com/view/ Entry/272264. For a vehement rejection of "globalization" as a useful analytical category, see Frederick Cooper's classic "What Is the Concept of Globalization Good For?"

26. For a more systematic treatment of "noosphere," see the works of the Russian cosmist Vladimir I. Vernadsky. Surprisingly little has been translated into English but see, for instance, Vernadsky, "Biosphere."

27. On Teilhard's so-called cosmic vision, see Samuel, "Le Corbusier"; Sesé, "Pierre Teilhard de Chardin"; and Osborne, "From Sputnik to Spaceship Earth"; see also Schmölders, "Heaven on Earth," 59–62.

28. Teilhard de Chardin, "Life and the Planets," 122. This brief remark pre- dates by half a decade the so-called Fermi paradox—"Where is everybody?"—as formulated by the Italian American physicist Enrico Fermi in the summer of 1950.

29. Teilhard de Chardin passed away in April 1955, two and a half years be- fore Sputnik. For an insightful, if eclectic reading of three dozen philosophers'

reactions to the early Space Age and its various milestones sometimes referred to as "Space Firsts," see Kreienbrock, *Sich im Weltall orientieren.*

30. Koselleck, *Vergangene Zukunft*, 350.

31. Once identified, any possible candidate for such an Archimedean point would instantly be territorialized and then give rise to yet another search. Thus, the process can never be completed, and it is for this reason that human geocentrism is impossible to overcome. Arendt, "Conquest of Space," 46; for a revised version, see Arendt, *Between Past and Future*, 265–80, 278. A comparison between Arendt's Archimedean point and Teilhard de Chardin's "Omega point" would be instructive but is beyond the scope of this chapter.

32. Mazlish, "Comparing Global History," 389 (emphasis in the original). To give but one counterexample: space, planetarity and extraterrestrial perspectives did not play any role in Geyer and Bright's otherwise programmatic essay "World History."

33. Spivak, "Planetarity."

34. Spivak, "Planetarity"; and Spivak, *Imperative zur Neuerfindung.*

35. Elias and Moraru, *Planetary Turn*; Bonneuil, "Der Historiker"; Clark and Szerszynski, *Planetary Social Thought*; and Chakrabarty, *Climate of History.*

36. Willy Ley, "The End of the Rocket Society," *Astounding Science-Fiction* 31/32, August/September 1943, part 2, 58.

EPILOGUE. AFTERWORLDS

1. See John Tresch's discussion of Ortega's artwork in Tresch, "Technological World-Pictures, 89–91.

2. K. Crawford and Joler, *Anatomy of an AI System*, 3.

3. K. Crawford and Joler, *Anatomy of an AI System*, 3.

4. As Bruno Latour explains: "We still don't know how to assemble, in a single, visually coherent space, all the entities necessary for a thing to become an object—Ortega's installation notwithstanding. When we have learned how to do that, we might finally get our (material) materialism back—and our cosmic things to boot." Latour, "Can We Get Our Materialism Back?" 143.

5. Siddiqi, Introduction to this book.

6. Siddiqi, Introduction to this book.

7. For discussion of cosmograms, see Tresch, "Technological World-Pictures," 92.

8. Wolfe, *Right Stuff*, 128. Cited in Reser, chapter 1 in this book. Reser has suggested that this construction of emptiness is a fundamental component of representations of space activities and of American technology projects more generally.

9. Southall, *Woomera*, 3. Cited in Gorman, chapter 4 in this book.

10. Tim Ingold, for instance, drawing on Deleuze and Guattari, has argued that the "scape" suggests the static rather than the dynamic and multidimen-

sional. See Ingold, *Being Alive*, chap. 10. On the ecocritical approach that also sees nature and environment as process rather than unchanging or static, see Buell, *Environmental Imagination*.

11. Gorman, chapter 4 in this book. Gorman also makes clear how the emptiness of space is part of apologias for the extension of colonial pasts. "The absence of humans who would suffer the ill effects of such colonialism has been recast as a justification rather than an obstacle. Such a response ignores the reality that it is as much the system of belief and action that constitutes colonialism: it is the process, not necessarily the target." Gorman, chapter 4 in this book.

12. See Reser, chapter 1 in this book.

13. Gorman, chapter 4 in this book.

14. Reser, chapter 1 in this book.

15. Armstrong, Fragment I in this book. On the more general belief of the Space Age as the end of an evolutionary trajectory that begins with the "Stone Age," see Gorman, chapter 4 in this book.

16. On the importance of technological and social progress narratives in space activities, see Gorman, "Cultural Landscape"; Gorman, *Dr Space Junk*; and Schwartz, "Myth-Free Space."

17. Siddiqi, chapter 3 in this book. For the example of Alcantara, Brazil, see S. Mitchell, *Constellations of Inequality*. On the Kennedy Space Center, see Reser, chapter 1 in this book.

18. Jones-Imhotep, "Paris-Montreal-Babylon."

19. On Vandenberg, see MacKenzie, "Missile Accuracy."

20. The postage stamps featuring Intelsat's decontextualized Managua Earth station, similarly served "to obscure, if not erase entirely, several salient facts about the Earth station's location . . . above a small lake in the middle of populous, urban Managua." Evans and Lundgren, chapter 2 in this book. When Gerald Bull, the university professor turned international arms dealer, sought to depict his envisioned supergun, he invoked a similar technique. See Jones-Imhotep, "Paris-Montreal-Babylon."

21. On mechanic orders, see Jones-Imhotep, *Unreliable Nation*.

22. On displaced empires, see Immerwahr, *How to Hide an Empire*; and Paul Kramer, "How Not to Write." On the relation to space activities, see, for example, Messeri, *Placing Outer Space*; Vertesi, *Seeing Like a Rover*; Olson, *Into the Extreme*; and Gorman, *Dr Space Junk*.

23. Satia, *Spies in Arabia*, 7.

24. Siddiqi, "Dispersed Sites."

25. Siddiqi, "Dispersed Sites."

26. Siddiqi, "Dispersed Sites," 175–76; and Tsing, *Mushroom*.

27. Siddiqi, "Dispersed Sites," 177.

28. Siddiqi, "Dispersed Sites," 180.

29. Siddiqi, "Dispersed Sites," 177.

30. Bijker, "Social Construction," 159.

31. Galison, *Image and Logic*, Introduction.

32. Homi Bhabha points to this estrangement as central to colonialism in Frantz Fanon's thought. See Bhabha, "Foreword," in Frantz Fanon, *The Wretched of the Earth*, trans. Richard Philcox (New York: Grove Press, 2004), xi.

33. Fanon, *Toward the African Revolution*, 53. Here, Fanon gestures toward the more general (and technical) concept of "alienation" as both a psychiatric condition and a technique of colonialism.

34. Oodgeroo Noonuccal, "We Are Going," cited in Gorman, chapter 4 in this book.

35. Durrani, chapter 5 in this book.

36. Petr Trotsenko, "Kritika ili Kleveta? Kak mery Baikonura sudilis s kazakhstanskim pravozashchitnikom," Radio Azattyk, April 11, 2019, https://rus.azattyq.org/a/kazakhstan-how-mayors-of-baikonur-sue-kazakh-human-rights-defender/29874952.html; cited in Bekus, chapter 6 in this book.

37. Shannon Stirone, "This Is Where the International Space Station Will Go to Die," *Popular Science Magazine*, June 13, 2016.

38. Wijeyeratne, chapter 9 in this book.

39. On this militarized geography of islands in both the Pacific and Atlantic, and its place in space research, see Jones-Imhotep, "Paris-Montreal-Babylon."

40. Jackson, "Rethinking Repair," cited in Gál, chapter 8 in this book.

41. European Space Agency, *Annual Environmental Report* GEN-DB-LOG-00288-OPS-SD, Darmstadt, ESA Space Debris Office, 2022, 3.

42. Heyck and Kaiser, "Introduction."

43. See Rand, Fragment III in this book.

44. See Klinger, chapter 7 in this book.

45. Gál, chapter 8 in this book.

46. Gal, chapter 8 in this book.

47. For an excellent examination of Operation Morning Light, see Power and Keeling, "Cleaning Up Cosmos.": Satellite Debris, Radioactive Risk, and the Politics of Knowledge in Operation Morning Light." *The Northern Review* (2018): 81–109.

48. Reynolds and Merges, *Outer Space*, 179–89.

49. Power and Keeling, "Cleaning Up Cosmos," 91.

50. Power, "Narratives of Radiation."

51. Power, "Narratives of Radiation."

52. See Volf (chapter 10), Cuenca (chapter 11), Majsova (chapter 12), and Geppert (Fragment IV) in this book.

53. Teilhard de Chardin, "Human Energy," 132–33.

54. Teilhard de Chardin, *Letters to Léontine Zanta*, 117.

55. Rankin, "Geography of Radionavigation."

56. For another example, see Siddiqi, chapter 3 in this book.

57. See Gafijczuk, "Dwelling Within"; and Rand, Fragment III in this book.

58. "Americana," by Ambrose Akinmusire, recorded February 16, 2017, track 3 on *Origami Harvest*, Blue Note Records, compact disc.

BIBLIOGRAPHY

Agrawal, Binod C., N. Sudhakar Rao, and P. C. Gurivi Reddy. *Yanadi Response to Change: An Attempt in Action Anthropology*. New Delhi: Concept, 1985.

Ahmed, Syed Ishtiaque, Steven J. Jackson, and Md. Rashidujjaman Rifat. "Learning to Fix: Knowledge, Collaboration and Mobile Phone Repair in Dhaka, Bangladesh." In *Proceedings of the Seventh International Conference on Information and Communication Technologies and Development*, 1–10. ICTD '15. New York: Association for Computing Machinery, 2015. https://doi.org/10.1145/2737856.2738018.

Akinmusire, Ambrose. "Americana / The Garden Waits for You to Match Her Wilderness." *Origami Harvest*. Blue Note Records, 2018.

Aldrich, Daniel P. *Site Fights: Divisive Facilities and Civil Society in Japan and the West*. Ithaca, NY: Cornell University Press, 2008.

Alonso, Pedro Ignacio, and Hugo Palmarola. "NASA in Chile: Technology and Branding of a Satellite-Tracking Station." *Design Issues* 33, no. 2 (2017): 31–42.

Altshuler, José. "From Shortwave and Scatter to Satellite: Cuba's International Communications." In Butrica, *Beyond the Ionosphere*, 243–50.

Anand, Nikhil, Akhil Gupta, and Hannah Appel, eds. *The Promise of Infrastructure*. Durham, NC: Duke University Press, 2018.

Anderson, Warwick. "Introduction: Postcolonial Technoscience." *Social Studies of Science* 32, nos. 5/6 (2002): 643–58.

Andrews, James T. *Red Cosmos: K. E. Tsiolkovskii, Grandfather of Soviet Rocketry.* College Station: Texas A&M University Press, 2009.

Andrews, James T., and Asif A. Siddiqi, eds. *Into the Cosmos: Space Exploration and Soviet Culture.* Pittsburgh, PA: University of Pittsburgh Press, 2011.

Aravamudan, R. "A Space Saga: The Thumba Years." *Space India* (October–November 2003): 13–19.

Aravamudan, R., and Gita Aravamudan. *ISRO: A Personal Journey.* Noida: HarperCollins, 2017.

Arendt, Hannah. *Between Past and Future: Eight Exercises in Political Thought.* New York: Viking Press, 1968.

Arendt, Hannah. "The Conquest of Space and the Stature of Man." In *A Symposium on Space: Has Man's Conquest of Space Increased or Diminished His Stature?* edited by William Benton, 34–47. Chicago: Encyclopedia Britannica, 1963.

Armiero, Marco. "The Case for the Wasteocene." *Environmental History* 26, no. 3 (2021): 425–30.

Armiero, Marco. *Wasteocene: Stories from the Global Dump.* Cambridge: Cambridge University Press, 2021.

Armstrong, Eleanor Sophie. "Exploring Space (s): Queer Feminist Approaches to Understanding Pedagogy in Science Museum Galleries." PhD diss., University College London, 2020.

Arnold, David. "Europe, Technology, and Colonialism in the 20th Century." *History and Technology* 21, no. 1 (2005): 85–106.

Arnold, David. "Nehruvian Science and Postcolonial India." *Isis* 104 (2013): 360–70.

Arnold, David. *The Problem of Nature: Environment, Culture and European Expansion.* New York: Wiley-Blackwell, 1996.

Arnold, David. *The Tropics and the Traveling Gaze: India, Landscape, and Science, 1800–1856.* Seattle: University of Washington Press, 2015.

Aronowsky, Leah V. "Of Astronauts and Algae: NASA and the Dream of Multispecies Spaceflight." *Environmental Humanities* 9, no. 2 (2017): 359–77. https://doi.org/10.1215/22011919–4215343.

Asad, Talal. "Introduction." In *Genealogies of Religion: Discipline and Reasons of Power in Christianity and Islam,* 1–24. Baltimore: Johns Hopkins University Press, 1993.

Astronautics and Aeronautics, 1964: Chronology on Science, Technology, and Policy. NASA SP4005. Washington, DC: NASA Historical Staff, Office of Policy, 1965.

Astronautics and Aeronautics, 1966: Chronology on Science, Technology, and Policy. NASA SP4007. Washington, DC: NASA Historical Staff, Office of Policy, 1967.

Baiocchi, Dave, and William Welser. "The Democratization of Space." *Foreign Affairs* 94, no. 3 (2015): 98–104.

Bakhtin, Mikhail M. "Discourse in the Novel." In Holquist, *Dialogic Imagination*, 269–422.

Bakhtin, Mikhail M. "Forms of Time and of the Chronotope in the Novel: Notes Toward a Historical Poetics." In Holquist, *Dialogic Imagination*, 48–254.

Bakytova, Leila, and Kulshat Medeuova. "'Cosmos' in the Museum-Memorial Landscape of Zhezkazgan. *Antropologicheskij forum* no. 57 (2023). https://anthropologie.kunstkamera.ru/en/06/2023_57/bakytova_medeuova.

Balagopal, K. "Land Unrest in Andhra Pradesh-III: Illegal Acquisition in Tribal Areas." *Economic and Political Weekly* 42, no. 40 (October 6–12, 2007): 4029–34.

Banks, Elisabeth, Robyn d'Avignon, and Asif Siddiqi. "Introduction: The African-Soviet Modern." *Comparative Studies of South Asia, Africa and the Middle East* 41, no. 1 (2021): 2–10.

Barentine, John C., Aparna Venkatesan, Jessica Heim, James Lowenthal, Miroslav Kocifaj, and Salvador Bará. "Aggregate Effects of Proliferating Low-Earth-Orbit Objects and Implications for Astronomical Data Lost in the Noise." *Nature Astronomy* 7, no. 3 (March 2023): 252–58.

Baudry, Gérard-Henry. *Dictionnaire Teilhard de Chardin*. Saint-Étienne: Aubin, 2009.

Bawaka Country, including A. Mitchell, S. Wright, S. Suchet-Pearson, K. Lloyd, L. Burarrwanga, R. Ganambarr, M. Ganambarr-Stubbs, B. Ganambarr, D. Maymuru, and R. Maymuru. "Dukarr Lakarama: Listening to Guwak, Talking Back to Space Colonization." *Political Geography* 81 (2020): 102218. https://doi.org/10.1016/j.polgeo.2020.102218.

Beery, Jason. "State, Capital and Spaceships: A Terrestrial Geography of Space Tourism." *Geoforum*, 43 no.1 (2012): 25–34.

Bekus, Nelly. "Outer Space Technopolitics and Postcolonial Modernity in Kazakhstan." *Central Asian Survey* 41, no. 2 (2022): 347–67.

Bekus, Nelly, and Zhomart Medeuov. "Aspirations and Challenges of Space Techno-Science in Global Semi-Peripheries: A View from Kazakhstan." *Science, Technology and Society* 29, no. 2 (2024): 322–42.

Beloglazova, Ekaterina. *Sovershenno sekretnyi general* [Top secret general]. Moscow: Geroi otechestva, 2005.

Benjamin, Marina. *Rocket Dreams: How the Space Age Shaped Our Vision of a World Beyond*. New York: Free Press, 2003.

Benson, Charles D., and William Barnaby Faherty, *Moonport: A History of Apollo Launch Facilities and Operations*. Washington, DC: NASA, 1978.

Beumers, Birgit. "Special/Spatial Effects in Soviet Cinema." In *Russian Aviation, Space Flight and Visual Culture*, edited by Vlad Strukov and Helena Goscilo, 169–88. Oxon: Routledge, 2016.

Bhabha, Homi. "Foreword: Remembering Fanon." In Fanon, *Black Skin, White Masks*, vii–xxvi.

Bhabha, Homi. "Science and the Problems of Development." *Science* 151, no. 3710 (February 4, 1966): 541–48.

Bijker, Wiebe E. "The Social Construction of Bakelite: Toward a Theory of Invention." In Bijker, Hughes, and Pinch, *Social Construction of Technological Systems*, 155–82.

Bijker, Wiebe E., Thomas P. Hughes, and Trevor Pinch, eds. *The Social Construction of Technological Systems*. Cambridge, MA: MIT Press, 2012.

Bijoy, C. R. "The Adivasis of India: A History of Discrimination, Conflict and Resistance." *Indigenous Affairs* no. 1 (2001): 54–61.

Bilstein, Roger E. *Stages to Saturn: A Technological History of the Apollo/Saturn Launch Vehicles*. Washington, DC: NASA, 1980.

Bimm, Jordan. "Andean Man and the Astronaut: Race and the 1958 Mount Evans Acclimatization Experiment." *Historical Studies in the Natural Sciences* 51, no. 3 (2021): 285–329.

Biolsi, Tom. *Power and Progress on the Prairie: Governing People on Rosebud Reservation*. Minneapolis: University of Minnesota Press, 2018.

Black, Megan. *The Global Interior: Mineral Frontiers and American Power*. Cambridge, MA: Harvard University Press, 2018.

Blenkinsopp, Joseph. "The Bible, Archaeology and Politics; or The Empty Land Revisited." *Journal for the Study of the Old Testament* 27, no. 2 (2002): 169–87. https://doi.org/10.1177/030908920202700202.

Boczkowska, Kornelia. *The Impact of American and Russian Cosmism on the Representation of Space Exploration in 20th Century American and Soviet Space Art*, 1st ed. Poznań: Wydawnictwo Naukowe UAM, 2016.

Boczkowska, Kornelia. "Spaceflight as the (Trans)National Spectacle: Transforming Technological Sublime and Panoramic Realism in Early IMAX Space Films." In *Multiculturalism, Multilingualism and the Self: Literature and Culture Studies; Second Language Learning and Teaching*, edited by Jacek Mydla, Małgorzata Poks, and Leszek Drong, 123–37. New York: Springer, 2017.

Bonneuil, Christophe. "Der Historiker und der Planet: Planetaritätsregimes an der Schnittstelle von Welt-Ökologien, ökologischen Reflexivitäten und Geo-Mächten." In *Gesellschaftstheorie im Anthropozän*, edited by Frank Adloff and Sighard Neckel, 55–92. Frankfurt am Main: Campus, 2020.

Boudignon, Patrice. *Pierre Teilhard de Chardin*. Paris: Editions du Cerf, 2008.

Bowen, Bleddyn. *Original Sin: Power, Technology and War in Outer Space*. London: Hurst, 2022.

Bowden, Mark. *Pitt Rivers: The Life and Archaeological Work of Lieutenant-General Augustus Henry Lane Fox Pitt Rivers*. Cambridge: Cambridge University Press, 1991.

Boym, Svetlana. "Kosmos: Remembrances of the Future." In *Kosmos: A Portrait of the Russian Space Age*, edited by Adam Bartos and Svetlana Boym, 82–99. New York: Princeton Architectural Press, 2001.

Boym, Svetlana. "Nostalgia and Its Discontents." *Hedgehog Review* 9, no. 2 (2007): 7–19.

Brenner, Neil, and Nikos Katsikis. "Operational Landscapes: Hinterlands of the Capitalocene." *Architectural Design* 90, no. 1 (2020): 22–31.

Brewster, Anne. "Oodgeroo: Orator, Poet, Storyteller." *Australian Literary Studies* 16, no. 4 (1994): 92–104. https://doi.org/10.20314/als.69adc84156.

Brockington, Dan. *Fortress Conservation: The Preservation of the Mkomazi Game Reserve, Tanzania*. Oxford: James Currey, 2002.

Bronfman, Alejandra. *Isles of Noise: Sonic Media in the Caribbean*. Chapel Hill: University of North Carolina Press, 2016.

Buck-Morss, Susan. *The Dialectics of Seeing: Walter Benjamin and the Arcades Project*. Cambridge, MA: MIT Press, 1989.

Buell, Lawrence. *The Environmental Imagination: Thoreau, Nature Writing, and the Formation of American Culture*. Cambridge, MA: Harvard University Press, 1995.

Bulkeley, Rip. *The Sputniks Crisis and Early United States Space Policy: A Critique of the Historiography of Space*. Bloomington: Indiana University Press, 1991.

Bullen, Ripley P. Adelaide K. Bullen, and William J. Bryant. *Archaeological Investigations at the Ross Hammock Site, Florida*. William J. Bryant Foundation, American Studies Report no. 7 (1967): 1. https://palmm.digital.flvc .org/islandora/object/ucf%3A15242#page/006/mode/2up

Bumpus, A. G., and D. Liverman. "Carbon Colonialism? Offsets, Greenhouse Gas Reductions, and Sustainable Development." In *Global Political Ecology*, edited by Richard Peet, Paul Robbins, and Michael Watts, 203–24. New York: Taylor and Francis, 2010.

Bürgel, Bruno H. *Der Mensch und die Sterne*. Berlin: Aufbau, 1946.

Burrell, Jenna. *Invisible Users: Youth in the Internet Cafés of Urban Ghana*. Cambridge, MA: MIT Press, 2012.

Bush, Vannevar. *Science the Endless Frontier: A Report to the President by Vannevar Bush, Director of the Office of Scientific Research and Development*. Washington, DC: US GPO, 1945.

Bushuev, Konstantin. *Soiuz i Apollon: rasskazyvaiut sovetskie uchenye, inzhenery i kosmonavty—uchastniki sovmestnykh rabot s amerikanskimi spetsialistami* [Soyuz and Apollo: Soviet Scientists, Engineers, and Astronauts: Participants in Joint Work with American Specialists Speak]. Moscow: Politizdat, 1976.

Buss, Jared S. *Willy Ley: Prophet of the Space Age*. Gainesville: University Press of Florida, 2017.

Butrica, Andrew, ed. *Beyond the Ionosphere: Fifty Years of Satellite Communication*. Washington, DC: NASA History Series, 1997.

Byrne, Denis R. "Counter-Mapping: New South Wales and Southeast Asia." *Transforming Cultures* 3, no. 1 (2008): 256–64.

Byrne, Denis R. "Nervous Landscapes: Race and Space in Australia." *Journal of Social Archaeology* 3, no. 2 (2003): 169–93. https://doi.org/10.1177/14696053 03003002003.

Byrnes, Mark E. *Politics and Space: Image Making by NASA.* Westport, CT: Praeger, 1994.

Cairns, Kate, and Josée Johnston. *Food and Femininity.* London: Bloomsbury, 2015.

Carnes, J. M., and Luis A. Frederick. "Deep Time and Vast Place: Visualizing Land/ Water Relations across Time and Space in Moonshot; The Indigenous Comics Collection." In *Graphic Indigeneity: Comics in the Americas and Australasia*, edited by Frederick Luis Aldama, 299–315. Jackson: University Press of Mississippi, 2020. https://doi.org/10.14325/mississippi/9781496828019.003.0015.

Carroll, R. P. "The Myth of the Empty Land." *Semeia* 59 (1992): 79–93.

Carroll, Siobhan. "Lost in Space: Surviving Globalization in *Gravity* and *The Martian*." *Science Fiction Studies* 46, no. 1 (2019): 127–42.

Castaneda, Carlos. *Journey to Ixtlan: The Lessons of Don Juan.* Harmondsworth: Penguin, 1981.

Catchpole, John E. *The International Space Station: Building for the Future.* Chichester: Praxis, 2008.

Cederlöf, Gunnel, and K. Sivaramakrishnan. *Ecological Nationalisms: Nature Livelihoods and Identities in South Asia.* Seattle: University of Washington Press, 2006.

Cezula, Ntozakhe Simon, and Leepo Modise. "The Empty Land Myth: A Biblical and Social-Historical Exploration." *Studia Historiae Ecclesiasticae* 46, no. 2 (2020): 15–17.

Chakrabarty, Dipesh. *The Climate of History in a Planetary Age.* Chicago: University of Chicago Press, 2021.

Chernykh, I. A., and A. E. Fominykh. "Diskursy iadernogo nerasprostraneniia v Tsentral'noi Asii." In *Iadernyi Mir: novye vyzovy rezhimu iadernogo nerasprostraneniia*, edited by E. B. Mikhailenko, 204–17. Yekaterinburg: Uralsk Universitet, 2017.

Child, Jack. *Miniature Messages: The Semiotics and Politics of Latin American Postage Stamps.* Durham, NC: Duke University Press, 2008.

Chowdhury, Indira, and Ananya Dasgupta. *A Masterful Spirit: Homi J. Bhabha, 1909–1966.* Delhi: Penguin Books, 2010.

Clark, Nigel, and Bronislaw Szerszynski. *Planetary Social Thought: The Anthropocene Challenge to the Social Sciences.* Cambridge: Polity Press, 2021.

Clarke, Arthur C. "Extra-Terrestrial Relays: Can Rocket Stations Give Worldwide Radio Coverage?" *Wireless World* 51, no. 10 (October 1945): 305–8.

Clarke, Arthur C. *The Promise of Space.* New York: Harper and Row, 1968.

Cockell, Charles, and Gerda Horneck. "A Planetary Park System for Mars." *Space Policy* 20, no. 4 (2004): 291–95.

Compton, W. David, and Charles D. Benson. *Living and Working in Space: A History of Skylab*. Washington, DC: NASA, 1983.

Condee, Nancy. *The Imperial Trace: Recent Russian Cinema*. Oxford: Oxford University Press, 2009.

Cooley, Alexander. "Imperial Wreckage: Property Rights, Sovereignty, and Security in the Post-Soviet Space." *International Security* 25, no. 3 (2001): 100–127.

Cooper, Frederick. "What Is the Concept of Globalization Good For? An African Historian's Perspective." *African Affairs* 100, no. 399 (April 2001): 189–213.

Cornum, L. C. "The Outer Space Future of Blackness and Indigeneity in *Midnight Robber* and *The Moons of Palmares*." PhD diss., University of British Columbia, 2015.

Cosgrove, Denis. "Contested Global Visions: One-World, Whole-Earth, and the Apollo Space Photographs." *Annals of the Association of American Geographers* 84, no. 2 (1994): 270–94.

Cox, Nicole C. "Selling Seduction: Women and Feminine Nature in 1920s Florida Advertising." *Florida Historical Quarterly* 89, no. 2 (2010): 186–209.

Crawford, Ian A., Parvathy Prem, Carle Peters, and Mahesh Anand. "Managing Activities at the Lunar Poles for Science." *Space Research Today* 215 (2022): 45–51. https://doi.org/10.48550/arxiv.2212.01363.

Crawford, Kate, and Vladan Joler. *Anatomy of an AI System: The Amazon Echo as an Anatomical Map of Human*. Novi Sad: Share Foundation, 2018.

Creager, Angela N. H. *Life Atomic: A History of Radioisotopes in Science and Medicine*. Chicago: University of Chicago Press, 2013.

Cresswell, Tim. *In Place/Out of Place: Geography, Ideology, and Transgression*. Minneapolis: University of Minnesota Press, 1996.

Crilley, Rhys, and Precious N. Chatterje-Doody. "From Russia with Lols: Humour, RT, and the Legitimation of Russian Foreign Policy." *Global Society* 35, no. 2 (2021): 269–88.

Cronon, William. *Changes in the Land: Indians, Colonists, and the Ecology of New England*. New York: Hill and Wang, 1983.

Crosby, Alexandra, and Jesse Adams Stein. "Repair." *Environmental Humanities* 12, no. 1 (2020): 179–85. https://doi.org/10.1215/22011919-8142275.

Cushing, Sumner W. "Sriharikota and the Yanadis: A Sandy Island off the East Coast of India and Its Inhabitants." *Annals of the Association of American Geographers* 7 (1917): 17–23.

Damjanov, Katarina. "Of Defunct Satellites and Other Space Debris: Media Waste in the Orbital Commons." *Science, Technology, and Human Values* 42, no. 1 (2017): 166–85.

Darrin, M. Ann Garrison, and Beth Laura O'Leary, eds. *Handbook of Space Engineering, Archaeology, and Heritage*. Boca Raton, FL: CRC Press, 2009.

Das Gupta, Sanjukta, and Raj Sekhar Basu, eds. *Narratives from the Margins: Aspects of Adivasi History in India*. Delhi: Primus Books, 2012.

Davies, Sarah. *Popular Opinion in Stalin's Russia: Terror, Propaganda and Dissent, 1934–1941*. Cambridge: Cambridge University Press, 1997.

d'Avignon, Robyn. *A Ritual Geology: Gold and Subterrenean Knowledge in Savanna West Africa*. Durham, NC: Duke University Press, 2022.

Dawson, Jane I. *Eco-Nationalism: Anti-Nuclear Activism and National Identity in Russia, Lithuania and Ukraine*. Durham, NC: Duke University Press, 1996.

Deetz, James. *In Small Things Forgotten: An Archaeology of Early American Life*. New York: Anchor Books/Doubleday, 1996.

Department of Atomic Energy, *Atomic Energy and Space Research: A Profile for the Decade, 1970–80*. Bombay: Publications Office, Dept. of Atomic Energy, 1970.

De Souza, Marcelo Lopes. "'Sacrifice Zone': The Environment–Territory–Place of Disposable Lives." *Community Development Journal* 56, no. 2 (2021): 220–43. https://doi.org/10.1093/cdj/bsaa042.

Deudney, Daniel. *Dark Skies: Space Expansionism, Planetary Geopolitics, and the Ends of Humanity*. Oxford: Oxford University Press, 2020.

Dick, Steven J. *Life on Other Worlds: The 20th-Century Extraterrestrial Life Debate*. Cambridge: Cambridge University Press, 1998.

Dick, Steven J. *Remembering the Space Age: Proceedings of the Fiftieth Anniversary Conference*. Washington, DC: NASA, 2008.

Dick, Steven J., and Roger D. Launius, eds. *Societal Impact of Spaceflight*. Washington, DC: NASA, 2007.

Dickens, Peter. "The Cosmos as Capitalism's Outside." In Parker and Bell, *Space Travel and Culture*, 66–82.

Dickens, Peter, and James S. Ormrod. *Cosmic Society: Towards a Sociology of the Universe*. London: Routledge, 2007.

Douglas, Mary. *Purity and Danger: An Analysis of Concepts of Pollution and Taboo*. New York: Routledge, 1966.

Downing, John. "International Communications and the Second World: Developments in Communication Strategies." *European Journal of Communication* 4, no. 1 (1989): 99–119.

Downing, John. "The Intersputnik System and Soviet Television." *Soviet Studies* 37, no. 4 (1985): 465–83.

Dubuisson, Eva-Marie. "Whose World? Discourses of Protection for Land, Environment, and Natural Resources in Kazakhstan." *Problems of Post Communism* 69, no. 4–5, (2020): 410–22.

Dudziak, Mary. *Cold War Civil Rights: Race and the Image of American Democracy*. Princeton, NJ: Princeton University Press, 2000.

Dufournaud, Daniel. "'Our Common Community': Third Way Cultural Work

in *Pleasantville* and *October Sky.*" *Film Criticism* 43, no. 1 (2019). https://doi
.org/10.3998/fc.13761232.0043.103.

Dusinberre, Martin. *Hard Times in the Hometown: A History of Community Survival in Modern Japan.* Honolulu: University of Hawai'i Press, 2012.

Early, L. B., L. Kumins, and J. Baer. "Business Forecasting for Communication Satellite Systems." In *Communication Satellite Systems Technology,* edited by Richard B. Marsten, 941–54. New York: Academic Press, 1966.

Edgerton, David. "From Innovation to Use: Ten Eclectic Theses on the Historiography of Technology." *History and Technology, an International Journal* 16, no. 2 (1999): 111–36.

Edgerton, David. "Innovation, Technology, or History: What Is the Historiography of Technology About." *Technology and Culture* 51, no. 3 (2010): 680–97.

Edgington, Ryan H. *Range Wars: The Environmental Contest for White Sands Missile Range.* Lincoln: University of Nebraska Press, 2014.

Elias, Amy J., and Christian Moraru, eds. *The Planetary Turn: Relationality and Geoaesthetics in the Twenty-first Century.* Evanston, IL: Northwestern University Press, 2015.

Elvis, Martin, and Tony Milligan. "How Much of the Solar System Should We Leave as Wilderness?" *Acta Astronomica* 162 (2019): 574–80.

Empson, William. *Some Versions of Pastoral.* New York: New Directions, 1960 (original 1935).

Engelhardt, Tom. *The End of Victory Culture: Cold War America and the Disillusioning of a Generation.* Amherst: University of Massachusetts Press, 2007.

Estes, Nick, Ruth Wilson Gilmore, and Christopher Loperena. "United in Struggle." *NACLA Report on the Americas* 53, no. 3 (2021): 255–67.

Evans, Christine E., and Lars Lundgren. *No Heavenly Bodies: A History of Satellite Communications Infrastructure.* Cambridge, MA: MIT Press, 2023.

Executive Office of the President, *Annual Report on Activities and Accomplishments under Communications Satellite Act of 1962.* Washington, DC: GPO, 1970.

Eyermann, Karl-Heinz. *Sojus-Apollo 1975.* Leipzig: Urania-Verl, 1975.

Ezell, Edward C., and Linda N. Ezell. *The Partnership: A History of the Apollo-Soyuz Test Project.* Washington, DC: NASA, 1978.

Ezell, Linda N. *NASA Historical Data Book: Vol.3, Programs and Projects 1969–1978.* Washington, DC: NASA Historical Series, 1988.

Faherty, William Barnaby. *Florida's Space Coast: The Impact of NASA on the Sunshine State.* Gainesville: University Press of Florida, 2002.

Fanon, Frantz. *Black Skin, White Masks.* London: Pluto Press, 1967.

Fanon, Frantz. *Toward the African Revolution: Political Essays.* London: Grove Press, 1988.

Ferdinando, Peter John. "Atlantic Ais in the Sixteenth and Seventeenth Centuries: Maritime Adaptation, Indigenous Wrecking, and Buccaneer Raids on

Florida's Central East Coast." PhD diss., Florida International University, 2015.

Fischer, Joachim, and Dierk Spreen, eds. *Soziologie der Weltraumfahrt*. Bielefeld: transcript, 2014.

Fischman, Robert L. "The National Wildlife Refuge System and the Hallmarks of Modern Organic Legislation." *Ecology Law Quarterly* 29, no. 3 (September, 2002): 458–622.

Fredengren, Christina. "Nature: Cultures—Heritage, Sustainability and Feminist Posthumanism." *Current Swedish Archaeology* 23, no. 1 (2015): 109–30. https://doi.org/10.37718/CSA.2015.09.

Freud, Sigmund. "The Uncanny." *Imago* 5 (1919): 297–324.

Fuller, Robert S., Duane W. Hamacher, and Ray P. Norris. "Astronomical Orientations of bora Ceremonial Grounds in Southeast Australia." *Australian Archaeology*, no. 77 (2013): 30–37.

Furaih, Ameer Chasib. "'Let No One Say the Past Is Dead': History Wars and the Poetry of Oodgeroo Noonuccal and Sonia Sanchez." *Queensland Review* (St. Lucia) 25, no. 1 (2018): 163–76. https://doi.org/10.1017/qre.2018.14.

Furniss, Elizabeth. "Indians, Odysseys and Vast, Empty Lands: The Myth of the Frontier in the Canadian Justice System." *Anthropologica* (Ottawa) 41, no. 2 (1999): 195–208.

Fürst, Elisabeth L'orange. "Cooking and Femininity." *Women's Studies International Forum* 20, no. 3 (1997): 41–49.

Gafijczuk, Dariusz. "Dwelling Within: The Inhabited Ruins of History." *History and Theory* 52, no. 2 (2013): 149–70.

Galison, Peter. *Image and Logic: A Material Culture of Microphysics*. Chicago: University of Chicago Press, 1997.

Gammage, Bill. *The Biggest Estate on Earth: How Aborigines Made Australia*. Crows Nest, NSW: Allen and Unwin, 2011.

Gao, Mobo. "Early Chinese Migrants to Australia: A Critique of the Sojourner Narrative on Nineteenth-Century Chinese Migration to British Colonies." *Asian Studies Review* 41, no. 3 (2017): 389–404. https://doi.org/10.1080/10 357823.2017.1336747.

Garber, Stephen J., and Lisa Ruth Rand. "A Montreal Protocol for Space Junk?" *Issues in Science and Technology* 38, no. 3 (2022): 20–22.

Garrard, Greg. *Ecocriticism*, 2nd ed. Abingdon: Routledge, 2011.

Garriott, Owen K. "Skylab Report: Man's Role in Space Research." *Science* 186, no. 4160 (October 18, 1974): 219–26. https://doi.org/10.1126/science.186.4160.219.

Gavaghan, Helen. *Something New under the Sun: Satellites and the Beginning of the Space Age*. New York: Springer, 1998.

George, Timothy S. *Minamata: Pollution and the Struggle for Democracy in Postwar Japan*. Cambridge, MA: Harvard University Asia Center, 2002.

Geppert, Alexander C. T. "Die Zeit des Weltraumzeitalters, 1942–1972." In *Obsession der Gegenwart: Zeit im 20. Jahrhundert*, edited by Alexander C. T. Geppert and Till Kössler, 218–50. Göttingen: Vandenhoeck & Ruprecht, 2015.

Geppert, Alexander C. T. "European Astrofuturism, Cosmic Provincialism: Historicizing the Space Age." In Geppert, *Imagining Outer Space*, 3–28.

Geppert, Alexander C. T., ed. *Imagining Outer Space: European Astroculture in the Twentieth Century*, 2nd ed. London: Palgrave Macmillan, 2018.

Geppert, Alexander C. T, ed. *Limiting Outer Space: Astroculture after Apollo*, 2nd ed. London: Palgrave Macmillan, 2020.

Geppert, Alexander C. T. "The Post-Apollo Paradox: Envisioning Limits during the Planetized 1970s." In Geppert, *Limiting Outer Space*, 3–26.

Geppert, Alexander C. T., ed. "Rocket Stars: Astrocultural Genealogies in the Global Space Age." Special issue, *British Journal for the History of Science*, forthcoming

Geppert, Alexander C.T., Daniel Brandau, and Tilmann Siebeneichner, eds. *Militarizing Outer Space: Astroculture, Dystopia and the Cold War*. London: Palgrave Macmillan, 2021.

Gerovitch, Slava. *From Newspeak to Cyberspeak: A History of Soviet Cybernetics*. Cambridge, MA: MIT Press, 2002.

Gerovitch, Slava. *Soviet Space Mythologies: Public Images, Private Memories and the Making of a Cultural Identity*. Pittsburgh, PA: Pittsburgh University Press, 2015.

Gerovitch, Slava. " 'Why Are We Telling Lies?' The Creation of Soviet Space History Myths." *Russian Review* 70, no. 3 (2011): 460–84.

Geyer, Michael, and Charles Bright. "World History in a Global Age." *American Historical Review* 100, no. 4 (October 1995): 1034–60.

Gilman, Nils. *Mandarins of the Future: Modernization Theory in Cold War America*. Baltimore: Johns Hopkins University Press, 2004.

Gindilis, L. M. "40 let SET v SSSR i Rossii." *Biulleten' spetsial'noe astrofizicheskoi observatorii* 60–61 (2007): 25–31.

Gindilis, Lev M., and Leonid I. Gurvits, "SET in Russia, USSR and the Post-Soviet Space: A Century of Research." *Acta Astronautica* 162 (September 2019): 1–13.

Gorbiel, Andrzej. "The Legal Status of Geostationary Orbit: Some Remarks." *Journal of Space Law* 6 (1978): 171–77.

Gordillo, Gastón R. *Rubble: The Afterlife of Destruction*. Durham, NC: Duke University Press, 2014.

Gordillo, Gastón R. "The Void: Invisible Ruins on the Edges of Empire." In Stoler, *Imperial Debris*, 227–51.

Gordon, Robert. "Critical Legal Histories." *Stanford Law Review* 36 (January 1984): 57–125.

Gorman, Alice. "The Anthropocene in the Solar System." *Journal of Contemporary Archaeology* 1, no. 1 (2014): 87–91.

Gorman, Alice. "The Archaeology of Space Exploration." In Parker and Bell, *Space Travel and Culture*, 129–42.

Gorman, Alice. "Beyond the Space Race: The Significance of Space Sites in a New Global Context." In *Contemporary Archaeologies: Excavating Now*, edited by Angela Piccini and Cornelius Holtorf, 161–80. Bern: Peter Lang: 2009.

Gorman, Alice. "The Cultural Landscape of Interplanetary Space." *Journal of Social Archaeology* 5, no. 1 (2005): 85–107. https://doi.org/10.1177/146960 5305050148.

Gorman, Alice. "The Cultural Landscape of Space." In Darrin and O'Leary, *Handbook of Space Engineering*, 331–42.

Gorman, Alice. *Dr Space Junk vs The Universe: Archaeology and the Future*. Cambridge, MA: MIT Press, 2019.

Gorman, Alice. "Ghosts in the Machine: Space Junk and the Future of Earth Orbit." *Architectural Design* 89, no. 1 (2019): 106–11.

Gorman, Alice. "How We Let the Moon Die, and Why It Isn't Dead." In *Mythologies of Outer Space*, edited by James Ellis and Noreen Humble. Calgary: University of Calgary Press, 2025.

Gorman, Alice. "La Terre et l'Espace: Rockets, Prisons, Protests and Heritage in Australia and French Guiana." *Archaeologies* 3, no. 2 (2007): 153–68. https://doi.org/10.1007/s11759-007-9017-9.

Gorman, Alice. "The Sky Is Falling: How Skylab Became an Australian Icon." *Journal of Australian Studies* 35, no. 4 (2011): 529–46. https://doi.org/10.108 0/14443058.2011.618507.

Gorman, Alice. "Space Cowboys: The Wild West and the Myth of the American Hero." *New England Review* (February 2005): 10–12.

Graham, Loren R. *Science in Russia and the Soviet Union: A Short History*. Cambridge: Cambridge University Press, 1993.

Graham, Loren R., and Irina Dezhina. *Science in the New Russia: Crisis, Aid, Reform*. Bloomington: Indiana University Press, 2008.

Graham, Stephen, and Nigel Thrift. "Out of Order: Understanding Repair and Maintenance." *Theory, Culture & Society* 24, no. 3 (2007): 1–25. https://doi .org/10.1177/0263276407075954.

Grampp, Sven. "Watching Television, Picturing Outer Space and Observing the Observer Beyond: The First Manned Moon Landing as Seen on East and West German Television." In *Television across and beyond the Iron Curtain*, edited by Kirsten Bönker, 55–93. Newcastle upon Tyne: Cambridge Scholars, 2016.

Green, C. M. *Vanguard: A History*. NASA SP-4202. Washington, DC: NASA, 1970.

Grisinger, Joanna. "'South Africa Is the Mississippi of the World': Anti-Apartheid Activism through Domestic Civil Rights Law." *Law and History Review* 38, no. 4 (2020): 843–81.

Gumbert, Heather L. "Cold War Theaters: Cosmonaut Titov at the Berlin Wall." In Andrews and Siddiqi, *Into the Cosmos*, 271–92.

Haney, Jennifer J., Claire Havice, and Jerry T. Mitchell. "Science or Fiction: The Persistence of Disaster Myths in Hollywood Films." *International Journal of Mass Emergencies and Disasters* 37, no. 3 (2019): 286–305.

Harland, David M. *The Story of Space Station Mir.* Chichester: Praxis, 2005.

Harper, Harry. *Dawn of the Space Age.* London: Sampson Low, Marston, 1946.

Harraway, Donna. "Anthropocene, Capitalocene, Plantationocene, Chthulucene: Making Kin." *Environmental Humanities* 6 (2015): 159–65.

Harris, Gordon L. *The Kennedy Space Center Story.* The Kennedy Space Center, NASA, 1968.

Harris, Stephen. "Dawn of the Soviet Jet Age Aeroflot Passengers and Aviation Culture under Nikita Khrushchev." *Kritika: Explorations in Russian and Eurasian History* 21, no. 3 (2020): 591–626.

Harrison, Rodney. "Beyond 'Natural' and 'Cultural' Heritage: Toward an Ontological Politics of Heritage in the Age of Anthropocene." *Heritage & Society* 8, no. 1 (2015): 24–42. https://doi.org/10.1179/2159032X15Z.00000000036.

Hart, Michael H. "An Explanation for the Absence of Extraterrestrials on Earth." *Quarterly Journal of the Royal Astronomical Society* 16 (1975): 128–35.

Hart, Philip Solomon, and Anthony A. Leiserowitz. "Finding the Teachable Moment: An Analysis of Information-Seeking Behavior on Global Warming Related Websites during the Release of *The Day After Tomorrow.*" *Environmental Communication* 3, no. 3 (2009): 355–66.

Harvey, Brian. *China in Space: The Great Leap Forward*, 2nd ed. New York: Springer, 2019.

Harvey, Brian, Henk H. F. Smid, and Theo Pirard. *Emerging Space Powers: The New Space Programs of Asia, the Middle East and South-America.* Chichester: Praxis, 2011.

Hecht, Gabrielle. *Being Nuclear: Africans and the Global Uranium Trade.* Cambridge, MA: MIT Press, 2012.

Hecht, Gabrielle, ed. *Entangled Geographies: Empire and Technopolitics in the Global Cold War.* Cambridge, MA: MIT Press, 2011.

Hecht, Gabrielle. *The Radiance of France: Nuclear Power and National Identity after World War II.* Cambridge, MA: MIT Press, 2009.

Hecht, Susanna, and Alexander Cockburn. *The Fate of the Forest: Developers, Destroyers, and Defenders of the Amazon.* Chicago: University of Chicago Press, 1990.

Hendrickx, Bart, and Bert Vis. *Energiya-Buran: The Soviet Space Shuttle.* New York: Praxis, 2007.

Henke, Christopher R. "The Mechanics of Workplace Order: Toward a Sociology of Repair." *Berkeley Journal of Sociology* 44 (1999): 55–81.

Hennaoui, Lenna, and Marzhan Nurzhan. "Dealing with a Nuclear Past." *International Spectator* 58, no. 4 (2023): 91–109.

Henry, Holly, and Amanda Taylor. "Re-thinking Apollo: Envisioning Environmentalism in Space." In Barker and Bell, *Space Travel and Culture*, 190–203.

Heppenheimer, T. A. *The Space Shuttle Decision: NASA's Search for a Reusable Space Vehicle*. Washington, DC: NASA, 1999.

Hernández, Diana. "Sacrifice along the Energy Continuum: A Call for Energy Justice." *Environmental Justice* 8, no. 4 (2015): 151–56.

Heyck, Hunter, and David Kaiser. "Introduction." *Isis* 101, no. 2 (2010): 362–66.

Hickam,Homer H. Jr. *Rocket Boys: A Memoir*. New York: Delacorte Press, 1998.

Hitt, David, Owen K Garriott, and Joe Kerwin. *Homesteading Space: The Skylab Story*. Lincoln: University of Nebraska Press, 2008.

Høgetveit, Åsne. 2018. "Female Aliens in (Post-) Soviet Sci-Fi Cinema: Technology, Sacrifice and Morality." *Studies in Russian, Eurasian and Central European New Media*, no. 19 (2018): 41–71.

Högselius, Per. "The Hidden Integration of Central Asia: The Making of a Region through Technical Infrastructure." *Central Asian Survey* 41, no. 2 (2022): 223–43.

Holquist, Michael, ed. *The Dialogic Imagination*. Austin: University of Texas Press, 2004.

Holt, Nathalia. *Rise of the Rocket Girls: The Women Who Propelled Us, from Missiles to the Moon to Mars*. New York: Little, Brown, 2016.

Hughes, Thomas P. *American Genesis: A History of the American Genius for Invention*. New York: Penguin Books, 1990.

Hughes, Thomas P. *Networks of Power: Electrification in Western Society, 1880–1930*. Baltimore: Johns Hopkins University Press, 1983.

Hughes, Thomas P. *Rescuing Prometheus*. New York: Pantheon Books, 1998.

Hurst, J. Willard. *Law and the Conditions of Freedom in the Nineteenth-Century United States*. Madison: University of Wisconsin Press, 1956.

Ikhsano, Andre, and Jakarudi Jakarudi. "Representation of Black Feminism in *Hidden Figures*." *Nyimak: Journal of Communication* 4, no. 2 (2020): 169–80.

Immerwahr, Daniel. *How to Hide an Empire: A History of the Greater United States*. New York: Farrar, Straus and Giroux, 2019.

Ingold, Timothy. *Being Alive: Essays on Movement, Knowledge and Description*. New York: Routledge, 2021.

Jackson, Steven. "Rethinking Repair." In *Media Technologies: Essays on Communication, Materiality, and Society*, edited by Tarleton Gillespie, Pablo J. Boczkowski, and Kirsten A. Foot, 221–40. Cambridge, MA: MIT Press, 2014.

Jaya Kumar, G. Stanley. *Tribals from Tradition to Transition: A Study of Yanadi Tribe of Andhra Pradesh*. New Delhi: MD, 1995.

Jenks, Andrew L. *Collaboration in Space and the Search for Peace on Earth.* London: Anthem Press, 2022.

Jenks, Andrew L. *The Cosmonaut Who Couldn't Stop Smiling: The Life and Legend of Yuri Gagarin.* DeKalb: Northern Illinois University Press, 2012.

Jenks, Andrew L "Securitization and Secrecy in the Late Cold War: The View from Space." *Kritika: Explorations in Russian and Eurasian History* 21, no. 3 (2020): 659–89.

Jessup, Philip C. "The Monroe Doctrine in 1940." *American Journal of International Law* 34, no. 3 (1940): 704–11.

Jessup, Philip C., and Howard J. Taubenfeld. *Controls for Outer Space and the Antarctic Analogy.* New York: Columbia University Press, 1959.

Johnson, Lyndon. "International Control of Outer Space." In *The Impact of Air Power: National Security and World Politics,* edited by Eugene M. Emme, 719–23. Princeton, NJ: D. van Nostrand, 1959.

Johnson, Stephen B. *The Secret of Apollo: Systems Management in American and European Space Programs.* Baltimore: Johns Hopkins University Press, 2002.

Jones-Imhotep, Edward. "Paris-Montreal-Babylon: The Modernist Genealogies of Gerald Bull." In *Made Modern: Science and Technology in Canadian History,* edited by Edward Jones-Imhotep and Tina Adcock, 185–215. Vancouver: UBC Press, 2018.

Jones-Imhotep, Edward. *The Unreliable Nation: Hostile Nature and Technological Failure in the Cold War.* Cambridge, MA: MIT Press, 2017.

Joravsky, David. *Soviet Marxism and Natural Science, 1917–1932.* London: Routledge, 1961.

Jorgenson, Darren. "Middle America, the Moon, the Sublime and the Uncanny." In Parker and Bell, *Space Travel and Culture,* 178–89.

Jorgenson, Darren. "States of Weightlessness: Cosmonauts in Film and Television." *Science Fiction Film and Television* 2, no. 2 (2009): 205–24.

Josephson, Paul R. "Rockets, Reactors, and Soviet Culture." In *Science and the Soviet Societal Order,* edited by Loren Graham, 168–91. Cambridge MA: Harvard University Press, 1990.

K stoletiiu so dnia rozhdeniia Alekseia Fedorovicha Bogomolova. Kniga 2. Ocherki razvitiiia OKB MEI v litsakh. Period 1965–1988. Moscow: AO OKB MEI, 2015.

Kabanova, Daria. "Mourning the Mimesis: Aleksei Fedorchenko's *First on the Moon* and the Post-Soviet Practice of Writing History." *Studies in Slavic Culture,* no. 10 (2012): 75–93.

Kakoudaki, Despina. "Representing Politics in Disaster Films." *International Journal of Media and Cultural Politics* 7, no. 3 (2011): 349–56.

Kakoudaki, Despina. "Spectacles of History: Race Relations, Melodrama, and the Science Fiction/Disaster Film." *Camera Obscura* 17, no. 2 (2002): 109–53.

Kalam, A. P. J. Abdul, and Arun Tiwari. *Wings of Fire: An Autobiography*. Hyderabad: Universities Press, 1999.

Karash, Yuri. *The Superpower Odyssey. A Russian Perspective on Space Cooperation*. Reston, VA: American Institute of Aeronautics and Astronautics, 1999.

Kardashev, N. S. "Peredacha informatsii vnezemnymi tsivilizatsiiami" [The Communication of Information by Civilizations on Other Worlds]. *Astronomicheskii zhurnal* 41 (1964).

Kauffman, James L. *Selling Outer Space: Kennedy, the Media, and Funding for Project Apollo, 1961–1963*. Tuscaloosa: University of Alabama Press, 1994.

Kavanaugh, James H. "'Son of a Bitch': Feminism, Humanism, and Science in 'Alien.'" *October* 13 (1980): 90–100.

Kelly, Scott, and Margaret Lazarus Dean. *Endurance: A Year in Space, a Lifetime of Discovery*. New York: Alfred A. Knopf, 2017.

Kennedy, Duncan. "The Three Globalizations of Law and Legal Thought: 1850–2000." In *The New Law and Economic Development*, edited by David Trubek and Alvaro Santos, 19–73. New York: Cambridge University Press, 2006.

Kessler, Elizabeth A. *Picturing the Cosmos: Hubble Space Telescope Images and the Astronomical Sublime*. Minneapolis: University of Minnesota Press, 2012.

Kilgore, De Witt Douglas. *Astrofuturism: Science, Race, and Visions of Utopia in Space*. Philadelphia: University of Pennsylvania Press, 2003.

King, Ursula. *Spirit of Fire: The Life and Vision of Teilhard de Chardin*, rev. ed. Maryknoll, NY: Orbis Books, 1996.

Kirchhof, Astrid Mignon, and Jan-Henrik Meyer. "Global Protest against Nuclear Power. Transfer and Transnational Exchange in the 1970s and 1980s." *Historical Social Research / Historische Sozialforschung* 39, no. 1 (2014): 165–90.

Klinger, Julie Michelle. "Environmental Geopolitics and Outer Space." *Geopolitics* 26, no. 3 (2021): 666–703.

Klinger, Julie Michelle. *Rare Earth Frontiers: From Terrestrial Subsoils to Lunar Landscapes*. Ithaca, NY: Cornell University Press, 2017.

Koch, Natalie. "Whose Apocalypse? Biosphere 2 and the Spectacle of Settler Science in the Desert." *Geoforum* no. 124 (2021): 36–45.

Kock, R. A., M. Orynbayev, S. Robinson, S. Zuther, N. J. Singh, W. Beauvais, E. R. Morgan et al. "Saigas on the Brink: Multidisciplinary Analysis of the Factors Influencing Mass Mortality Events." *Science Advances* 4, no. 1 (2018): 1–10.

Kohonen, Iina. *Picturing the Cosmos: A Visual History of Early Soviet Space Endeavor*. Bristol: Intellect Books, 2017.

Kohonen, Iina. "The Space Race and Soviet Utopian Thinking." *Sociological Review* 57, no. 1 (2009): 114–31.

Kojevnikov, Alexei. *Stalin's Great Science: The Times and Adventures of Soviet Physicists*. London: Imperial College Press, 2004.

Koot, Stasja, Bram Büscher, and Lerato Thakholi. "The New Green Apartheid? Race, Capital and Logics of Enclosure in South Africa's Wildlife Economy." *Environment and Planning E: Nature and Space* (2022). https://doi.org/10.1177/2514848622111.

Kopack, Robert A. "Baikonur 2.0: 'Inland-Offshore' Space Economies in post-Soviet Kazakhstan." *Culture, Theory and Critique* 62, no. 1–2 (2021): 96–112. https://doi.org/10.1080/14735784.2021.1929363.

Kopack, Robert A. "Rocket Wastelands in Kazakhstan: Scientific Authoritarianism and the Baikonur Cosmodrome." *Annals of the American Association of Geographers* 109, no. 2 (2019): 556–67.

Koselleck, Reinhart. *Vergangene Zukunft: Zur Semantik geschichtlicher Zeiten*. Frankfurt am Main: Suhrkamp, 1979.

Kramer, Paul. "How Not to Write the History of U.S. Empire." *Diplomatic History* 42, no. 5 (2018): 911–31.

Kramer, Peter. "'Want to Take a Ride?': Reflections on the Blockbuster Experience in *Contact* (1997)." In *Movie Blockbusters*, edited by Julian Stringer, 128–40. London: Routledge, 2003.

Krasnyak, Olga. "The Apollo–Soyuz Test Project: Construction of an Ideal Type of Science Diplomacy." *Hague Journal of Diplomacy* 13, no. 4 (2018): 410–31.

Kreienbrock, Jörg. *Sich im Weltall orientieren: Philosophieren im Kosmos 1950–1970*. Vienna: Turia + Kant, 2020.

Krige, John. *Fifty Years of European Cooperation in Space: Building on Its Past, ESA Shapes the Future*. Paris: Beauchesne, 2014.

Krige, John. "NASA as an Instrument of U.S. Foreign Policy." In Dick and Launius, *Societal Impact of Spaceflight*, 207–18.

Kristeva, Julia. *Powers of Horror: An Essay on Abjection*. New York: Columbia University Press, 1982.

Lahusen, Thomas. "Decay or Endurance? The Ruins of Socialism." *Slavic Review* 65, no. 4 (2006): 736–46.

Landis, Geoffrey A. "Spaceflight and Science Fiction." *Journal of Astrosociology* 1 (2015): 57–67.

Larkin, Brian. "The Politics and Poetics of Infrastructure." *Annual Review of Anthropology* 42, no. 1 (2013): 327–43.

Latham, Michael E. *Modernization as Ideology: American Social Science and "Nation Building" in the Kennedy Era*. Chapel Hill: University of North Carolina Press, 2000.

Latour, Bruno. "Can We Get Our Materialism Back, Please?" *Isis* 98, no. 1 (2007): 138–42.

Latour, Bruno. *Down to Earth, Politics in the New Climatic Regime*. Translated by Catherine Porter. Cambridge: Polity Press, 2018.

Latour, Bruno. "From Realpolitik to Dingpolitik or How to Make Things Public." In *Making Things Public-Atmospheres of Democracy*, edited by Bruno Latour et al., 515–39. Cambridge, MA: MIT Press, 2005.

Latour, Bruno. "On Actor-Network Theory: A Few Clarifications Plus a Few More Complications." *Soziale Welt* 15 (1996): 369–81.

Latour, Bruno. "Technology Is Society Made Durable." In *A Sociology of Monsters: Essays on Power, Technology, and Domination*, edited by John Law, 103–30. New York: Routledge, 1991.

Latour, Bruno. "Telling Friends from Foes at the Time of the Anthropocene." In *The Anthropocene and the Global Environmental Crisis: Rethinking Modernity in a New Epoch*, edited by Clive Hamilton, Francois Gemenne, and Christophe Bonneuil, 145–55. New York: Routledge, 2015.

Launius, Roger D. *Apollo's Legacy: Perspectives on the Moon Landings*. Washington, DC: Smithsonian Books, 2019.

Launius, Roger D. "The Historical Dimension of Space Exploration: Reflections and Possibilities." *Space Policy* 16 (2000): 23–38.

Launius, Roger D. *NASA: A History of the U.S. Civil Space Program*. Malabar, FL: Krieger, 1994.

Launius, Roger D. "Public Opinion Polls and Perceptions of US Human Spaceflight." *Space Policy* 19 (2003): 163–75.

Launius, Roger D. "Underlying Assumptions of Human Spaceflight in the United States." *Acta Astronautica* 62, no 6–7 (2008): 341–356.

Launius, Roger D. "What Are the Turning Points in History, and What Were They for the Space Age?" In Dick and Launius, *Societal Impact of Spaceflight*, 19–39.

Launius, Roger D., John M. Logsdon, and Robert W. Smith, eds. *Reconsidering Sputnik: Forty Years since the Soviet Satellite*. Amsterdam: Harwood, 2000.

Launius, Roger D., and Howard E. McCurdy, eds. *Spaceflight and the Myth of Presidential Leadership*. Urbana: University of Illinois Press, 1997.

Lawler, Andrew. "Accident Clouds US Future on Mir." *Science* 276, no. 5322 (1997): 26–28.

Lee, David Johnson. "De-Centring Managua: Post-Earthquake Reconstruction and Revolution in Nicaragua." *Urban History* 42, no. 4 (2015): 663–85.

Lenin, V. I. *Imperialism, the Highest Stage of Capitalism: A Popular Outline*. New York: International, 1969.

Lerner, Steve. *Sacrifice Zones: The Front Lines of Toxic Chemical Exposure in the United States*. Cambridge, MA: MIT University Press, 2010.

"Let There Be (Natural) Light." *Nature Astronomy* 7, no. 3 (March 2023): 235.

Levin, Jamie, Gustavo de Carvalho, Kristin Cavoukian, and Ross Cuthbert. "Before and after Borders: The Nomadic Challenge to Sovereign Territoriality." In *Nomad-State Relationships in International Relations: Before and after Borders*, edited by Jamie Levin, 63–76. Cham: Springer, 2020.

Lewis, Cathleen S. *Cosmonaut: A Cultural History*. Gainesville: University Press of Florida, 2023.

Liboiron, Max. *Pollution Is Colonialism*. Durham, NC: Duke University Press, 2021.

Liboiron, Max, and Josh Lepawsky. *Discard Studies: Wasting, Systems, and Power*. Cambridge, MA: MIT Press, 2022.

Linenger, Jerry M. *Off the Planet: Surviving Five Perilous Months aboard the Space Station MIR*. New York: McGraw-Hill, 2000.

Lipartito, Kenneth, and Orville R. Butler, *A History of the Kennedy Space Center*. Gainesville: University Press of Florida, 2007.

Liu, Hongbo, and Xiang Li. "How Travel Earns Us Bragging Rights." *Journal of Travel Research* 60, no. 8 (2021): 1635–53.

Livesey, Sophie. "Climate Change, Capitalism, 9/11, and *The Day after Tomorrow*." *Film Matters* 5, no. 1 (2014): 71–75.

Llinares, Dario. "Idealized Heroes of 'Retrotopia': History, Identity and the Postmodern in *Apollo 13*." *Sociological Review* 57, no. 1 (2009): 164–77.

Logsdon, John M. *After Apollo? Richard Nixon and the American Space Program*. New York: Palgrave Macmillan, 2015.

Logsdon, John M. *John F. Kennedy and the Race to the Moon*. New York: Palgrave Macmillan, 2010.

Logsdon, John M. "U.S.-Japanese Space Relations at a Crossroads." *Science*, New Series 255, no. 5042 (January 17, 1992): 294–300.

Lousma, Jack "The Role of Man in Space." In *Space Station Policy Planning and Utilization*. Arlington, VA: American Institute of Aeronautics and Astronautics, 1983. https://doi.org/10.2514/6.1983-3104.

Low, Morris F. "Accelerators and Politics in Postwar Japan." *Historical Studies in the Physical and Biological Sciences* 36, no. 2 (March 2006): 276–96.

Lozano, Henry Knight. "Race, Mobility, and Fantasy: Afromobiling in Tropical Florida." *Journal of American Studies* 51, no. 3 (2017): 805–83.

Lozano, Henry Knight. *Tropic of Hopes: California, Florida, and the Selling of American Paradise, 1869–1929*. Gainesville: University Press of Florida, 2013.

Lu, Juliet, and Oliver Schönweger. "Great Expectations: Chinese Investment in Laos and the Myth of Empty Land." *Territory, Politics, Governance* 7, no. 1 (2019): 61–78. https://doi.org/10.1080/21622671.2017.1360195.

Lucier, Paul. "Capitalism and Science." In *The Routledge History of American Science*, edited by Timothy W. Kneeland. New York: Routledge, 2022.

Lucke, Bernhard, Mohammed Shunnaq, Bethany Walker, Atef Shiyab, Zeidoun al-Muheisen, Hussein al-Sababha, Rupert Bäumler, and Michael Schmidt. "Questioning Transjordan's Historic Desertification: A Critical

Review of the Paradigm of 'Empty Lands.'" *Levant* (London) 44, no. 1 (2012): 101–26. https://doi.org/10.1179/175638012X13285409187955.

Lyle, Charles Dillard. "Highlights of the History of Triana." *Historic Huntsville Quarterly of Local Architecture and Preservation* (Summer 1997): 114–24.

Macaulay, Rose. *Pleasure of Ruins*. London: Weidenfeld and Nicholson, 1953.

MacDonald, Fraser. *Escape from Earth: A Secret History of the Space Rocket*. New York: Public Affairs, 2019.

MacKenzie, Donald A. "Missile Accuracy: A Case Study in the Social Processes of Technological Change." In Bijker, Hughes, and Pinch, *Social Construction of Technological Systems*, 195–222.

MacLeod, Roy, ed. "Nature and Empire: Science and the Colonial Enterprise." Special issue, *Osiris* 15 (2000).

Maharaj, Ashok. "Space for Development: US-Indian Space Relations 1955–1976." PhD diss., Georgia Institute of Technology, 2011.

Maher, Neil M. *Apollo in the Age of Aquarius*. Cambridge, MA: Harvard University Press, 2017.

Mailer, Norman. *Of a Fire on the Moon*. New York: Little, Brown, 1970.

Majsova, Natalija. "Neither Rupture nor Continuity: Memorializing the Dawn of the Space Age in Contemporary Russian Cinematography." In *The Twentieth Century in European Memory: Transcultural Mediation and Reception*, edited by Tea Sindbæk Andersen and Barbara Törnquist-Plewa, 198–219. Leiden: Brill, 2017.

Majsova, Natalija. *Soviet Science Fiction Cinema and the Space Age: Memorable Futures*. Lanham, MD: Lexington Books, 2021.

Malone, Peter. "Catholic Movies." In *The Continuum Companion to Religion and Film*, edited by William L. Blizek, 137–47. London: Bloomsbury, 2009.

Mann, G. A, J. D. A. Clarke, and V. A. Gostin. "Surveying for Mars Analogue Research Sites in the Central Australian Deserts." *Australian Geographical Studies* 42, no. 1 (2004): 116–24. https://doi.org/10.1111/j.1467–8470.2004 .00247.x.

Mark, James, Artemy M. Kalinovsky, and Steffi Marung, eds. *Alternative Globalizations: Eastern Europe and the Postcolonial World*. Bloomington: Indiana University Press, 2020.

Martin, Anne-Sophie. "Kourou: The European Spaceport and Its Impact on the French Guyana Economy." In *Space Fostering Latin American Society: Developing the Latin American Continent through Space*, edited by Annette Froehlich, 73–82. Cham: Springer, 2022.

Martin, Joel W. "Anti-Feminism in Recent Apocalyptic Film." *Journal of Religion & Film* 4, no. 1. https://digitalcommons.unomaha.edu/jrf/vol4/iss 1/1.

Martin, Terry. *The Affirmative Action Empire: Nations and Nationalism in the USSR, 1923–1939*. Ithaca, NY: Cornell University Press, 2001.

Marx, Leo. *The Machine in the Garden: Technology and the Pastoral Ideal in America*. Oxford: Oxford University Press, 1964.

Masao, Kihara. "Production of Weapons in Postwar Japan and Its Characteristics: Production of Rockets in Particular." *Kyoto University Economic Review* 47, no. 1–2 (April–October 1977): 1–26.

Masco, Joseph. "'Survival Is Your Business': Engineering Ruins and Affect in Nuclear America." *Cultural Anthropology* 23, no. 2 (2008): 361–98.

Matogawa, Yasunori. "'Pencil' Rocket and Itokawa Hideo: A Pioneering Work of Japanese Rocketry." In *History of Rocketry and Astronautics: Proceedings of the History Symposia of the International Academy of Astronautics (IAA), 1994–1995, Vol. 23*, edited by Donald C. Elder and Christophe Rothmund, 121–32. San Diego, CA: American Astronautical Society, 2001.

Matogawa, Yasunori. "A Survey of Rocketry for Space Science in Japan." In *History of Rocketry and Astronautics: Proceedings of the History Symposia of the International Academy of Astronautics (IAA), 1993, Vol. 22*, edited by Philippe Jung, 203–25. San Diego, CA: American Astronautical Society, 1998.

Mattern, Shannon. "Maintenance and Care." *Places Journal* 20 (November 2018). https://doi.org/10.22269/181120.

Maurer, Eva, Julia Richers, Monica Rüthers, and Carmen Scheide, eds. *Soviet Space Culture: Cosmic Enthusiasm in Socialist Societies*. New York: Palgrave Macmillan, 2011.

Mazlish, Bruce. "Comparing Global History to World History." *Journal of Interdisciplinary History* 28, no. 3 (Winter 1998): 385–95.

Mazlish, Bruce, ed. *The Railroad and the Space Program: An Exploration in the Historical Analogy*. Cambridge, MA: MIT Press, 1965.

McCarthy, Joseph M. *Pierre Teilhard de Chardin: A Comprehensive Bibliography*. New York: Garland, 1981.

McCray, Patrick. *The Visioneers: How a Group of Elite Scientists Pursued Space Colonies, Nanotechnologies, and a Limitless Future*. Princeton, NJ: Princeton University Press, 2013.

McCurdy, Howard E. *Space and the American Imagination*, 2nd ed. Baltimore: Johns Hopkins University Press, 2011.

McDougall, Walter. *The Heavens and the Earth: A Political History of the Space Age*. Baltimore: Johns Hopkins University Press, 1985.

McGregor, Russell. *Imagined Destinies: Aboriginal Australians and the Doomed Race Theory, 1880–1939*. Carlton South, Victoria: Melbourne University Press, 1997.

McMahon Adam M. "To the Moon and Back: Reexamining Presidential Decision-Making and the Apollo Program." *Space Policy* 62 (2022). https://doi.org/10.1016/j.spacepol.2022.101516.

Merchant, Carolyn. "Shades of Darkness: Race and Environmental History." *Environmental History* 8, no. 3 (2003): 380–94.

Messeri, Lisa. *Placing Outer Space: An Earthly Ethnography of Other Worlds.* Durham, NC: Duke University Press, 2016.

Middleton, Sallie. "Space Rush: Local Impact of Federal Aerospace Programs on Brevard and Surrounding Counties." *Florida Historical Quarterly* 87, no. 2 (Fall 2008): 258–89.

Miller, Carolyn. *Extraction Ecologies and the Literature of the Long Exhaustion.* Princeton, NJ: Princeton University Press, 2021.

Miller, Thomas Kent. *Mars in the Movies: A History.* Jefferson, NC: McFarland, 2016.

Mindell, David A. *Digital Apollo: Human and Machine in Spaceflight.* Cambridge, MA: MIT Press, 2011.

Mitchell, Jeffrey T., Deborah S. K. Thomas, Arleen A. Hill, and Susan L. Cutter. "Catastrophe in Reel Life versus Real Life: Perpetuating Disaster Myth through Hollywood Films." *International Journal of Mass Emergencies and Disasters* 18, no. 3 (2000): 383–402.

Mitchell, Mary X. "Land, Culture, and Marshall Islanders' Struggles for Self-Determination during the 1970s." *Environmental History* 22 (2017): 209–34.

Mitchell, Sean T. *Constellations of Inequality: Space, Race & Utopia in Brazil.* Chicago: University of Chicago Press, 2017.

Mitchell, Timothy. "The Limits of the State: Beyond Statist Approaches and Their Critics." *Political Science Review* 85, no. 1 (March 1991): 77–96.

Morgan, Lewis H. *Ancient Society or, Researches in the Lines of Human Progress from Savagery through Barbarism to Civilization.* London: Macmillan, 1877.

Morris, Narrelle. *Japan-Bashing: Anti-Japanism since the 1980s.* London: Routledge, 2010.

Morrison, Alexander. "Introduction: Killing the Cotton Canard and Getting Rid of the Great Game: Rewriting the Russian Conquest of Central Asia, 1814–1895." *Central Asian Survey* 33, no. 2 (2014): 131–42.

Morton, Peter. *Fire across the Desert: Woomera and the Anglo-Australian Joint Project 1946–1980.* Canberra: AGPS Press, 1989.

Moss-Wellington, Wyatt. "Individual and Collaborative Labour in the Space Crisis Movie: From *Apollo 13* to *The Martian*." *Quarterly Review of Film and Video* 37, no. 7 (2020): 634–65.

Moyn, Samuel. "Imaginary Intellectual History." In *Rethinking Modern European Intellectual History*, edited by Darrin M. McMahon and Samuel Moyn, 112–30. New York: Oxford University Press, 2014.

Moynagh, Maureen. *Political Tourism and Its Texts.* Toronto: University of Toronto Press, 2008.

Mozingo, Louise A. *Pastoral Capitalism: A History of Suburban Corporate Landscapes.* Cambridge, MA: MIT Press, 2014.

Mudrooroo. "The Poetemics of Oodgeroo of the Tribe Noonuccal." *Australian Literary Studies* 16, no. 4 (1994): 57–62.

Muir-Harmony, Teasel. *Apollo to the Moon: A History in 50 Objects*. Washington, DC: National Geographic, 2018.

Muir-Harmony, Teasel. *Operation Moonglow: A Political History of Project Apollo*. New York: Basic Books, 2020.

Narayan, S. Nambi, and Arun Ram. *Ready to Fire: How India and I Survived the ISRO Spy Case*. London: Bloomsbury, 2018.

Navakas, Michele Currie. *Liquid Landscape: Geography and Settlement at the Edge of Early America*. Philadelphia: University of Pennsylvania Press, 2007.

Nazarbayev, Nursultan. *Kazakhstan's Way*. Karaganda: Stacey International, 2008.

Neufeld, Michael J. *Von Braun: Dreamer of Space, Engineer of War*. New York: Alfred A. Knopf, 2007.

Newell, Catherine L. *Destined for the Stars: Faith, the Future, and America's Final Frontier*. Pittsburgh, PA: University of Pittsburgh Press, 2019.

Nichols, Tiffany. "Finding Stillness: Navigating Conflicting Land Interests during the Site Selection of the Laser Interfermoteter Gravitational-Wave Observatory." Paper presented at the New York History of Science Lecture Series, December 7, 2022.

Nilsen, Alf Gunvald. *Adivasis and the State: Subalternity and Citizenship in India's Bhil Heartland*. Cambridge: Cambridge University Press, 2018.

Nixon, Rob. "The Anthropocene: The Promise and Pitfalls of an Epochal Idea." In *Future Remains: A Cabinet of Curiosities for the Anthropocene*, edited by Gregg Mitman, Marco Armiero, and Robert S. Emmett, 1–18. Chicago: University of Chicago Press, 2018.

Nixon, Rob. "The Unequal Anthropocene." In *Living in the Anthropocene: Earth in the Age of Humans*, edited by John W. Kress and Jeffrey K. Stine, 149–60. Washington, DC: Smithsonian Books, 2017.

Nkrumah, Tara. "Problems of Portrayal: *Hidden Figures* in the Development of Science Educators." *Cultural Studies of Science Education* (2021). https://doi.org/10.1007/s11422-021-10021-3.

Nobuyoshi, Fugono. "Eisei yowa: Motomichi shin sōgō kenkyū shochō— Kashima 50 shūnen (to watashi)." *Space Japan Review*, no. 90 (October–January 2015/2016): 2–4.

Nochlin, Linda. *The Body in Pieces: The Fragment as a Metaphor of Modernity*. New York: Thames and Hudson, 1994.

Noonuccal, Oodgeroo. *My People*, 5th ed. Milton, Queensland: Wiley, 2021.

Noonuccal, Oodgeroo. *We Are Going: Poems*. Brisbane: Jacaranda Press, 1964.

Novak, William, "The 'Myth' of the Weak American State." *American Historical Review* 113, no. 3 (2008): 752–72.

Oda Minoru, *Uchū kūkan kansoku 30-nen shi*. Tokyo: Uchū kūkan kansoku 30-nen shi henshū iinkai, 1987.

Oldenziel, Ruth. "Islands: The United States as a Networked Empire." In Hecht, *Entangled Geographies* 13–42.

Olson, Valerie. *Into the Extreme: U.S. Environmental Systems and Politics beyond Earth.* Minneapolis: University of Minnesota Press, 2018.

Olson, Valerie. "Political Ecology in the Extreme: Asteroid Activism and the Making of an Environmental Solar System." *Anthropological Quarterly* 58, no. 4 (2012): 1027–44.

Olson, Valerie, and Lisa Messeri. "Beyond the Anthropocene: Un-Earthing an Epoch." *Environment and Society: Advances in Research* 6, no. 1 (2015): 28–47.

O'Neill, Gerard K. *The High Frontier.* London: Jonathan Cape, 1977.

Ormrod, James S. "Beyond World Risk Society? A Critique of Ulrich Beck's World Risk Society Thesis as a Framework for Understanding Risk Associated with Human Activity in Outer Space." *Environment and Planning D: Society and Space* 31, no. 4 (2013): 727–44.

Orr, Julian E. *Talking about Machines: An Ethnography of a Modern Job.* Ithaca, NY: Cornell University Press, 1996.

Osborne, Catherine R. "From Sputnik to Spaceship Earth: American Catholics and the Space Age." *Religion and American Culture* 25, no. 2 (Summer 2015): 218–63.

Osborne, Michael A., ed. "The Social History of Science, Technoscience and Imperialism." Special issue, *Science, Technology & Society* 4, no. 2 (1999).

Osokina, Elena. *Stalin's Quest for Gold: The Torgsin Hard-Currency Shops and Soviet Industrialization.* Ithaca, NY: Cornell University Press, 2021.

Owen, John R., Deanna Kemp, Alex M. Lechner, Jill Harris, Ruilian Zhang, and Élénore Lèbre. "Energy Transition Minerals and Their Intersection with Land-Connected Peoples." *Nature Sustainability* 6, no. 2 (2023): 203–11.

Paat-Dahlstrom, Emeline, and Eric Dahlstrom. "Democratizing Access to Space: Cocreating a Star Trek Universe." *New Space* 8, no. 4 (December 1, 2020): 166–70.

Pacner, Karel. *Sojuz volá Apollo* [Soyuz Calls Apollo]. Prague: Albatros, 1976.

Padel, Felix, and Samarendra Das. *Out of This Earth: East India Adivasis and the Aluminium Cartel.* New Delhi: Orient Blackswan, 2010.

Pahl, Berenice. "Pussy Riot's Humour and the Social Media: Self-Irony, Subversion, and Solidarity." *European Journal of Humour Research* 4, no. 4 (2017): 67–104.

Pak, Chris. *Terraforming: Ecopolitical Transformations and Environmentalism in Science fiction.* Liverpool: Liverpool University Press, 2016.

Palmer, Lorrie. "Untethered Technology in *Gravity*: Gender and Spaceflight from Science Fact to Fiction." *Science Fiction Film and Television* 12, no. 1 (2019): 29–51.

Parazynski, Scott, and Susy Flory. *The Sky Below: A True Story of Summits, Space, and Speed.* New York: Little A, 2017.

Parillo, Nicholas. "Leviathan and Interpretive Revolution: The Administrative State, the Judiciary, and the Rise of Legislative History, 1890–1950." *Yale Law Journal* 123, no. 2 (2013): 266–411.

Parker, Martin. "Capitalists in Space." In Parker and Bell, *Space Travel and Culture*, 83–97.

Parker, Martin, and David Bell, eds., *Space Travel and Culture: From Apollo to Space Tourism*. Malden, MA: Wiley-Blackwell, 2009.

Parks, Lisa. "Global Networking and the Contrapuntal Node: The Project Mercury Earth Station in Zanzibar, 1959–64." *Zeitschrift Für Medien- Und Kulturforschung* 11, no. 1 (2020): 40–57.

Parks, Lisa, and Nicole Starosielski. "Introduction." In *Signal Traffic: Critical Studies of Media Infrastructure*, edited by Lisa Parks and Nicole Starosielski. Urbana: University of Illinois Press, 2015.

Paul, Richard, and Steven Moss. *We Could Not Fail: The First African Americans in the Space Program*. Austin: University of Texas Press, 2016.

Pearson, Wendy G., Veronica Hollinger, and Joan Gordon, eds. *Queer Universes: Sexualities in Science Fiction*. Liverpool: Liverpool University Press, 2008.

Peldszus, Regina. "Architectures of Command: The Dual-Use Legacy of Mission Control Centers." In Geppert, Brandau, and Siebeneichner, *Militarizing Outer Space*, 285–312.

Perdue, Peter. *China Marches West: The Qing Conquest of Central Eurasia*. Cambridge, MA: Harvard University Press, 2005.

Peterson, Maya. *Pipe Dreams: Water and Empire in Central Asia's Aral Sea Basin*. Cambridge: Cambridge University Press, 2019.

Pfarrer, Chuck, Howard Cobb, and Jimmy Palmiotti, *Virus*. Milwaukie, OR: Dark Horse Comics, 1998.

Pierce, J. R., and R. Kompfner. "Transoceanic Communication by Means of Satellites." *Proceedings of the Institute of Radio Engineers* (March 1959): 372–80.

Pierson, David P. "Speculative Finance and Network Temporality in Duncan Jones's *Moon* and *Source Code*." *CR: The New Centennial Review* 19, no. 1 (2019): 255–76.

Postoutenko, Kirill, Alexey Tikhomirov, and Dmitri Zakharine, eds. *Media and Communication in the Soviet Union (1917–1953): General Perspectives*. Cham: Palgrave Macmillan, 2022.

Power, Ellen. "Narratives of Radiation, Risk and Uncertainty in the Clean-up of Satellite Cosmos 954." Paper presented at the annual meeting of the Society for the History of Technology, Saint Louis, MO, October 11–14, 2018.

Power, Ellen, and Arn Keeling. "Cleaning Up Cosmos: Satellite Debris, Radioactive Risk, and the Politics of Knowledge in Operation Morning Light." *Northern Review* (2018): 81–109.

Poznakhirko, Sergei. "Spravochnik: kosmicheskaia era v *Filatelii*." Supplement to *Filateliia* 39, no. 3 (2009): 1–79.

Prakash, Gyan. *Another Reason: Science and the Imagination of Modern India.* Princeton, NJ: Princeton University Press, 1999.

Prelinger, Megan Shaw. *Another Science Fiction: Advertising the Space Race 1957–1962.* New York: Blast Books, 2010.

Pretes, Michael. "Touring Mines and Mining Tourists." *Annals of Tourism Research* 29, no. 2 (April 2002): 439–56.

Pritchard, Sara B., and Carl A. Zimring. *Technology and the Environment in History.* Baltimore: Johns Hopkins University Press, 2020.

Purse, Lisa. "Working Space: *Gravity* (Alfonso Cuarón 2013) and the Digital Long Take." In *The Long Take: Palgrave Close Readings in Film and Television*, edited by J. Gibbs and D. Pye, 221–37. London: Palgrave Macmillan, 2017.

Quarantelli, E. L. "Realities and Mythologies in Disaster Films." *Communications* 11, no. 1 (1985): 31–44.

Quast, Paul E. "A Profile of Humanity: The Cultural Signature of Earth's Inhabitants beyond the Atmosphere." *International Journal of Astrobiology* 20, no. 3 (2021): 194–214.

Radhakrishna, Meena, ed. *First Citizens: Studies on Adivasis, Tribals, and Indigenous Peoples in India.* New York: Oxford University Press, 2016.

Raj, Gopal. *Reach for the Stars: The Evolution of India's Rocket Programme.* New Delhi: Viking, 2000.

Ramaswamy, Sumanthi. *Passions of the Tongue: Language Devotion in Tamil India, 1891–1970.* Berkeley: University of California Press, 1997.

Ramirez, Paul Edward Montgomery. "Indigenous Latino Heritage: Destruction, Invisibility, Appropriation, Revival, Survivance." In *Critical Perspectives on Cultural Memory and Heritage: Construction, Transformation and Destruction*, edited by Veysel Apaydin, 155–68. London: UCL Press, 2020.

Rand, Lisa Ruth. "Falling Cosmos: Nuclear Reentry and the Environmental History of Earth Orbit." *Environmental History* 24, no. 1 (January 2019): 78–103.

Rand, Lisa Ruth. "Orbital Decay: Space Junk and the Environmental History of Earth's Planetary Borderlands." PhD diss., University of Pennsylvania, 2016.

Rand, Lisa Ruth. *Space Junk: A History of Waste in Orbit.* Cambridge, MA: Harvard University Press, forthcoming.

Rankin, William. "The Geography of Radionavigation and the Politics of Intangible Artifacts." *Technology and Culture* 55, no. 3 (2014): 622–74.

Rao, P. V. Manoranjan, ed. *From Fishing Hamlet to Red Planet: India's Space Journey.* Noida: HarperCollins, 2015.

Rao, P. V. Manoranjan, and P Radhakrishnan. *A Brief History of Rocketry in ISRO.* Hyderabad: Universities Press, 2012.

Rao, S. Sudhakar. *Ethnography of a Nomadic Tribe: A Study of Yanadi.* New Delhi: Concept, 2002.

Rao, U. R. *India's Rise as a Space Power.* Delhi: Foundation Books, 2014.

Ravikumar, G. "Ethnographic Profile of the Yanadis in Nellore District of Andhra Pradesh." *Review of Research Journal* 6, no. 10 (July 2017): 1–9.

Reddy, S. Sudhakara. *Displaced Populations and Socio-Cultural Change.* New Delhi: Commonwealth, 1995.

Redfield, Peter. "Beneath a Modern Sky: Space Technology and Its Place on the Ground." *Science, Technology and Human Values* 23, no. 3 (1996): 251–74.

Redfield, Peter. "The Half-Life of Empire in Outer Space." *Social Studies of Science* 31, no. 5/6 (2002): 791–825.

Redfield, Peter. *Space in the Tropics: From Convicts to Rockets in French Guiana.* Berkeley: University of California Press, 2000.

Reich, Charles. "The New Property." *Yale Law Journal* 73, no. 5 (1964): 733–87.

Reid, Donald M. "The Symbolism of Postage Stamps: A Source for the Historian." *Journal of Contemporary History* 19, no. 2 (1984): 223–49.

Reusswig, Fritz, and Anthony A. Leiserowitz. "The International Impact of the *Day after Tomorrow.*" *Environment: Science and Policy for Sustainable Development* 47, no. 3 (2005): 41–44.

Reynolds, Glenn H., and Robert P. Merges. *Outer Space: Problems of Law and Policy.* New York: Routledge, 1998.

Richers, Julia. "Himmelssturm, Raumfahrt und 'kosmische' Symbolik in der visuellen Kultur der Sowjetunion." In *Die Spur des Sputnik*, edited by Igor J. Polianski and Matthias Schwartz, 181–209. Frankfurt am Main: Campus, 2009.

Rieppel, Lukas, Eugenia Lean, and William Derringer, eds. "Science and Capitalism: Entangled Histories." Special issue, *Osiris* 33 (2018).

Rifat, Mohammad Rashidujjaman, Hasan Mahmud Prottoy, and Syed Ishtiaque Ahmed. "The Breaking Hand: Skills, Care, and Sufferings of the Hands of an Electronic Waste Worker in Bangladesh." In *Proceedings of the 2019 CHI Conference on Human Factors in Computing Systems*, 1–14. CHI '19. New York: Association for Computing Machinery, 2019. https://doi.org/10.1145/3290605.3300253.

Rifkin, Mark. *Fictions of Land and Flesh: Blackness, Indigeneity, Speculation.* Durham, NC: Duke University Press, 2019.

Ring, Natalie J. "Inventing the Tropical South: Race, Region, and the Colonial Model." *Mississippi Quarterly* (2003): 619–31.

Rizzoni, Eitel M. "An Overview of Latin American Telecommunications, Past, Present, and Future." *IEEE Transactions in Communications* COM-24, no. 3 (1976): 290–305.

Roberts, Peder, Adrian Howkins, and Lize-Marié van der Watt. "Antarctica: A Continent for the Humanities." In *Antarctica and the Humanities*, edited by Roberts Peder, Lize-Marié van der Watt, and Adrian Howkins, 1–23. London: Palgrave, 2016.

Robey, Sarah. *Atomic Americans: Citizens in a Nuclear State*. Ithaca, NY: Cornell University Press, 2022.

Robinson, Cedric J. *Black Marxism: The Making of the Black Radical Tradition*. Chapel Hill: University of North Carolina Press, 2000.

Rockwell, Trevor S. "They May Remake Our Image of Mankind: Representations of Cosmonauts and Astronauts in Soviet and American Propaganda Magazines, 1961–1981." In *Spacefarers: Images of Astronauts and Cosmonauts in the Heroic Era of Spaceflight*, edited by Michael J. Neufeld, 125–47. Washington, DC: Smithsonian Institution Press, 2013.

Rodgers, Dennis, and Bruce O'Neill. "Infrastructural Violence: Introduction to the Special Issue." *Ethnography* 13, no. 4 (2012): 401–12.

Rose, Kenneth D. *One Nation Underground: The Fallout Shelter in American Culture*. New York: New York University Press, 2004.

Rostow, Walter W. *The Stages of Economic Growth: A Non-Communist Manifesto*. Cambridge: Cambridge University Press, 1960.

Roy, Stephanie A., Elaine C. Gresham, and Carissa Bryce Christensen. "The Complex Fabric of Public Opinion on Space." *Acta Astronautica* 47 (2000): 665–75.

Rozsa, George Gregory. "The Nevada Movement: A Model of Trans-Indigenous Antinuclear Solidarity." *Journal of Transnational American Studies* 11, no. 2 (2020): 99–123.

Rush, Kathryn A. "Astronauts' Sensemaking of Dangerous Beauty." *Journal of Applied Behavioral Science* (2022). https://doi.org/10.1177/0021886322 1136289.

Rushdie, Salman. *The Jaguar Smile: A Nicaraguan Journey*. New York: Penguin Books, 1988.

Russell, Andrew L., and Lee Vinsel. "Make Maintainers: Engineering Education and an Ethics of Care." In *Does America Need More Innovators?* edited by Matthew H. Wisnioski, Eric S. Hintz, and Marie Stettler Kleine, 249–69. Cambridge, MA: MIT Press, 2019.

Russell, James F., and David M. Klaus. "Maintenance, Reliability and Policies for Orbital Space Station Life Support Systems." *Reliability Engineering & System Safety* 92, no. 6 (June 1, 2007): 808–20. https://doi.org/10.1016/j .ress.2006.04.020.

Ryan, Robert G., Eloise A. Marais, Chloe J. Balhatchet, and Sebastian D. Eastham. "Impact of Rocket Launch and Space Debris Air Pollutant Emissions on Stratospheric Ozone and Global Climate." *Earth's Future* 10, no. 6 (2022). https://doi.org/10.1029/2021EF002612.

Sabitov, Daniyar. *Kazakhstanskii Kosmos: Realnost' i perspectivy*. Astana: Institut Mirovoi Ekomiki i Politiki, 2016.

Sackley, Nicole. "Passage to Modernity: American Social Scientists, India, and the Pursuit of Development, 1945–1961." PhD diss., Princeton University, 2004.

Sagan, Carl. *Contact.* New York: Simon and Schuster, 1985.

Sagan, Carl. *Pale Blue Dot: A Vision of the Human Future in Space.* New York: Random House, 1994.

Sagdeev, Roald. *The Making of a Soviet Scientist: My Adventures in Nuclear Fusion and Space from Stalin to Star Wars.* New York: Wiley, 1994.

Samuel, Flora. "Le Corbusier, Teilhard de Chardin and 'The Planetisation of Mankind.'" *Journal of Architecture* 4, no. 2 (1999): 149–65.

Sanchez-Sibony, Oscar. *Red Globalization: The Political Economy of the Soviet Cold War from Stalin to Khrushchev.* Cambridge: Cambridge University Press, 2014.

Sandlas, Ved Prakash. *The Leapfroggers: An Insider's Account of ISRO.* Noida: HarperCollins, 2018.

Sandner, David. "Shooting For the Moon: Méliès, Verne, Wells, and the Imperial Satire." *Extrapolation* 39, no. 1 (1998): 5–25.

Sankar, U. *The Economics of India's Space Programme: An Exploratory Analysis.* New Delhi: Oxford University Press, 2007.

Saperstein, Julie. "Opening the Black Box of Outer Space: The Case of Jason-3." *Geopolitics* 26, no. 3 (2021): 729–46.

Sarabhai, Vikram. *Science Policy and National Development*, edited by Kamla Chowdhry. Delhi: Macmillan India, 1974.

Satia, Priya. *Spies in Arabia.* New York: Oxford University Press, 2008.

Sato, Yasushi. "A Contested Gift of Power: American Assistance to Japan's Space Launch Vehicle Technology 1965–1975." *Historia Scientiarum* 11, no.2 (2001): 176–204.

Schama, Simon. *Landscape and Memory.* New York: Vintage, 1995.

Scharmen, Fred. *Space Forces: A Critical History of Life in Outer Space.* Brooklyn: Verso Books, 2021.

Scharmen, Fred. *Space Settlements.* New York: Columbia Books on Architecture and the City, 2019.

Scheiber, Harry. "Review of *The People's Welfare: Law and Regulation in Nineteenth-Century America*, by William Novak: 'Private Rights and Public Power: American Law, Capitalism, and the Republican Polity in Nineteenth Century America.'" *Yale Law Journal* 107, no. 3 (December 1997): 823–61.

Schiebinger, Londa, ed. "Focus: Colonial Science." Special issue, *Isis* 96, no. 1 (2005).

Schiwy, Günther. *Teilhard de Chardin: sein Leben und seine Zeit.* 2 vols. Munich: Kösel, 1981.

Schmalzer, Sigrid. *Red Revolution, Green Revolution: Scientific Farming in Socialist China.* Chicago: University of Chicago Press, 2015.

Schmitt, Carl. *Land und Meer: Eine weltgeschichtliche Betrachtung.* Stuttgart: Reclam, 1942.

Schmitt, Carl, *The* Nomos *of the Earth.* Translated by G.L. Ulmen. New York: Telos Press, 2003 (original 1950).

Schmölders, Claudia. "Heaven on Earth: Tunguska, 30 June 1908." In Geppert, *Imagining Outer Space,* 51–71.

Schönle, Andreas. "Ruins and History: Observations on Russian Approaches to Destruction and Decay." *Slavic Review* 65, no. 4 (2006): 649–69.

Schwartz, James S. J. "Myth-Free Space Advocacy Part II: The Myth of the Space Frontier." *Astropolitics* 15, no. 2 (2017): 167–84. https://doi.org/10.10 80/14777622.2017.1339255.

Schwartz, Matthias. *Expeditionen in andere Welten Sowjetische Abenteuerliteratur und Science-Fiction von der Oktoberrevolution bis zum Ende der Stalinzeit.* Cologne: Böhlau Verlag, 2014.

Schwoch, James. *Wired into Nature: The Telegraph and the North American Frontier.* Urbana: University of Illinois Press, 2018.

Scott, James C. *Domination and the Art of Resistance: Hidden Transcripts.* New Haven, CT: Yale University Press, 1990.

Sculos, Bryant W. "*The Martian*: A NASA-tionalist Utopia." *Class, Race and Corporate Power* 3, no. 2 (2015). http://digitalcommons.iu.edu/classrace corporatepower/vol3/iss2/6.

Seaton, A. V. "Guided by the Dark: From Thanatopsis to Thanatourism." *International Journal of Heritage Studies* 2, no. 4 (1996): 234–44.

Sen, Asoka Kumar. *Indigeneity, Landscape and History: Adivasi Self-fashioning in India.* Abingdon: Routledge, 2018.

Seo, Sarah. "The New Public." *Yale Law Journal* 125, no. 6 (2016): 1616–71.

Sesé, Bernard. "Pierre Teilhard de Chardin, prophète de la mondialisation?" *Études* 396, no. 4 (2002): 483–94.

Seth, Suman. "Putting Knowledge in Its Place: Science, Colonialism, and the Postcolonial." *Postcolonial Studies* 12, no. 4 (2009): 373–88.

Shah, Amrita. *Vikram Sarabhai: A Life.* New Delhi: Penguin, 2007.

Shapin, Steven, and Simon Schaffer. *Leviathan and the Air-Pump: Hobbes, Boyle, and the Experimental Life.* Princeton, NJ: Princeton University Press, 1985.

Shetterly, Margot Lee. *Hidden Figures: The American Dream and the Untold Story of the Black Women Mathematicians Who Helped Win the Space Race.* New York: HarperCollins, 2016.

Shrivastava, Aseem, and Ashish Kothari. *Churning the Earth: The Making of Global India.* Delhi: Penguin, 2014.

Siddiqi, Asif A. "American Space History: Legacies, Questions, and Opportunities for Future Research." In *Critical Issues in the History of Spaceflight,* edited by Steven J. Dick and Roger D. Launius, 433–80. Washington, DC: NASA History Division, 2006.

Siddiqi, Asif A. "Another Space: Global Science and the Cosmic Detritus of the Cold War." In *Space Race Archaeologies: Photographs, Biographies, and Design*, edited by Pedro Ignacio Alonso, 21–38. Berlin: DOM, 2016.

Siddiqi, Asif A. *Beyond Earth: A Chronicle of Deep Space Exploration, 1958–2016.* Washington, DC: NASA History Division, 2018.

Siddiqi, Asif A. *Challenge to Apollo: The Soviet Union and the Space Race, 1945–1974.* Washington, DC: NASA, 2000.

Siddiqi, Asif A. "Competing Technologies, National(ist) Narratives, and Universal Claims." *Technology and Culture* 51, no. 2 (2010): 425–43.

Siddiqi, Asif A. "Dispersed Sites: San Marco and the Launch from Kenya." In *How Knowledge Moves: Writing the Transnational History of Science and Technology*, edited by John Krige, 175–200. Oxford: Oxford University Press, 2019.

Siddiqi, Asif A. "From Cosmic Enthusiasm to Nostalgia for the Future. A Tale of Soviet Space Culture." In Maurer et al., *Soviet Space Culture*, 283–306.

Siddiqi, Asif A. "Imagining the Cosmos: Utopians, Mystics, and the Popular Culture of Spaceflight in Revolutionary Russia." *Osiris* 23 (2008): 260–88.

Siddiqi, Asif A. "Iskusstvenny sputnik zemli: The Launch of the World's First Artificial Satellite 50 Years Ago." *Spaceflight* 49 (2007): 426–47.

Siddiqi, Asif A. *The Red Rockets' Glare: Spaceflight and the Soviet Imagination, 1857–1957.* New York: Cambridge University Press, 2010.

Siddiqi, Asif A. "Science, Geography, and Nation: The Global Creation of Thumba." *History and Technology* 31, no. 4 (2015): 420–51.

Siddiqi, Asif A. "Shaping the World: Soviet-African Technologies from the Sahel to the Cosmos." *Comparative Studies of South Asia, Africa and the Middle East* 41, no. 1 (2021): 41–55.

Siddiqi, Asif A. "Sputnik 50 Years Later: New Evidence on Its Origins." *Acta Astronautica* 63: 529–39.

Siddiqi, Asif A. "Whose India? SITE and the Origins of Satellite Television in India." *History and Technology* 36, no. 3–4 (2020): 452–74.

Simmons, Henry, Edwin Diamond, and Henry W. Hubbard. *Language of a New World: Words of the Space Age; An Abridged Glossary*, 2nd ed. New York: Newsweek, 1962.

Singh, Gurbir. *The Indian Space Programme: India's Incredible Journey from the Third World to the First.* London: Astrotalkuk, 2017.

Slotten, Hugh. "The International Telecommunications Union, Space Radio Communications, and U.S. Cold War Diplomacy, 1957–63." *Diplomatic History* 37, no. 2 (2013): 313–71.

Slotten, Hugh. "Satellite Communications, Globalization, and the Cold War." *Technology and Culture* 43, no. 2 (2002): 315–50.

Small, Lindsay Marlies. "Space Museums: Technical and Cultural Considerations." PhD diss., University of Toronto, 2017.

Smiles, Deondre. "The Settler Logics of (Outer) Space." *Society and Space*, October 26, 2020. https://www.societyandspace.org/articles/the-settler-log ics-of-outer-space.

Smith, Jenny Leigh. *Works in Progress: Plans and Realities on Soviet Farms, 1930–1963*. New Haven, CT: Yale University Press, 2014.

Smith, Keith T. et al. "Losing the Darkness." *Science* 380, no. 6650 (June 16, 2023): 1116–17.

Smith, Michael L. "Selling the Moon: The U.S. Manned Space Program and the Triumph of Commodity Scientism." In *The Culture of Consumption: Critical Essays in American History, 1880–1980*, edited by Richard Wightman Fox and T. J. Jackson Lears, 175–209. New York: Pantheon Books, 1983.

Smith, Sierra E. "'A Cosmic Rorschach Test': The Origins and Development of the Search for Extraterrestrial Intelligence, 1959–1971." PhD diss., James Madison University, 2012.

Solomon, Matthew. *Fantastic Voyages of the Cinematic Imagination: George Méliès's Trip to the Moon*. Albany: State University of New York Press, 2012.

Solomon, Matthew. "Up-to-Date Magic: Theatrical Conjuring and the Trick Film." *Theatre Journal* 5, no. 4 (2006): 595–615.

Somsen, Geert J. "A History of Universalism: Conceptions of the Internationality of Science from the Enlightenment to the Cold War." *Minerva* 46 (2008): 361–79.

Southall, Ivan. *Woomera*. Sydney: Angus and Robertson, 1962.

Space in Africa, *Global Space Budgets: A Country-Level Analysis*. Lagos, 2021.

Speaight, Robert. *Teilhard de Chardin: A Biography*. London: Collins, 1967.

Spector, Sam, and James E. S. Higham. "Space Tourism in the Anthropocene." *Annals of Tourism Research* 79 (2019). https://doi.org/10.1016/j .annals.2019.102772.

Spence, Mark David. *Dispossessing the Wilderness: Indian Removal and the Making of the National Parks*. Oxford: Oxford University Press, 2000.

Sperb, Jason. "Specters of Film: New Nostalgia Movies and Hollywood's Digital Transition." *Jump Cut: A Review of Contemporary Media* 56 (2014–2015). https://www.ejumpcut.org/archive/jc56.2014–2015/SperbDigital-nostal gia/index.html.

Spivak, Gayatri Chakravorty. *Imperative zur Neuerfindung des Planeten/Imperatives to Re-Imagine the Planet*. Vienna: Passagen Verlag, 1999.

Spivak, Gayatri Chakravorty. "Planetarity." In *Dictionary of Untranslatables: A Philosophical Lexicon*, edited by Barbara Cassin, Emily Apter, Jacques Lezra, and Michael Wood, 1223. Princeton, NJ: Princeton University Press, 2014.

Sprengler, Christine. *Screening Nostalgia: Populuxe Props and Technicolor Aesthetics in Contemporary American Film*. New York: Berghahn Books, 2009.

Squeri, Lawrence. *Waiting for Contact: The Search for Extraterrestrial Intelligence*. Gainesville: University Press of Florida, 2016.

Stanek, Lukasz. *Architecture in Global Socialism: Eastern Europe, West Africa, and the Middle East in the Cold War*. Princeton, NJ: Princeton University Press, 2020.

Stanley, O'Brien, Nicki L. Michalski, Lane "Doc" Roth, and Steven J. Zani. *Martian Pictures: Analyzing the Cinema of the Red Planet*. Jefferson, NC: McFarland, 2018.

Star, Susan Leigh. "The Ethnography of Infrastructure." *American Behavioral Scientist* 43, no. 3 (1999): 377–91.

Steinke, Jocelyn. "Women Scientist Role Models on Screen: A Case Study of *Contact*." *Science Communication* 21, no. 2 (1999): 111–36.

Steinmetz, Sara. *Democratic Transition and Human Rights: Perspectives on U.S. Foreign Policy*. Albany: State University of New York Press, 1994.

Stepan, Nancy Leys. *Picturing Tropical Nature*. Ithaca, NY: Cornell University Press, 2001.

Stoler, Ann Laura, ed. *Imperial Debris: On Ruins and Ruination*. Durham, NC: Duke University Press, 2013.

Stoler, Ann Laura. "'The Rot Remains': From Ruins to Ruination." In Stoler, *Imperial Debris*, 1–35.

Stoler, Ann Laura, and David Bond. "Refractions Off Empire: Untimely Comparisons in Harsh Times." *Radical History Review* 2006, no. 95 (2006): 93–107.

Stovbun, V. D. "Osobennosti Mezhdunarodnoi organizatsii kosmicheskoi sviazi 'Intersputnik' v kachestve operativnoi (ekspluatsionnoi) organizatsii." *Istoricheskie, filosofskie, politicheskie i iuridicheskie nauki, kul'turologiia i iskusstvovedenie: Voprosy teorii i praktiki* 3, no. 4 (2009): 192–94.

Strebel, Ignaz, Alain Bovet, and Philippe Sormani. *Repair Work Ethnographies: Revisiting Breakdown, Relocating Materiality*. Basingstoke: Palgrave Macmillan, 2018.

Sundvall, Scott. "Clonetrolling the Future: Body, Space, and Ontology in Duncan Jones' *Moon* and Mark Romanek's *Never Let Me Go*." *Politics of Place* 2 (2015): 20–37.

Swanson, Kara. *Banking on the Body: The Market in Blood, Milk, and Sperm in Modern America*. Cambridge, MA: Harvard University Press, 2014.

Świtała-Trybek, Dorota. "Memorial Tourism: A Turn toward Locality." In *Social Sciences of Sport: Achievements and Perspectives*, edited by Jerzy Kosiewicz, Teresa Drozdek-Małolepsza, and Eligiusz Małolepszy, 229–39. Częstochowa: Wydawnictwo im. Stanisława Podobińskiego Akademii im. Jana Długosza, 2017.

Tani, Karen M. *States of Dependency: Welfare, Rights, and American Governance, 1935–1972*. Cambridge: Cambridge University Press, 2016.

Tarr, Joel. *The Search for the Ultimate Sink: Urban Pollution in Historical Perspective*. Akron, OH: University of Akron Press, 1996.

Teilhard de Chardin, Pierre. *The Future of Man*. London: Collins, 1964.

Teilhard de Chardin. "A Great Event Foreshadowed: The Planetisation of Mankind." In *Future of Man*, 124–39.

Teilhard de Chardin, Pierre. "Human Energy." In *Human Energy*, 132–33. New York: Harcourt Brace Jovanovich, 1971.

Teilhard de Chardin, Pierre. *L'Avenir de l'homme*. Paris: Éditions du Seuil, 1959.

Teilhard de Chardin, Pierre. *Le Phénomène humain*. Paris: Éditions du Seuil, 1955.

Teilhard de Chardin, Pierre. *Letters to Léontine Zanta*. Translated by Bernard Wall. London: Collins, 1969.

Teilhard de Chardin, Pierre. *Letters to Two Friends, 1926–1952*. New York: New American Library, 1968.

Teilhard de Chardin, Pierre. "Life and the Planets: What Is Happening at this Moment on Earth?" In *Future of Man*, 97–123.

Teilhard de Chardin, Pierre. "Un grand événement qui se dessine: la planétisation humaine." *Cahiers du Monde Nouveau* 2, no. 7 (August–September 1946): 1–13.

Teilhard de Chardin, Pierre. "Vie et planètes: que se passe-t-il en ce moment sur la Terre?" *Études* 248 (May 1946): 145–69.

Thompson, Courtney L. "'See What She Becomes': Black Women's Resistance in *Hidden Figures*." *Feminist Media Studies* (2020): 1–17.

Tichi, Cecelia. *Shifting Gears: Technology, Literature, Culture in Modernist America*. Chapel Hill: University of North Carolina Press, 1987.

Torras i Segura, Daniel. "'First Man': The Meaning and Use of Three Very Human Silences in an Inner Spatial Struggle." *Quarterly Review of Film and Video* (2021): 1–31.

Tresch, John. "Technological World-Pictures: Cosmic Things and Cosmograms." *Isis* 98, no. 1 (2007): 84–99.

Treviño, Natalie B. "The Cosmos Is Not Finished." PhD diss., University of Western Ontario, 2020.

Tribbe, Matthew D. *No Requiem for the Space Age*. New York: Oxford University Press, 2014.

Trischler, Helmuth. "The Anthropocene: A Challenge for the History of Science, Technology, and the Environment." *NTM Zeitschrift für Geschichte der Wissenschaftern, Technik und Medizin* 24 (2016): 309–35.

Tronzo, William, ed. *The Fragment: An Incomplete History*. Los Angeles, CA: Getty Research Institute, 2009.

Trouillot, Michel-Ralph. *Silencing the Past Power and the Production of History*. Boston: Beacon Press, 1995.

Trubek, David, and Mark Galanter. "Scholars in Self-Estrangement: Some Reflections on the Crisis in Law and Development Studies in the United States." *Law and Society* 4 (1974): 1062–1103.

Tsing, Anna Lowenhaupt. *In the Realm of the Diamond Queen Marginality in an Out-of-the-Way Place*. Princeton, NJ: Princeton University Press, 1993. https://doi.org/10.1515/9781400843473.

Tsing, Anna Lowenhaupt. *The Mushroom at the End of the World*. Princeton, NJ: Princeton University Press, 2015.

Tsing, Anna Lowenhaupt, Heather Swanson, Elaine Gan, and Nils Bubandt, eds. *Arts of Living on a Damaged Planet*. Minneapolis: University of Minnesota Press, 2017.

Tutton, Richard. "Sociotechnical Imaginaries and Techno-Optimism: Examining Outer Space Utopias of Silicon Valley." *Science as Culture* 30, no. 3 (July 3, 2020): 416–39. https://doi.org/10.1080/09505431.2020.1841151.

Uddin, Sayed Mohammad Nazim, and Jutta Gutberlet. "Livelihoods and Health Status of Informal Recyclers in Mongolia." *Resources, Conservation and Recycling* 134 (July 2018): 1–9.

Urey, Harold. *The Planets: Their Origin and Development*. New Haven, CT: Yale University Press, 1952.

Valentine, David. "Exit Strategy: Profit, Cosmology, and the Future of Humans in Space." *Anthropological Quarterly* 85, no. 4 (2012): 1045–67.

Valentine, David. "For the Machine." *History and Anthropology* 28, no. 3 (2017): 302–7.

Valenzuela, Mattias. "Policy-Making as an Articulation and Communication Process: Nicaragua's Telecommunications Reforms." PhD diss., University of Washington, 2000.

van Eeden, Jeanne. "Surveying the 'Empty Land' in Selected South African Landscape Postcards." *International Journal of Tourism Research* 13, no. 6 (2011): 600–612. https://doi.org/10.1002/jtr.832.

Vernadsky, Vladimir I. "The Biosphere and the Noösphere." *American Scientist* 33, no. 1 (January 1945): xxii–12.

Vertesi, Janet. *Seeing Like a Rover: How Robots, Teams, and Images Craft Knowledge of Mars*. Chicago: University of Chicago Press, 2014.

Viney, William. *Waste: A Philosophy of Things*. New York: Bloomsbury Academic, 2014.

Volf, Darina. "Between Cooperation and Competition: Cold War Imaginaries and Representations of U.S.-Soviet Encounters in Space." In *Science on Screen and Paper: Media Cultures of Knowledge Production in Cold War Europe*, edited by Mariana Ivanova and Juliane Scholz, 95–117. New York: Berghahn Books, 2024.

Volnarovski, Petar. "*Deep Impact* vs. *Armageddon*: Both Sides of the Same Story." *Blesok / Shine—Literature & Other Arts* 7 (1999): 1–3.

Von Burg, Ron. "Decades Away or *The Day After Tomorrow*?: Rhetoric, Film, and the Global Warming Debate." *Critical Studies in Media Communication* 29, no. 1 (2012): 7–26.

von Mossner, Alexa Weik. "Facing *The Day After Tomorrow*: Filmed Disaster, Emotional Engagement, and Climate Risk Perception." In *American Environments: Climate, Cultures, Catastrophe*, edited by Christof Mauch and Sylvia Mayer, 97–115. Heidelberg: Universitätsverlag Winter, 2012.

Walsh, Justin St. P. "Protection of Humanity's Cultural and Historic Heritage in Space." *Space Policy* 28, no. 4 (November 2012): 234–43.

Walsh, Justin St. P., and Alice C. Gorman. "A Method for Space Archaeology Research: The International Space Station Archaeological Project." *Antiquity* 95, no. 383 (October 2021): 1331–43.

Wang, Zuoyue. "Science and the State in Modern China." *Isis* 98, no. 3 (2007): 558–70.

Warren, Kate. "'Moon': Clones Are Human Too." *Screen Education* 60 (2010): 119–25.

Watanabe, Hirotaka. "Japan-U.S. Space Relations during the 1970s: After the Exchange of Notes." 55th International Astronautical Congress, 2004, Vancouver. https://arc.aiaa.org/doi/10.2514/6.IAC-04-IAA.6.15.2.08.

Watson, K., B. Murray, and H. Brown. "On the Possible Presence of Ice on the Moon." *Journal of Geophysical Research* 66 (1961): 1598–1600.

Weaver, David C., and James R. Anderson. "Some Aspects of Metropolitan Development in the Cape Kennedy Sphere of Influence." *Tijdschrift voor economische en sociale geografie*, May/June (1969): 187–92.

Webber, Derek. "Space Tourism: Its History, Future and Importance." *Acta Astronautica* 92, no. 2 (2013): 138–43.

Weinthal, Erika, and Kate Watters. "Transnational Environmental Activism in Central Asia: The Coupling of Domestic Law and International Conventions." *Environmental Politics* 19, no. 5, (2010): 782–807.

Weir, Andy. *The Martian: A Novel*. New York: Crown, 2014.

Weitekamp, Margaret A. *Space Craze: America's Enduring Fascination with Real and Imagined Spaceflight*. Washington, DC: Smithsonian Institutuion Press, 2022.

Werron, Tobias. "On Public Forms of Competition." *Cultural Studies ↔ Critical Methodologies* 14 (2014): 62–76.

Westad, Odd Arne. *The Global Cold War: Third World Interventions and the Making of Our Times*. Cambridge: Cambridge University Press, 2007.

Westling, Louise H. *The Green Breast of the New World: Landscape, Gender, and American Fiction*. Athens: University of Georgia Press, 1998.

Whalen, David. "Billion Dollar Technology: A Short Historical Overview of the Origins of Communications Satellite Technology, 1945–1965." In Butrica, *Beyond the Ionosphere*, 95–131.

Whalen, David. *The Origins of Satellite Communications, 1945–1965*. New York: Soho Press, 2014.

White, Frank. *The Overview Effect: Space Exploration and Human Evolution.* Reston, VA: American Institute of Aeronautics and Astronautics, 1998.

Wijeyeratne, Subodhana. "Between the Rocket and the Deep Blue Sea: Space Facilities and the 'Fishing Problem' in Southern Japan, 1950–1990." *Historia Scientiarum* 31, no. 22 (2022).

Wilde, Richard C. et al. "One Hundred US EVAs: A Perspective on Spacewalks." *Acta Astronautica* 51, no. 1 (1 July 2002): 579–90. https://doi .org/10.1016/S0094-5765(02)00096-6.

Willems, Brian. "The Potential of the Past: *First on the Moon.*" *Science Fiction Film and Television* 9, no. 2 (2016): 159–79.

Wilmes, Justin. "Empire Reloaded: Sacred Power in a Postmodern Era." In *Cinemasaurus*, edited by Nancy Condee, Alexander Prokhorov, and Elena Prokhorova, 48–64. Boston: Academic Studies Press, 2020.

Wilson, Sandra. "The 'New Paradise': Japanese Emigration to Manchuria in the 1930s and 1940s." *International History Review* 17, no. 2 (1995): 249–86.

Wolfe, Thomas C. *Governing Soviet Journalism: The Press and the Socialist Person after Stalin.* Bloomington: Indiana University Press, 2005.

Wolfe, Tom. *The Right Stuff.* New York: Farrar, Straus and Giroux, 1979.

Wolk, Coryn. "Invisible Damage, Imaginary Injuries, and the Political Ecology of Redress in a Quebec Mining and Smelter Town." MA thesis, University of Delaware, 2023.

Wolverton, Mark. *Burning the Sky: Operation Argus and the Untold Story of the Cold War Nuclear Tests in Outer Space.* New York: Overlook Press, 2018.

Woodburn, Amber Victoria. "Pushback in the Jet Age: Investigating Neighborhood Change, Environmental Justice, and Planning Process in Airport-Adjacent Communities." PhD diss., University of Pennsylvania, 2016.

Wormbs, Nina, and Lisa Ruth Rand. "Techno-Diplomacy of the Planetary Periphery, 1960s–1970s." In *History of the International Telecommunication Union: Transnational Techno-Diplomacy from the Telegraph to the Internet*, edited by Andreas Fickers and Gabriele Balbi. Berlin: De Gruyter Oldenbourg, 2020.

Wright, Alexis. "A Weapon of Poetry [The Poetry of Oodgeroo Noonuccal]." *Overland* 193, no. 193 (2008): 19–24.

Wukelic, G. E., ed. *Handbook of Soviet Space-Science Research.* New York: Gordon and Breach, 1968.

Yessenova, Saulesh. "The Tengiz Oil Enclave: Labour Business and the State." *Polar-Political and Legal Anthropology Review* 35, no. 1 (2012): 94–114.

Zimmerman, Robert. *Leaving Earth: Space Stations, Rival Superpowers, and the Quest for Interplanetary Travel.* Washington, DC: Joseph Henry, 2006.

CONTRIBUTORS

ELEANOR S. ARMSTRONG is a Space Research Fellow at the University of Leicester's Institute of Space Research. She obtained her PhD from University College London, and has held posts at the University of Delaware and Stockholm University, and visiting positions at, among others, the University of Cambridge, Ingenium Canada's Museums of Science and Innovation, New York University, and University of Vienna. Her research focuses on queer feminist approaches to social studies of outer space, particularly the presentation of femininities, feminisms, and femmes in public discourses about outer space, published in journals such as *Queer-Feminist Science and Technology Studies Forum; Quest: History of Space Flight Quarterly,* and in edited volumes including *Space Feminisms, Queering Science Communication* and *Routledge Handbook on Critical Social Studies of Outer Space.*

NELLY BEKUS, PhD (2007) is an assistant professor at the University of Exeter. She previously held research posts at Harvard University, the Institute of Human Sciences in Vienna, and the University of Warsaw. Her academic interests lie in Eastern European and Eurasian studies, and she focuses on Soviet and post-Soviet nation-building, postcolonialism, techno-politics, and memory studies. She is the author of two monographs and over thirty chapter and journal articles published, among others, in *Theory and Society, British Journal of*

Sociology, Slavic Review, Europe-Asia Studies, and *Nationalities Papers,* and the editor of several special issues, including *Slavic Review, Communist and Postcommunist Studies,* and *International Journal of Heritage Studies.*

REBECCA CHARBONNEAU is a historian of science with expertise in radio astronomy and the search for extraterrestrial intelligence (SETI). She holds a PhD in history and philosophy of science from the University of Cambridge. She has served as the historian-in-residence at the Harvard|Smithsonian Center for Astrophysics and as a Jansky Fellow at the National Radio Astronomy Observatory. She is currently a historian at the American Institute of Physics. Her first book, *Mixed Signals: Alien Communication across the Iron Curtain* (2024), explores the history of SETI in the US and USSR during the Cold War period.

ESTHER LIBERMAN CUENCA is an assistant professor of history at the University of Houston-Victoria. She has published on film, music, and television in the journals *Popular Music* and *Open Library of Humanities* and in the edited volume *One-Track Mind: Capitalism, Technology, and the Art of the Pop Song.* She has also coedited a coursebook examining medievalist film, *Law, Justice, and Society in the Medieval World: An Introduction through Film* (Fordham University Press, 2025).

HARIS A. DURRANI is a scholar of law, technology, and globalization. He holds a PhD from the Department of History at Princeton University (Program in History of Science). He holds a JD from Columbia Law School, an MPhil in history and philosophy of science from University of Cambridge, and a BS in applied physics from Columbia Engineering. His scholarship and essays appear in the *Columbia Journal of Transnational Law, Quest: The History of Spaceflight, The Nation,* and *The Washington Post.* His fiction includes *Technologies of the Self* and short stories in *McSweeney's, Analog Science Fiction and Fact,* and elsewhere.

CHRISTINE E. EVANS is an associate professor of history at the University of Wisconsin-Milwaukee. Her first book, *Between Truth and Time: A History of Soviet Central Television* (Yale University Press, 2016), received honorable mention for the 2017 USC Book Prize in Literary and Cultural Studies from the Association for Slavic, Eastern European, and Eurasian Studies. Together with Lars Lundgren, she recently published *No Heavenly Bodies: A History of Satellite Communications Infrastructure* (MIT Press, 2023), which explores the transnational history of satellite communications infrastructure during the Cold War.

RÉKA PATRICIA GÁL is a feminist and decolonial technoscience scholar. She is a postdoctoral researcher at the Technische Universität München's Department of Science, Technology and Society. She is the coeditor with Petra Löffler of *Earth and Beyond in Tumultuous Times: A Critical Atlas of the Anthropocene* (meson press, 2021). She completed her dissertation at the University of Toronto's Faculty of Information. Her doctoral dissertation mapped the genealogies of technological maintenance and care labors connected to human survival on space stations, focusing on the implications of human–machine interdependence in outer space as it relates to issues of environmental and labor justice.

ALEXANDER C. T. GEPPERT is an associate professor of history and European studies at New York University (NYU), jointly appointed by NYU New York and NYU Shanghai. He held the Charles A. Lindbergh Chair at the Smithsonian National Air and Space Museum and served as Eleanor Searle Visiting Professor at Caltech. His book publications include a trilogy on European astroculture, consisting of the edited volumes *Imagining Outer Space: European Astroculture in the Twentieth Century* (2018) and *Limiting Outer Space: Astroculture after Apollo* (2020), and a coedited volume (with Daniel Bandau and Tilmann Siebeneichner), *Militarizing Outer Space: Astroculture, Dystopia and the Cold War* (2021). He is at work on two monographs, *Astroculture: Europe in the Age of Space*, and a sequel, *Planetizing Earth: An Extra-Terrestrial History of the Global Present*. He also runs the lecture series "NYU Space Talks: History, Politics, Astroculture" (www.space-talks.com).

ALICE GORMAN is a space archaeologist and author of the award-winning book *Dr Space Junk vs The Universe* (MIT Press, 2019). Her research focuses on the archaeology and heritage of space junk in Earth orbit, planetary landing sites, and deep space probes. She is an associate professor at Flinders University, Adelaide, and a heritage consultant with over thirty years' experience working with Indigenous communities in Australia. In 2021 she was a member of a collective that wrote the first Declaration of the Rights of the Moon. She is a Vice Chair of the Global Expert Group for Sustainable Lunar Activity. Asteroid 551014 Gorman is named after her.

EDWARD JONES-IMHOTEP is a professor and the director of the Institute for the History and Philosophy of Science and Technology at the University of Toronto. His first book, *The Unreliable Nation: Hostile Nature and Technological Failure in the Cold War* (MIT Press, 2017), won the Sidney Edelstein Prize for best scholarly book in the field of history of technology. His new project—The Black Androids—explores Black technological selfhood in nineteenth- and twentieth-century New York through a history of the "Black androids"—automata in the form of Black humans.

JULIE MICHELLE KLINGER is an associate professor in the Department of Geography and Spatial Sciences at the University of Delaware. She has published numerous articles on rare earth elements, natural resource use, environmental politics, and outer space, including the award-winning book *Rare Earth Frontiers: From Terrestrial Subsoils to Lunar Landscapes* (Cornell University Press, 2018).

LARS LUNDGREN is a professor of media and communication studies at Södertörn University, Sweden. His research explores transnational television, cooperation and integration of television networks across Cold War divides, and the emergence of a global communications satellite system. With Christine Evans he recently published *No Heavenly Bodies: A History of Satellite Communications Infrastructure* (MIT Press, 2023). His work also appears in journals such as *Media History, International Journal of Communication*, and the *European Journal of Cultural Studies*.

NATALIJA MAJSOVA is an associate professor and researcher in the Department of Cultural Studies at the Faculty of Social Sciences, University of Ljubljana. Her research interests include memory studies, cultural theories, film studies, and media archaeology. She has authored numerous articles in foreign and Slovenian scientific journals and two books on space, film, and utopias. Her latest books are *Soviet SF Cinema and the Space Age: Memorable Futures* (Lexington Books, 2021) and *Faith in a Beam of Light: Magic Lantern and Belief in Western Europe, 1860–1940*, coedited with Sabine Lenk (Brepols, 2022).

LISA RUTH RAND is a historian of science, technology, and the environment, currently an assistant professor of history and William H. Hurt Scholar at the California Institute of Technology. Her research explores histories of waste and decay, particularly in extreme environments. An upcoming book examines the material, political, and cultural significance of orbital debris in the transformation of outer space into a contested natural resource. She has also researched and published on maintenance and repair in the space industry, the scientific politics of planetary analog habitats, and gendered narratives of reproductive space futurity.

ANNA RESER is a historian of American spaceflight technology and culture. She is the coauthor of *Forces of Nature: The Women Who Changed Science* (2021) and the cofounder and former coeditor in chief of *Lady Science*, a magazine for the history of women in science. Her writing has appeared in the *Atlantic, Technology's Stories*, and StarTrek.com.

ASIF A. SIDDIQI is a professor of history at Fordham University in New York. His early writing focused on Soviet efforts to imagine and explore space, described in *The Red Rockets' Glare: Spaceflight and the Soviet Imagination, 1857–*

1957 (Cambridge University Press, 2010). More recently, his interests have gravitated to new directions, the subjects of two books: *Soviet Science and the Making of Stalin's Gulag* (Oxford University Press, forthcoming) and *Departure Gates: Global Histories of Space on Earth* (MIT Press, forthcoming). He was the recipient of a Guggenheim Fellowship in 2015.

DARINA VOLF is a postdoctoral researcher in the Cooperation and Competition in the Sciences research group at the University of Munich, funded by the German Research Foundation. Her two research projects within the group focus on the history of Soviet–American cooperation in manned spaceflight in the mid-1970s (Apollo–Soyuz) and the history of space cooperation between socialist countries (Interkosmos). Her further research interests include cultural history, history of science, history of socialism, and memory studies. Her article "Evolution of the Apollo–Soyuz Test Project: The Effects of the 'Third' on the Interplay between Cooperation and Competition" was published in *Minerva* in 2021.

SUBODHANA WIJEYERATNE is an assistant professor of history in the history department at Purdue University. His work focuses on the relationship between technology, society, and culture. His forthcoming book, *Of Rockets and the Rising Sun*, charts the development of Japan's civilian space programs between 1920 and 2003, with a focus on the human and institutional elements of the story. He is also a writer of fiction, with over twenty short stories and three book-length publications in print. His second novel, *Triangulum: An Epic of the Nine Worlds of Surya* (Rosarium, 2024), was described by *Publisher's Weekly* as an "elegant and layered space opera."

INDEX

INDEX